The Philosophy Behind Physics

Thomas Brody The
Philosophy
Behind
Physics

Edited by
Luis de la Peña and
Peter E. Hodgson

Springer-Verlag
Berlin Heidelberg New York
London Paris Tokyo
Hong Kong Barcelona
Budapest

Professor Thomas A. Brody †

Editors

Professor Luis de la Peña

Universidad Autónoma de México
Instituto de Física
Apartado postal 20-364
01000 México DF
México

Dr. Peter E. Hodgson

University of Oxford
Department of Physics
Keble Road
Oxford OX1 3RH
Great Britain

Second Printing 1994

ISBN 3-540-57952-4 Springer-Verlag Berlin Heidelberg New York
ISBN 0-387-57952-4 Springer-Verlag New York Berlin Heidelberg

ISBN 3-540-55914-0 1. Auflage (Gebunden)
Springer-Verlag Berlin Heidelberg New York
ISBN 0-387-55914-0 1st Edition (Hardcover)
Springer-Verlag New York Berlin Heidelberg

CIP data applied for

The use of general descriptive names, registered names, trademarks, etc. in this publication does not imply, even in the absence of a specific statement, that such names are exempt from the relevant protective laws and regulations and therefore free for general use.

Typesetting: Springer $T_{E}X$ in-house system
Printing: Druckhaus Beltz, Hemsbach, Germany
Binding: Buchbinderei Schäffer, Grünstadt, Germany
SPIN 10466290   55/3140-5 4 3 2 1 0 – Printed on acid-free paper

Editors' Preface

Thomas Brody had one of the most powerful and wide-ranging intellects of his generation. Although primarily a physicist who worked on statistical problems in nuclear physics, on probability theory and on computational physics he had an extensive knowledge of the philosophy of science and of philosophy, and was fluent in many languages. He is well-known among physicists for the Brody-Moshinsky transformation but his extensive work on probability and on the philosophy of science remained almost unknown. This was because the originality of his ideas entailed many lengthy battles with uncomprehending referees, and he frequently published in Mexican journals of limited circulation. In addition, his strongly critical spirit inhibited his willingness to publish his ideas.

He was always most concerned by the very unsatisfactory situation in the philosophy of physics, that is largely due to the generally poor knowledge that physicists and philosophers have of each other's disciplines. Philosophers of science write at length about physics without any detailed first-hand knowledge of how research is actually carried out. Physicists, for their part, often implicitly assume naive or erroneous philosophical ideas, and this often hinders their scientific work, besides spreading further confusion if they try to give an account of what they are doing. Many of his writings are devoted to various aspects of this subject, and after his appointment as emeritus researcher he began to collect together some of his ideas into a book on the philosophy behind physics. He had already completed much of this work when it was interrupted by his untimely death.

Enough of the book was written for it to be published, and it was considered desirable to complete it with a selection of his many articles on the theory of probability, the philosophy of quantum mechanics and artificial intelligence. He frequently returned to the same subject, and so there is inevitably some repetition in our selection. We considered, however, that this is preferable to our trying to combine several articles into one. There is often an advantage in reading treatments of the same subject from somewhat different angles and with different degrees of sophistication. Thus, apart from some very minor editorial changes, all the words here are his own, except that five articles have been translated from the Spanish by his sons Julian and Carlos, who are themselves working in philosophy and in artificial intelligence. The result is not the book that Thomas would have produced had he lived to complete it, but it will give some idea of the importance and range of his work. We trust that will be of value, even in its present form.

What it cannot convey is the personality of Thomas Brody, one of the wisest, kindest and gentlest of all the men we have been privileged to know.

1992

L. de la Peña
P.E. Hodgson

Explanations and Acknowledgements

Of the material included in this book, Chaps. 1–4, 9 and 11 were originally written for it, and Chaps. 5, 10 and 19 come from manuscripts written at the same time and very probably as a preparation for it. Chapter 23 was also written at the same time. The remaining chapters were selected by the editors from his publications.

The editors and publisher are indebted to the original journals and editors for permission to publish the remainder of the papers, and are grateful to Julian and Carlos Brody for translating Chaps. 6, 8, 21, 22, 24. The references to the originals of these chapters are included in the list of publications.

Preface

Thirty years spent doing research in various fields of experimental and theoretical physics have gradually brought me to the conviction that present-day philosophy of science is far too widely divorced from the problems which develop in the course of scientific practice. There are, of course, several different schools of thought; but none of them appear to be of any help to the working scientist in resolving the difficulties that crop up in what he is doing; and it may not be too far off the mark to say that some of them actively hinder their resolution.

Yet these difficulties are real. They must be faced by any scientist whose work is not simple routine and who is thus forced to think clearly about what he is doing and the implications of his work. In this book I attempt to spell out certain basic concepts and their consequences in the philosophy of physics, concepts which are very much at variance with what is taught in the philosophy schools and unfortunately often also with what is taught in physics schools. I have nevertheless been forced into the views I here expound by actual experience; I have perhaps been luckier than many in that I have had opportunity to work in several different areas of physics, and also in having some experience in the applications as well as in the basic physical sciences.

Three areas of disagreement with the standard views have conditioned most of my thinking:

Firstly, the experimental side of physics, which is central to it, is misrepresented as consisting chiefly of "observation". In fact, a real experiment is very much more than that, and it is not a set of observed results but their interrelation with the experimental preparation and their evaluation in terms of the deliberately constructed set-up, the equally deliberately established calibrations of the instruments, and the experimental error limits found from the statistical work-up of the systematic repetitions of the experiment that really provides a basis for new knowledge. Properly understood, these facts not only explain (or help to explain) why experimentation is so central to physics, a fact largely ignored by philosophy of science, but they also shed a wholly new light on some age-old basic problems of general philosophy. Furthermore, they set the stage for the next step in creating a suitable philosophical framework for understanding science as an ongoing activity.

This second step concerns the construction of theories in physics. These can no longer be seen as individual and isolated deductive schemes. Rather, the more formal elements, i.e. the various mathematical formalisms associated with a theory, are seen to be embedded in a number of non-formalisable elements which nevertheless are essential to the theory and make the difference between a mathematical elaboration and a physical theory. The mathematical structures themselves are seen to be transformed by these non-formal parts of the theoretical picture. Moreover, the theories are no longer independent; they form a tightly knit but very flexible network, interconnected by common concepts, shared basic assumptions, the opposing processes of derivation and generalisation, and numerous common experimental results, either in an immediate form or as "universal" constants. Theories also form a kind of hierarchy, in which a lower-level structure rounds out the picture of a higher-level one by including much extraneous special information, until, at the lowest theoretical level, numerical values are assigned to initial or boundary conditions and actual predictions become possible. Such conceptual structures cannot be assimilated to (logical) propositions and hence it is meaningless to assign them truth values. Nor is the connection between theory and experiment so simple; not only is the dividing line relative and shifting, according to circumstances, but the information from experiment that the theory uses is very much more than that the experiment succeeded or not. And as is clear from actual practice, physicists do not either accept (provisionally accept) or reject theoretical pictures, they determine their scope and adapt them and use them in combination, often building a lower-level theory from several incompatible higher-level ones. In these processes as at many other points of the whole enterprise, the different approximation techniques play a decisive rôle (wholly ignored by philosophy of science) as the "lubricant" that maintains the entire edifice sufficiently flexible to fit the extraordinary richness of reality.

The third area of disagreement with the usual conceptions that I must note here arises from probability and its applications in physics. The relevant philosophical problems form a small and closed-off field of specialisation in philosophy; but in physics they are all-pervasive and hence a more satisfactory account of the nature of random phenomena is needed. Here I take the standpoint that the generality of a physical theory obliges us to consider the whole set of corresponding possible more specialised models that we can build from it, rather than only one particular case; this gives rise at once to the notion of an ensemble (in the sense used in statistical mechanics), which does not need the probability concept for its adequate description. Taking averages over the ensemble then provides the basis – not only for the notion of probability – but also for the construction of a different level of theory, with essentially new concepts at its core. Thus light is shed on the relation between a statistical theory and its underlying non-statistical one. When the notion of determinism is examined in this light, it is seen that, once again, philosophy of science misrepresents it. Determinism is a property which characterises, not the reality, but our description of it; not the world, but some of our theories about it. But since all theories have finite scope, so does their determinism, whatever its type may be. And there are several different varieties of determinism in physical theories. The Laplacean kind, allowing both prediction and retrodiction, turns out to be incompatible with random phenomena; but a determinism unilateral in time does accept them. Here we find the explanation of why computers generate random numbers, and of the possibility of random behaviour in deterministic systems, both of which have given rise to extraordinarily interesting developments in physics of late. Causality, on the other hand – which must carefully be distinguished from determinism, in spite of the presently common confusion – does have direct relevance to the underlying reality in that it describes an essential element in the interrelation between different physical systems; and in as much as it requires distinguising between causes and non-causes, between effects of these causes and non-effects, it is incompatible with the kind of conception that would generalise the Laplacean form of determinism to the entire world, seen as a single system. Finally, these views about probabilistic (or statistical) theories and their relations with other theories find an application in considering the case of quantum mechanics; here we find a situation which is unique: a statistical theory whose nature is hidden by the associated mathematical formalism and deformed by a philosophically conditioned interpretation, disguised by a wealth of irrelevant or even misleading formalisations. These matters I discuss in a separate chapter, in which I attempt to show that a common-sense point of view opens a number of interesting new areas to research, some of which are beginning to yield results.

In the chapters that follow I try to work out the consequences of these views. But it must be stressed that while they will sound extraordinary, perhaps even shocking, to the traditionally minded philosopher of science, there is very little new that I have contributed. Nearly every one of the points I make in the following discussions belongs in one way or another to the "folk lore" of the physics community – which is by no means unanimous, of course, but does contain a central stream which I have tried to sample. I can thus lay no great claim to originality; my aim has been, rather, to present in a reasonably coherent fashion, a consensus of common views that a very large proportion of experienced physicists would agree to. I have presented large fractions of this material, in the form of papers, seminars, and informal talks, to many of my colleagues, and have been both greatly helped by their critical comments and stimulated by their encouragement. I have also have had much fruitful contact with philosophers of science; and their acute comments and frequent disagreement with my ideas has likewise been of enormous help, a help which was made the greater by their friendly reception of my often highly critical remarks and their active seeking out of whatever enlightenment I, as a practicing physicist, might be able to offer them.

In order to make what I have to say more accessible and useful, I have largely abstained from discussing other views; to do this exhaustively would have lengthened unbearably an already overburdened text, while a selective approach would necessarily have been arbitrary and give an incorrect picture. Only at one point do I make such an attempt, namely concerning earlier views of the nature of probability. In this one case I believe that this will make my approach more intelligible.

I have tried to make my discussion comprehensible both to the professional physicist and to the professional philosopher, for if my starting point is not too wide of the mark we have a great deal to learn from each other; the present separation between the two disciplines does much damage to both.

But if such collaboration is to be possible, I believe both sides must recognise a general principle underlying both areas. This is that the search for the absolute, for a wholly unshakeable foundation of our theories, is a snare and an illusion. The natural sciences already recognise this; but at present a very dangerous tendency, which I discuss under the name of formalism, has arisen which attempts to reverse this basic point of view. It is dangerous because it subverts the possible aims of the scientific enterprise, and by misleading both the scientific community and the general public about what science could achieve may generate the impression that science is useless since its purpose – as stated by formalism – cannot be realised. Formalism is a consequence of an offshoot of the search for the absolute starting point of all science and all philosophy, and that is the identification of the rational with the merely logical. And this also is a matter I take up below.

I should wish to acknowledge my debt to the many colleagues who suggested one or another of the ideas I have here assembled; but I cannot do so, for these suggestions accumulated and changed over more than three decades, long before I ever considered writing them down and before I therefore thought of noting their originators. Let this word then express my gratitude to them, which I cannot discharge individually. I must also add the numerous students whom I taught and from whose often acute questionings I surely learnt more than they from me. Last but far from least, my thanks go to my colleague Luis de la Peña who helped me disentangle many a thread of thought presented here, as well as many more that are not.

Thomas Brody

Table of Contents

Part IV.   General

Introduction

The relation between philosophy and the sciences is peculiar. Physics, for instance, originated by splitting off from philosophy, and to this day is called 'natural philosophy' in a few universities. The split was carried through by each side ignoring the other's doings. On the side of physics, the justification given was the irrelevance of traditional philosophical conceptions to what the new science was doing, a point well made by C.R. Weld in his *History of the Royal Society*. The philosophical view of the goings on among the new breed of practical philosophers was never so clearly spelt out; but it permeates Berkeley's remarkable critique of Newton's new calculus – correct in every detail and wholly wrongheaded in its totality.

Since then physics, along with the other sciences, has grown and matured and above all has become an effective force in social change, as philosophy never was, because it was able to provide an increasing store of useful knowledge. Not that philosophy has remained static; among other new things it has created a specialised field of the philosophy of science, intended to examine critically the foundations on which the sciences are built. Such an effort should in principle be of enormous interest to the scientist, inasmuch as it could help him to improve his methods of work and understand more clearly what it is that he is achieving. But the fact is that scientists in general and physicists in particular mostly pay very little attention to what the philosophers of science have to say. It is true, nevertheless, that when a physicist becomes too old to do good physics any longer he turns to philosophy. Numerous physicists – the names of Mach, Einstein, Planck, Eddington, and Bridgman come to mind, though a full list is much longer – have indeed written on the philosophy of their craft.

Unfortunately much of this literature, instead of building on the physicists' professional experience in order to enrich philosophy, simply reinterprets physics from the point of view of one or another philosophical school. Sometimes – as in the case of Bridgman's operationalism or Mach's formidable reworking of phenomenalism – it does contribute a valuable new idea to its chosen school of thought. In a few cases it does scant justice to its leading philosophical thought; a case in point is Jeans' oversimplification of neopositivism. Seldom does it reach the height of Planck's searching questioning of the concepts underlying physics (although he did not come to any very firm conclusions himself, his statement of the problems cannot be ignored), let alone the acute and original rethinking of some of the foundations of physics that we owe to Einstein. It is to be deplored that philosophy of science largely ignores these mountains that rise from among the shallow seas of what one might call "standard" physicists' philosophising.

Thus physicists and philosophers on the whole turn their backs on each other. The results of this longstanding attitude are anything but beneficial for either side. As regards the physicists, their lack of sound knowledge of certain philosophical questions exposes them to the influence of the prevailing philosophical notions; these will not generally be recognised as such, yet their pervasive presence, as we shall have occasion to comment below, can often exert a most misleading influence on thinking in physics. It is not in fact true that physics can proceed in complete independence of philosophy; true, if one calculates a fourth-order Born approximation in the same way that the three preceding corrections have already been determined, no profound new principles are involved; but when one is breaking new ground, whether experimentally or theoretically, philosophical assumptions are always tacitly made and had better be brought out explicitly, to make sure of their adequacy. On their side the philosophers, by ignoring our century's main process for acquiring, interpreting and making use of new knowledge, or even worse by taking a partial and distorted view of it, cut themselves off from their best source of study material. Present-day practice of physics, together with its history, should really form the philosopher's laboratory. Any who refuse to enter it run the risks of irrelevance and sterility. And the amount of work done on what must be recognised as pseudoproblems, such as that of so-called induction, or Nelson Goodman's "grue" paradox, is witness to this.

Behind these differences, there is also a more basic question. Most philosophers see themselves as searching for fundamental principles, concepts that would once and for all answer fundamental questions. In the search for such principles, they develop astonishing ingenuity and subtlety which, however, are quite lost on the physicist. For in physics questions receive only approximate answers; physics is exact only insofar as it can quantitatively determine how inexact it is; and the fundamental questions have a way of either answering themselves or shifting their ground as a result of the bit-by-bit piling up of seemingly irrelevant results.

If their separation does harm to both physicists and philosophers, their collaboration can nevertheless not be fruitful until these underlying discrepancies are faced and at least to some extent resolved. In this book, I hope to make a start in this process; it will of course be long and involved, and I could not expect to achieve more than to initiate discussions that, instead of being merely conciliatory, attempt to dig out from each of the two disciplines what it can usefully offer the other. Over the last three centuries, physics has shown itself to be fruitful; the answers to most questions must therefore come from examining how it has managed to be so. But the questions themselves must largely come from philosophy, where the habit of critical self-examination is still alive in a way that should put physics to shame, even if it mostly exhausts itself in side issues. This, then, is the spirit I have attempted to establish in what follows: to lay the basis of a philosophy of science by examining what physics actually does.

The most fundamental question that clearly has to be answered here is why experimentation plays so central and pervasive a role in physics (as also in other sciences, of course; our study will however be mostly limited to physics, because of the author's incompetence in other fields). Philosophers of the traditional sort offer no plausible answer; indeed, their description of how theory and experiment interact generally runs counter to the physicist's experience by reducing it to a single question: does the theory correctly predict experimental outcome or not? The book begins by attempting to find a more satisfactory answer through an epistemology in which experiment, i.e. the planned interference with some part of nature, plays a central role. Getting to know things is an active process, not a mere passive perception. In a sense the book is no more than the working out of the consequences of this one idea.

It is of course not enough to attempt to spell out these consequences. Even in these matters on the borderline between philosophy and physics, it is necessary to understand what is actually done; a number of further necessary notions spring from physicists' practice.

In this sense, then, Bachelard's strictures on traditional philosophy's criticism of "naive" materialism are surely correct: this kind of criticism is directed at a variety of naive materialism that has long been left behind by three centuries' development of the sciences. We may call it naive; but it has become rich and subtle, and has proved itself to be effective over and over again. It should not be confused with mechanical materialism, which is largely a philosophers' creation. And inasmuch as it underlies all the successes physicists score, whatever their publicly stated belief, the aim pursued here is simply to make explicit the philosophy that physicists use when they are at work.

The presentation of the material in this book needs some remarks. It is addressed chiefly to those who have some knowledge of physics, whether as professional physicists or as philosophers of physics. But since very many fields of technical knowledge are drawn upon, and not all of them will be familiar to every reader, we have organised the main line of argument so as to be relatively free of technicalities.

Since the simple exposition of the points of view developed here already occupies so much space, the discussion of other points of view has been kept to a minimum. Only in the chapters on probability (Part II) and quantum mechanics (Part III) have I departed from this decision, chiefly because the contrast with other views here simplifies the exposition. It is planned to remedy this lack of critical comments on other conceptions in future work; but precisely in order to provide a suitable reference frame for that work, a simple exposition of what I have come to see as the fundamental questions in the process of working out a philosophy that appropriately satisfies the needs of working physicists appeared to be necessary and opportune.

I need hardly stress that everything I say here, insofar as it departs from well-established ideas, is to be taken as tentative and subject to review, correction and possibly even rejection, as further work may dictate. But such further work should always be guided by the principle that for the philosophy of physics the actual process of active research in the field must be the testing bed of all our philosophical conceptions. This must not, of course, be taken in the absurd sense that everything a physicist does is well done. Physicists, being as human as others, are just as prone to mistakes, to persist in error, and even to surprisingly violent protest if these failures are pointed out to them. The philosophy of physics must evidently have two aspects. On the one hand, it is (or rather, should be: as noted above, we are still far from reaching this ideal situation) an instrument of critical evaluation of what physicists do. And on the other, it must itself grow and develop by examining what physicists do, *and by finding the appropriate explanations for these actions.* A scientifically – and philosophically – justifiable action must be shown to proceed from understood (or at least understandable) conceptions; must, in other words, be shown to be rational; a conventional action – which might, without affecting the meaning of the results, have been decided differently – must be recognised as such. And whatever falls outside rationality – whether due to prejudice, confusion, or simply inadequacy – must also be marked as such. And in the process of effecting this very demanding analysis, the philosophical conceptions themselves on which the analysis is based are subject to constant reexamination.

The philosophy of physics, in other words, must itself become a scientific discipline. And just insofar as it achieves this, it will not merely escape from the contempt, at present all too often justified, with which physicists tend to regard it; it will become useful to physics and, as we shall argue below, find significant applications even in the practical sphere of the government of human societies. And in escaping from the confines of a stultifying academicism, it will reach maturity and a new dignity. It is in the hope of initiating the long labours that will lead to this end that I have written the present book.

Part I

The Philosophy of Physics

1. The Active Epistemology

(i)

Physics is presented to the university student as a vast accumulation of facts and theories, as a tremendous edifice of erudition which in the course of a few years he must learn to use as a problem-solving tool.

But a closer look – not unfortunately possible for the student, who does not yet possess the required knowledge – shows that this image is not altogether satisfactory. In the first place, the edifice lacks any very definite guiding principle. Physics might be defined the study of motion and structure, of how the structure of things determines their motions and how the laws of motion determine structure in their turn. But not only is this far too wide: the most obviously moving things in our environment, living beings, are studied in biology, not physics, while the most important structure, human society, is considered altogether outside the range of the natural scientist; what is more, much of what the student learns concerns the solution of equations. And where is motion or structure in Ohm's law? In fact, what we include in physics and what is seen in other sciences is only partly decided by reference to any abstract principle; historical factors are often more influential, and among them mere academic convenience is not the least. Nor is the division between the sciences fixed once and for all; some view astronomy as part of physics, others – on equally good grounds – call it an independent science; the frontiers with chemistry on the one hand and mathematics on the other are constantly shifting; and where to draw the line between physics and technology is one of the age's unresolved problems.

Secondly – and as we shall see, more significantly – the very idea of an "edifice of erudition", of a simple and steadily progressing accumulation of definite truth, is misleading. Once the student receives his degree he is free to initiate himself into the practice of physics; he then discovers that physics is only secondarily the sum of received wisdom it is commonly said to be. Rather is it a constant striving for improving, completing and in the process thoroughly remaking our previous ideas. This involves endless diversity of opinion and conception, interminable discussions and discrepancies, where the appearance of orthodoxy could almost herald is impending overturn. Physics – and science in general – is not the finished product but the process of creating it; not an established body of thought, coherent and unified by basic principles from which everything else derives, but the struggle and toil of the piecemeal acquisition of this knowledge. And at every step much that went before has to be entirely reworked. Physics is an ongoing endeavour, not a tradition-laden corpus of doctrine; or at least it should not be.

If, then, physics is essentially the process of acquiring physical knowledge, an analysis of this process must certainly form at least the starting point of any philosophy of physics. We therefore devote the rest of this chapter to a preliminary account, leaving a number of points to be elaborated later on, as needs arise. In contradistinction to the more usual approach, we shall develop an epistemological basis suited to explaining what physics actually does by attempting to find the common elements in various processes of knowledge acquisition. The results will also throw light on several other philosophical problems, not so directly related to physics, and will at the same time point to very practical implications of philosophy; often considered to be at best an ivory-tower pursuit, it turns out to have much relevance to ordering our public affairs.

(ii)

A beginning may be made by observing a human being – preferably a small child – when faced with an object whose nature and purpose he does not recognise. After a second or two, he will stretch out a hand, gingerly touch it with a finger, then grasp it and try to lift it. If this succeeds, he will turn it around and press it or pull it, just to see if it will 'do' something. He might smell it or try to bite it; some children will drop it on the floor, to see if it will break or make a nice noise. What no child will do is to sit contemplating the object, in the hope that he can organise his sense impressions into a meaningful conception of it.

An adult's reactions are less uniquely directed into this channel. In part, no doubt, because he has a much wider experience to draw upon, from which he commonly is able to derive at least a reasonable guess at the nature of the object ("Oh, it's just some newfangled kind of cigarette lighter, I'm sure!"). But also this same experience has taught him that the novelty of unknown objects wears off and does not often lead to interesting developments; an adult, in other words, has learnt to canalise his curiosity into directions that promise more lasting satisfactions, – whether these be the activities of his fellow human beings, the interactions of shapes of paint blotches on canvas, or the misbehaviour of electrons in a vacuum.

Such observations suggest that at least one basic way of acquiring knowledge about a segment of the world is to act on it and see what happens. This formulation is very imprecise; we will refine, complete and revise it as we go along; yet in its very imprecision there is a very useful aspect: the specific ways in which this notion can be embodied have themselves grown and multiplied and become much more powerful in the course of human history, and there seems no reason to doubt that this process will continue. It is therefore most undesirable that we should prematurely limit ourselves merely to achieve a spurious precision that offers little advantage.

But can we at least spell out in more detail what is meant by this "acting on a segment of the world and seeing what happens"? Insight is provided by examining the analogous problem in computer intelligence: a robot has to interpret the world it "sees" (through a television camera, for instance). The crude data it receives take the form of point-by-point information on light intensity (to simplify the discussion, we ignore colour). A first step is to recognise areas, delimited by lines where the illumination changes rather abruptly. The second step is to decide, firstly, which of these areas are simply shadows and should be joined to neighbouring areas, and secondly, how these areas combine to form images of three-dimensional bodies. In a very simple "world", containing a few regular solids, this can be done by examining all possible combinations and evaluating the results against a list of known body types (e.g. cubes, pyramids, cones, and so on), as well as using general information about laws to be satisfied. But if the number of solid objects grows, the number of possible combinations grows exponentially until the problem exceeds the capacity of any conceivable computer. And things become much worse if we allow irregular shapes; we must also remember that bodies can partially hide one another, and this makes it much more difficult to recognise them. Methods of ever increasing sophistication have been tried, without much success. The solution turned out to be surprising for the traditional epistemologist: it was not to study the data in exhaustive detail, but to make a few simple assumptions, on the basis of which the robot could be told to take an action that would modify his little world. For instance, only if certain lines really were the edges of shadows, then shifting the light source would make them move; and the directions and amounts they would move would give clear indications of the orientations of the surfaces the shadows lay on, and so enormously reduce the number of possible interpretations that still had to be examined. Such a programme of action is easy to implement and does not require huge computing resources. It will not by itself fully solve the problem of recognising the objects in a world of more or less regular solids; but it will reduce it to a problem of much smaller complexity. So the next steps just repeat this strategy until the problem has shrunk to manageable proportions. Thus we may move the light source again, in a different way; or we may move the camera; and in order to act on its little "world", the robot has an arm which may pick up some of the objects that have – tentatively – been recognised, among other things with the aim of revealing what lies behind them. So far, these techniques, though promising, still have not been elaborated into a complete system for the acquisition of knowledge; there appear to be not only technical but also certain conceptual problems remaining.

The epistemological implications of such work are clear. First of all, the problem of interpreting what an observer sees is computationally complex, in the following sense: if there are only a few elements in the scene, fairly straightforward comparison with known forms will suffice to eliminate the unwanted possibilites; but the required computer resources, execution time and memory space, grow exponentially with the number of elements. (Whether the problem is intractable, i.e. NP (non-deterministic polynomial time) complete or worse, in the sense of complexity theory [Aho et al. 1974] is not here a relevant question, since heuristic methods are employed to solve it.) Therefore attempts at complete interpretation of individual images will not in general be satisfactory. There are some exceptions, in particular for images of two-dimensional scenes; the interpretation of satellite photographs, though still enormously complex, is possible through the use of information obtained from other sources. Secondly, a feasible procedure for reducing the extraordinary richness and complexity of the scene that is being faced is to make a 'guess' at a partial interpretation, using some initial "model" of the scene; this guessed-at interpretation will imply that action $A1$ will make change $C1$ in the scene, action $A2$ will make change $C2$, and so on. One of these actions is carried out (the choice being guided by suitable heuristics); if the new image exhibits the expected change, the guess is to that extent confirmed. The prediction of the change may have contained undefined parameters, e.g. a shadow edge is expected to move when the light source is shifted, but the extent of motion is not predicted; values for these parameters may be derived from the observed change, so that the interpretation receives further detail. The cycle of guessing at interpretational elements on the basis of what has already been obtained, making predictions of what possible actions will yield, choosing and executing one of them, and using the changes as revealed by the new scene in order to improve the interpretation, – this cycle is repeated as often as may be needed to establish an interpretation with adequate detail.

This cycle (which for short will be called the epistemic cycle) is a more definite form of the notion set out above, of "doing something to see what happens". We proceed to examine a number of points our description has not dealt with.

(iii)

The epistemic cycle starts from an initial model; iterations of the cycle then improve the model until an adequate fit is achieved. The initial model must evidently be derived from other sources. In a computer controlling a robot, probably the most significant source is what the programmer has furnished as possible starting points; the human equivalent, information derived from interaction (conscious or unconscious) with one's fellow human beings, is equally important. Another source derives from experience with situations judged to be similar; a rich store of such knowledge can therefore provide us with models that hardly need any improving at all. Certain implications of this will be discussed in later chapters. Finally, a further source of initial models comes from a part of the mechanism not yet discussed, namely what is needed when the epistemic cycle misfires.

There is no guarantee that iterating the epistemic cycle will lead to a satisfactory model of the segment of nature under study; indeed, there cannot be. The process may not converge and could conceivably oscillate; and even if it converges, it might arrive at a representation that in a relevant respect is misleading. Should such a situation be detected, the model arrived at must be discarded and the epistemic cycle restarted from a different initial model.

A new initial model may be found in several different ways, including those already indicated for the first initial model. But there are others; among them stands out the one that concludes from how the failure being corrected arose to the change to be made in the first initial model. In other words, the mechanism of the epistemic cycle is being applied, but at a higher level: the action is not carried out directly on the segment of nature being studied, it touches the starting point of the lower-level cycle; and it enters when this lower-level cycle does not work correctly. For this purpose the higher-level cycle needs to have its own model, with an initial form and gradual movement towards a more adequate one. This higher-level model will be in some way an abstraction from the type of models used at the lower level; for our robot in a geometrical world, it will contain certain generalisations, e.g. that the objects are approximately within certain size limits, that they have regular shapes, are not reentrant, and are delimited by surfaces whose geometry can be described by functions of at most the second degree.

Such a higher-level epistemic cycle seems indispensable; but once the machinery for realising it exists, it will be recognised that it may be used recursively. Or in other words, once second-level epistemic cycles will work, we may have third-level cycles, and even higher, up to whatever may be needed in order to be able to create model that make adequate action possible at all levels. We do not yet know how to implement such ideas in a satisfactory and flexible way on the computer; but the philosophical implications of this idea are far-reaching. Some of them will be spelt out in what follows.

One such point is that the systematic following-out of this kind of epistemology will raise, in some minds, the spectre of the old questions of 'innate ideas'. What appears worrisome here is not, in itself, the concept of inborn predispositions; indeed, observation of newborns quite strongly suggests that indeed they possess, from the moment of birth and perhaps even before, certain reactions to their surroundings that function as conceivable starting points for epistemic cycles as described here. Rather, there are two problems: (i) it seems difficult to attribute to a newborn baby any concept that should mirror the world surrounding him well enough to give him some understanding; to posit any such concept merely pushes back the problem to the question where it springs from in its turn. And (ii) if the baby starts from such concepts, will they not colour and shape his notions about the world, will they not in fact hide the 'real' world from him? Will we not simply see those things that we are born to see? The answer to both problems is to be found in the extraordinary flexibility of the epistemic process. The resemblance of the initial to the last model it forms may be so slight as to be negligible; provided the initial model permits convergence to a satisfactory final one, it hardly matters what initial model the process starts from, the final one will be pretty much the same. At each execution of the cycle, so to say, the remnants of the initial 'content' of the model decrease in relative importance, while the contribution derived from adaptation to the real character of the segment being modelled increases correspondingly. And once this is understood, the first problem also disappears; for the starting model need not be a conceptual one at all, it suffices if it consists of no more than the impulse to "work the machinery" of the epistemic cycle with almost random actions in response to whatever feeling reigns at any moment. Much of the machinery of the body already works properly; and its interactions with the world then provide the needed starting point. Concepts come later.

Concept formation, indeed, requires that several higher levels of epistemic cycles are already working. The concept of "rattle" for instance – the name adults give to the notion of something you can grasp and that makes a nice noise when you shake it – cannot arise until several distinct first-level models have been built and recognised to be essentially the same: for only then can the expectation be set up – and this is a high-level model – that another rattle might appear; and the "rattle" concept is then completed when it is recognised that some rattle appearances have much in common, while others share only the central property of making a noise when shaking, otherwise being quite different.

(iv)

We return to the problem of computer interpretation of an observed scene, in order to note that the stepwise improvement of the model is not simply a uniform, over-all improvement. This is particularly evident when there are several objects in the 'world' to be understood. For they are generally recognised one by one, each object that is recognised reducing the number of open possibilities for the total picture. But the recognition of each individual object proceeds in much the same fashion whatever other objects may be present; in other words, a submodel is being generated for it. The model, then, is a collection of submodels, held together by the relations assumed to exist between the submodels. However, this relational framework must possess sufficient flexibility to accommodate whatever variation the submodels undergo as they are, each in its turn, refined; and since the different submodels can go different ways in this process, this means that they do not need to be wholly consistent. We shall study the possible range of inconsistency below; here we only note that it is essential, for two reasons: firstly, because it makes it possible to use already existing models as submodels in another one; and secondly, because it allows us (the computer, that is to say, or the individual human being) to adapt each submodel without reference to the whole model. Were this not possible, did one always have to adjust the entire set of superordinate models whenever one model at a certain level is being improved, the task of achieving understanding would be wholly beyond us since no partial understanding would be admissible, let alone useful. Thus the acceptable inconsistencies between submodels in a model are, so to say, the oil in the epistemic machinery.

(v)

If the ideas outlined in the preceding paragraphs are correct, it is obvious that after a finite number of executions of epistemic cycles of varying but finite depth a model may have been achieved which is good enough as a working tool; but it can only be a partial model, both in the sense that it cannot cover all the aspects of the segment of nature it represents and in the sense that its representation is not exact. Indeed, there will be some aspects that are grossly distorted. That this is useful will be obvious from a consideration of the analogy with maps. An Underground map of London, for instance, misrepresents distances between the stations, straightens and curves the lines almost at will, and paints small circles for stations with quite different shapes; yet this permits a clear presentation of the topology of the Underground network, which is what is needed in view of the specific purpose of the map. We have introduced a decisive consideration. Maps – like models – have a purpose. Different maps of the same region – and different models of the same segment of nature – will be incompatible because they distort different factors in order to suit distinct purposes. Nevertheless, we can combine them in order to achieve a more complete understanding of the region or segment. Thus the inconsistencies we have seen to be necessary arise because epistemic models have different and generally incompatible purposes.

(vi)

The notion of the purpose with which an epistemic model is built turns out to be relevant also to another point: the heuristics used to choose which of the possible actions in an epistemic cycle is executed, as well as those used to decide when to stop. The action in a cycle is chosen so that its expected result is the kind of thing we want to be able to plan on the basis of the model. And the cycle is repeated until the models provides us with plans that work to within whatever precision we need for the purpose we have in mind. Specifying some needed precision is commonly a very vague notion; but as we shall see later on, in physics it has to made quite detailed and clear – precise, in its turn.

Relatively low-level models are built with fairly specific aims; a higher-level model is built so that with its help many different lower-level models can easily be built to suit quite a wide range of different aims. Its purpose can therefore not be very exactly circumscribed. As the level and hence generality of the model increases, its purpose becomes more wide-ranging. But no model lacks a purpose; even the most general of all is intended, quite simply but quite definitely, to further our understanding. This is relevant to a purely philosophical matter: all the models we have so far considered apply to finite segments of our universe, indeed to rather small ones. Even cosmology treats only of certain average properties of the universe, and leaves aside the vastnesses of detail that a complete description of the universe would require. This rather suggests that the notion of a wholly exact theory of the universe, one that could predict every possible aspect of everything that goes on, is not tenable within the epistemology described here. Such a theory would constitute a universal model, in the terminology used here, that would never require revision; it might thus seem desirable to work towards it. But quite apart from the awkward circularity involved in a model that must also predict all the actions of those engaged in building it – i.e. in modifying it – before final convergence is achieved, this perfect and universal model is quite useless. For it must, in each and every aspect behave exactly like the universe it models; it will, therefore, be just as difficult to handle and understand as the universe itself; indeed, how could we even tell the universe and its model apart? According to Leibniz' criterion, they must be identical since they share every conceivable property. Thus the search for a universal and perfect model is self-stultifying.

Such a conclusion must not be taken to imply that there can be no such thing as truth. Indeed there can. But not absolute truth; only partial and relative truth. Nor do we need any other, since "too much truth" would fill our minds with endless irrelevancies, so that we could no longer see the wood for the trees.

This kind of truth is partial and limited in that the kind of model the epistemic process provides us with misrepresents and distorts some aspects of the segment of nature being modelled. Two implications of this statement must be faced:

(a) We have tacitly assumed a wholly realist – in the would-be derogatory term of modern academic philosophy, a "naively realist" – ontology. What makes such an ontology inescapable is that active interference with this segment of the world is an essential and indispensable part of the epistemic cycle; the (logically) independent existence of this world has thus to be presumed from the beginning. The solipsist answer, that we merely think we have acted but have no independent evidence that we really did so except through the evidence of our senses, is inadequate in two respects: one, that then the active epistemology here expounded, with its realist presupposition, appears constructed on the basis of an exactly opposite epistemology and ontology – a self-contradictory procedure; and two, that it would remain wholly inexplicable why, at least at first, our models make such mistaken predictions and why, by revising them, we can improve their predictive capabilities.

(b) The kind of truth concept used here is of the correspondence type. The usual difficulties faced by this conception nevertheless do not arise here; in part because of the sort of correspondence that is needed, and in part because the correspondence is, as we have seen, necessarily partial and inexact. The usual correspondence theories suppose (tacitly, in most cases) a photographic kind of correspondence, – perhaps a 3D colour picture. Neither the human brain nor the computer memory contain anything of the sort, nor would it be helpful. What is needed is a dynamic correspondence, a correspondence of behaviour; if in the real world a glass will shatter on the floor if we let it drop, then the model representation of the glass need not shatter itself, of course, but as the model distance between model glass and model floor reaches zero the model must show us the glass concept becoming inapplicable, and that of many shards having to be substituted for it. In other respects the model may differ; indeed, it must, so that it should be easy to use. In other words, the model reproduces the structural relations among the relevant aspects of the segment being modelled; but such structural relations may exist among many different kinds of things. In physical terms the model obeys the same (or at least similar) equations of evolution as the real thing does.

These two implications presuppose not merely that the epistemic cycle we have described *works*, in the sense of producing valid knowledge, but that it is the only method of acquiring knowledge available to us. This has not been proved so far; if it is the only method, no proof should be possible, as will be seen when the philosophical status of logic is examined. What can be done, however, is to argue that as a matter of experience this is the only way to acquire original new knowledge. (The condition of originality is merely intended to eliminate the irrelevant acquisition of knowledge from other people who already possess it.) Only two serious contenders seem to be in the field, inspiration and induction (in the sense of John Stuart Mill). Inspiration in its meaning of intuition or unconscious cerebration, is not a contender at all, for it is just what provides the improvement on the previous model that each execution of the epistemic cycle achieves; its fundamentally creative nature – so that acquiring knowledge is in a very real sense a process of re-*creating* the world – will be discussed later on. Inspiration, when separated from the epistemic cycle, suffers the defect of not being verifiable; and it is very commonly the case that where someone has had an inspiration, someone else has had another and incompatible one. Induction will be discussed in connection with its conceptual basis in probability, and shown not to be a workable method.

We note a final philosophical point, further comment upon which will be postponed until induction is discussed. The active epistemology provides us, as shown above, with a reasonable basis for a realist ontology. To be more precise, it provides us with reasonable grounds for assuming that those segments of nature on which we act – successfully or not – in the epistemic cycle have actual existence, independently of what we think about them; note that they are independent of what we think, but not independent of what we do. More than that, to the extent that the iteration of the epistemic cycle converges to a workable model, we have achieved knowledge – comprehension – of them. Thus Hume's argument does not apply; this convergence is proof of a kind that no *inductive generalisation*, if it were possible, could ever achieve.

(vii)

The discussion has been based on the observation of human beings in the process of acquiring knowledge, on the individual level; use has also been made of computer programs that carry out analogous tasks. Computers are capable – in principle, at least – of behaving in ways that can only be called intelligent. We thus have two entirely different kinds of systems that acquire knowledge; both use the epistemic cycle, and in fact, as our discussion shows, it is possible to observe one of these kinds of system and draw conclusions that illuminate the question of how the other works. This is very useful, because what we know of each can complement what we know of the other: the human process reaches high levels of abstraction in its knowledge but its inner workings are not easily accessible to examination (except through necessarily limited techniques of observation, limited by the need to remain non-invasive); the computer process, on the other hand, is open to very detailed study but has not as yet been developed very far.

But there is a third kind of system that exploits the epistemic cycle, and it is the subject of much of what is to be said in this book: scientific research. Although it is a human endeavour, it can justifiably be considered different from the individual's acquisition of knowledge, for in it the cycle is not often completed by a singe person and certainly never with the speed and almost un-awareness of its working in the individual. It is the social form of the epistemic cycle. A single execution of the cycle in research may take weeks, months, even years; it is commonly though not necessarily carried out by several researchers working together; and, especially, in physics, the building and modification of the model is often carried out by one group, the interference with a segment of nature by another.

We see here something that from the viewpoint of active epistemology is natural and immediate, but within any of the epistemologies of the armchair variety is difficult or impossible to explain: why does scientific research have to do experiments? It may be observed – with some sadness – that this question is often not addressed by the academic philosopher of science; indeed, the whole experimental side – which makes up half and perhaps more than half of physics – is barely touched upon. (There are honourable exceptions, to whom I hope to give adequate recognition in later chapters; but they are few.)

This third epistemic system offers much further illumination for the un-derstanding of the epistemic process. We can examine in some detail such questions as the structure and types of theoretical model, something not eas-ily accessible in the other two systems. It will be part of the method of study developed in the remainder of the book to refer to any of the other two kinds of epistemic system in order to understand the details of what goes on in the third.

This implies that we cannot agree that there is a fundamental and sharp distinction between scientific thinking and other (pre-scientific?) thought. The difference is, rather, one of degree. One is the consciously socialised form, or-ganised to work among groups, of the other; there is a qualitative difference between them, but also a gradual transition, with many intermediate situa-tions. What is more, when we come to examine the details of the scientific form of the epistemic cycle, we will see that they require and presuppose the individual form. The search for a logically formulatable criterion of demarca-tion of scientific theorising from other forms of thought must therefore remain futile; and – in spite of much ingenious effort expended on it – this is just what has resulted. There is a clear distinction between a theory developed by way of the epistemic cycle, so that interaction with the relevant segment of nature has appropriately shaped the theory, and a simple speculation. The speculation is not to be rejected out of hand; it may furnish us with useful hints; it may even be incorporated into an epistemic cycle and so become a theory properly integrated with its experimental part; but it may – and here lies the danger – come to be an obstacle for further epistemic cycles. This occurs when from speculation we elevate – or debase – it into a dogma.

This chapter offers no more than a summary description of what the ac-tive epistemology is and implies. In the following chapter some further details will be found concerning the higher-level cycles, and other comments appear in later chapter. However, the reader should remember that fleshing out the bare bones to be found here will require much further research; and one of the significant positive features of the active epistemology is that it is indeed susceptible of research – experimental research – because the process of knowledge acquisition it describes, containing as it does the vital element of active interference with a segment of the world, can be both observed as an ongoing process and also usefully affected by our actions. This is most evidently the case where its realisation in the computer is concerned; it is also the case where scientific research is concerned, for this preeminently the task of history of science; and a little thought will convince the reader that it is also true of the individual person's case, – without our having to trepan his braincase so as to implant numerous electrodes.

Reference

Aho, A., Hopcroft, J. (1974): *The Design and Analysis of Computer Algorithms.* Addison-Wesley, New York

2. Higher-Level Epistemic Cycles

(i)

In Chap. 1 we saw that the basic epistemic cycle includes in an essential way an action on the segment of the world being studied. This action is predicted, on the basis of the previous version of the epistemic model, to produce a certain effect; the fundamental item providing further knowledge is then the correlation of the effect actually found with that predicted. We also noted that it is by no means certain that iteration of the basic epistemic cycle will converge to a satisfactorily functioning model; it is therefore necessary to be able to build up a further epistemic cycle, whose action is now carried on, not directly on the segment being examined, but on the model in the lower-level cycle. This new cycle can also fail to converge satisfactorily, of course: perhaps it does not converge at all, or it converges to something that later proves not to work as had been hoped. Thus there is need for yet further levels of epistemic cycles. Each cycle in such a hierarchy must evidently involve a model, and when the cycle is activated, the model is appropriately revised, as in the basic cycle, whenever needed.

Such a hierarchy can in principle have an indefinite number of levels. Given the machinery for setting up fully functioning epistemic cycles of the first two levels, recursive-function theory (Rozsa 1981) suggests that any number of further levels can be constructed. If a recursive function is implemented on a computer, the activation of each new level only requires saving whatever is needed on the stack and bringing the relevant data for the new level into the function evaluator (the central processing unit in the computer); when everything necessary has been done, the preceding level of recursion is restored from the stack, the results at the level just concluded are used to modify the operating conditions, and the cycle continues. In actual exectution, the level may go quite deeply into the hierarchy or go up and down several times before returning to the first level; there is not limit to the complexity – or apparent complexity – of the computation that can be achieved.

It would be rash to affirm that the human central nervous system possesses a recursive processing unit able to carry out precisely the same operations that we know from computer theory; but that it has analogous capabilities seems rather well confirmed. In scientific research, it is likewise easy to observe that work shifts rather easily between the direct experimental level, the data-interpretation level, the (scientific) model level, and those of higher-order theories, in a way quite analogous to recursive processing. We shall therefore assume in this chapter that recursion – or an analogous type of processing – characterises the functioning of the higher-level epistemic cycles, as well as their interactions with the basic cycles.

The limit to the depth of recursive hierarchies is not set by any conceptual consideration; indeed, abstractly, infinite hierarchies are conceivable. Is the bugbear of infinite regress real? Fortunately no. There is a computational limit, for each new level requires space on the stack to store away the information to be held over from the previous level, and the stack is finite; there is some evidence that analogous limits of possibly physiological nature exist for the human brain; as for scientific research, no limit is yet in sight.

However, this type of limit is generally not even relevant. A deeper level of the hierarchy is required whenever at a given level no satisfactory convergence is achieved; but if the convergence is reasonable (in a sense we examine later) or has been reached in earlier epistemic processes, no further levels are required. (It is also possible – and happens all too commonly – that either a spurious convergence at some level is accepted or failure to construct an adequate model is registered.) In a computer, excessive depth or even infinite recursion is a sign of a mistake; the solution must be to start all over again, after thoroughly revising one's assumptions and methods. Does the same apply to human thinking? It seems not implausible that both too shallow an approach – an oversimplification, in other words – and too deep a one – looking for hidden depths that are not there – will lead astray.

(ii)

The need for only a finite (and commonly even small) depth to adjust an epistemic model has its roots in another very fundamental aspect: the model need not be more than approximate. Indeed, it is easy to see that it cannot be absolutely exact. We expect it to represent adequately those factors in the segment of nature it applies to which are relevant to the purpose we pursue, and to eliminate from consideration all others. Since this implies a distortion of reality, 'adequate' must here mean to within such an approximation as will still allow us to use the model to plan our actions with success. Any less good approximation will lead to our making undesirable mistakes; and better approximation will be superfluous and may, as we shall see below, prove misleading later on.

The approximate nature of all epistemic models furnished by the active epistemology outlined here is a very fundamental point. To clarify it, we reconsider the case of the map of the London Underground – which is, in fact, a collectively produced epistemic model of certain static aspects of that transport system. Firstly, the greater part of the geographical elements of London are ignored; even of the Underground itself, only certain features are indicated. Secondly, some of the features that are represented appear in a grossly distorted form – for instance, the Underground stations are neither squares nor single or double circles; and the straight lines and curves drawn on the map have very little relation to the actual geometry of the railway lines; the colours on the map are not those of the tunnels, the cars or the rails, and each line has at least two sets of rails (and sometimes more) where only one narrow coloured curve appears on the map. What, however, does closely correspond to reality is the topology of the stations and their interconnections: Chalk Farm is the stop just before Belsize Park on the Edgware branch of the Northern line, and you can change from the Metropolitan line to the Central line at Liverpool Street but not at Shepherds Bush. The map even has certain dynamical properties that the reality it represents does not possess: we can turn the map around in various directions, we can even fold it so that Bethnal Green falls on top of Ealing, while Edgware is next door to Tooting Bec. Yet all these distortions of reality – and this is a very important point to notice – are not simply irrelevant: most of them help us in bringing out those features we most need when we use the map. A more detailed mapping of the actual course run by the Underground trains would not help us to see which lines to take and where to change when we want to get from one place to another; and when would we want to know the exact layout of the stations? It is precisely the probability and ease of folding that makes the map useful to us, even though the Underground itself cannot be folded up and carted away. Insistence on exactness would here mean a map that is as huge and as unwieldy as the Underground transport system – and therefore just as difficult to find one's way around in. Indeed, if the map were absolutely exact, down to the last and most irrelevant little detail, how could we tell it apart from the reality it is supposed to map? The reality includes trains that move and can take us from one side of London to the other in a short time; if the map is exact as regards this, too, would the difference between map and reality even matter?

(iii)

The approximate nature of the epistemic model – which we have just seen to be essential – has a consequence of fundamental importance. The model is initially constructed for one specific segment of the world; but since it is only approximate, it will fit also other segments that share those features which are deemed relevant of the purpose in hand are therefore represented correctly in the model ("Fit", as should be evident from the discussion of Chap. 1, is to be interpreted here in the sense of having a dynamic evolution which is structurally similar to that of the modelled segment, so that predictions made from the model will be verified within the scope as defined below.) The map of London's Underground holds good not only for the weekdays when it was made, but also on Sundays; indeed, the behaviour of the modelled segment in this case is such that the map need be modified only quite exceptionally – perhaps once or twice in every decade. Other maps are much more widely applicable; the ancient Romans, for instance, built their military camps according to a single pattern, so that the same map would fit a great many camps. Higher-level models also apply to several different cases, but now in an indirect fashion: directly they apply to several lower-level epistemic cycles, which in turn each apply to a different set of real-world segments.

The concept of applicability requires some comment. It must be taken in the sense of *providing predictions that correspond to what actually happens within a margin of approximation compatible with the purpose in hand*. This specification implies that there are two ways in which applicability can reach its limit. We may use the model on a different segment of the world which we hoped had sufficient similarity to the original one but in fact does not; or we may use the model on the same segment (to be pedantically but sometimes relevantly precise, on the same segment at another time) but for a sufficiently different purpose for which the model no longer serves. Thus the map of the London Underground is no longer very useful either when we are in New York or when, while still in London, we propose to travel by bus.

Actual use of a map implies the creation of several higher-level models by means of which we can establish the parallels we need between the map (which, without these models, is simply another object) and also eliminate the irrelevancies of the map; we also make use of information derived from different sources, for instance knowledge about how to enter Underground stations and thereafter the trains, knowledge of fares and running times, and so on, which the map does not specify. These models can often work also on different lower-level models. If we know how to use the map of the London Underground, it is much easier to interpret a map of the New York Subway; having learnt how to program one computer by means of its manuals, the manuals written for another machine are much more easily understood. We emphasise the central role played in the concept of acceptability by the purpose or purposes to satisfy which the model was created: this is a feature of the active epistemology which is of great practical significance and to which we will have occasion to return.

(iv)

The range of segments of the real world to which a model applies (in the above sense) will from here on be called its *scope*. The scope is immediate only for the lowest-level models; at higher levels it is mediated by all the intervening levels of modelling. The scope is, as has been discussed, a function of the approximation required of the model. As the approximation improves, the scope decreases. In the limit of absolute exactness, the scope vanishes: no model can work to this degree of "approximation". At the other extreme, complete error tolerance, the model will apply to anything whatever. The scopes for different degrees of approximation are nested, as is obvious from the fact that if a model is satisfactory at one approximation for a given application, it will also be satisfactory for this application at a lower level of approximation.

The scope of a model need not be in any sense a continuous function of the approximation; indeed, in most cases the approximation cannot even be specified numerically, let alone expressed by a continuous variable. Generally, we can talk only about a rather restricted set of different approximations. No detailed study of this set and its structure has yet been made.

The scope, likewise, is not susceptible of numerical expression except in certain particular cases. As we shall see later on, in connection with the concepts of probability and induction, this may be formulated by stating that the scope is not measureable. We can say that the scope of one model is larger than that of another when one scope lies wholly inside the other; we can also say – as indeed often happens – that the scope of one model extends beyond that of another in one direction, while the opposite occurs in another direction. But when the uses covered by the scopes of two models are disjoint, no comparison is possible. These facts have significant consequences which will be discussed in their appropriate place. Here we mention only one: since the epistemic models are approximate and lose their value outside their – finite – scope, a purely logical analysis of their structure has little justification. Contradictions whose consequences make themselves felt outside the scope are quite acceptable, and as we shall see are even useful where they facilitate the interconnection of otherwise incompatible models. Contrary to present practice, we shall therefore have to distinguish between the rational and the logical, the former being characterised by its linkage to purpose and planning where the latter is essentially static; borrowing a terminology from measure theory, we shall say that the rational is almost everywhere locally logical.

(v)

The higher-level epistemic cycles and the associated models are central to at least four types of activities in which mental processes play a decisive role:

i) Their primary usefulness lies, as discussed above, in the readjustment of the basic epistemic cycles whenever these do not satisfactorily converge. We also stress their role in creating new basic cycles or cancelling old ones as need arises.

ii) Higher-level cycles, as noted above, attend to entire groups of basic (or at least lower-level) cycles. This creates the possibility of their *generalising* the models in the lower-level cycles, by combining cycles that have sufficient common elements into cycles of broader scope. Whether any "economy of thought" – along the lines proposed by Mach – motivates the trend to generalisation may be doubted; the genuine advantage of generalisation lies in the fact that the scope of a more general model will often extend beyond the combined scope of the models it supplants. This makes extrapolation possible, and the organism owning the model is enabled to face a range of new and unforeseen circumstances where the original models would have failed.

iii) If the degree of generality achieved in this way extends far enough, another significant advantage appears: for a wide range of activities the set of models (both basic and higher-level) is already adequate. Time is therefore not lost in what under some circumstances could be a lengthy process of acquiring the needed knowledge by activating epistemic cycles at appropriate levels. Note that higher-level models and their attendant generality are indispensable here, to ensure the flexibility required for facing the constantly changing demands of a world that never repeats itself. Note also that the evolutionary advantages of developing sufficient levels of higher epistemic cycles are obvious; the resultant ability to recognise new dangers and flee in fractions of a second would by itself enormously increase survival chances, and there are many other advantages.

iv) A further stage is reached when the set of higher-level cycles enables the organisms to model at least a significant fraction of the world it lives in. In this world the organisms itself plays a major role; the models of it must therefore include models of the organism itself – or at least such parts of it as are accessible to its own actions. Since these parts possess significant interactions and dependencies among themselves, a still higher-level model of the organism as a whole becomes necessary. At this stage of evolution the organism has become *aware* of itself; further levels of epistemic cycles that allow it to model and therefore control the processes of its own awareness now make possible, not simply the psychologically interesting but otherwise not vital process of self-consciousness, but the much more fundamental activity of planning and *foresight*: the time span over which organisms plan ahead might indeed be taken as an index of their intelligence that is meaningful in a way the too much touted IQ is not.

Planning and foresight become possible when two conditions are fulfilled: a) the network of basic epistemic cycles must be sufficiently dense and inter-linked by cycles at not much higher levels so that together they constitute a reasonably adequate representation of significant portions of the world we live in; and b) the cycles at levels above these can therefore to a considerable extent use this representation in order to discover what *would* happen *if* ... We have stated these conditions in a very vague and simpleminded way. Necessarily so. At this point it is not only not possible to be more definite since our understanding of the active epistemology is still very limited, it is not even desirable; a qualitative grasp of the idea at this point is to be preferred. Note also that the transition is not from no planning ability to a full capacity for it, but rather a gradual passage from, say, a worm's extremely limited ability to foresee, through a mammal's slight but easily observable capacity to think ahead, up to what human beings are in principle capable of – though regrettably often do not do.

(vi)

The description of the epistemic cycles, both basic and higher-level, in these two chapters has been schematic, in order to exhibit their structure and the generality of the concept. We have mentioned that there exist three real-isations of the epistemic process; one is the individual's process of acquiring personal knowledge by acting on his immediate world (which includes himself, of course); another is the process simulated on computers by means of artificial-intelligence techniques; and the third is that of scientific investigation. Here we will consider the first and third of these epistemic processes. But in one fundamental respect the schematic presentation must be extended, to become somewhat more realistic.

Human beings do not exist in isolation; they form societies, of varying complexity, size and internal harmony. Even when trying to understand why his roses do not bloom, a gardener, though all by himself, still uses all the resources he has derived from the society he belongs to. The individual epistemic cycle is formed with the constant help, first of his parents, then also of his teachers and childhood friends, and then of many others, both living and dead. These facts hardly need arguing; recent years have brought us endless publications stressing them, though too many of them stress the social factor to the exclusion of the individual's contribution: each of us not only receives a vast heritage from the society he is born into, he modifies that heritage to fit it to his particular needs. This process may be only small-scale; but insofar as his needs have shifted along lines fairly similar to those of others, his modifications may combine with those the others have developed and in their turn contribute to this heritage for the next generation. More than that; to a greater or lesser degree he has the ability to examine his social heritage critically and consciously to work towards its modification, by precept or teaching or merely by asking pertinent questions. Society makes us what we are; but we in turn make society what it is.

(vii)

Of the many mechanisms of social interaction – from the mother's touch onwards – by far the most important in modern societies is language. A detailed discussion of how language fits into the scheme of an active epistemology would exceed the scope of this book, and indeed the needed research has mostly not been done yet. Only a few disjointed comments are possible in the circumstances.

Reference

Rozsa, P. (1981): *Recursive Functions in Computer Theory*. Verlag der Ungarischen Akademie der Wissenschaften, Budapest

3. Systems and Experiments

In the preceding chapter we introduced the concept of a "system", i.e. of that part of the physical world which is represented by the model in an epistemic cycle, but we did not discuss how it is to be chosen nor how we interact with it; this will be done here.

We begin by observing that the concept has both ontological and epistemological aspects. It is ontological in that it implies that the world we live in is structured in such a way that we can hive off (conceptually and physically) a part of it and treat it, at least to a good approximation, as independent. It is epistemological in that this separation is not given to us, once and for all, by the structure of the world; it is something that we ourselves, in our search for knowledge, must carry out again and again. But because of this we have the flexibility of changing our system, expanding or contracting it as needed, using several different and even overlapping systems, dividing a system into subsystems or amalgamating several systems into a grand system. We can thus adapt the system to the problem in hand. Observe that the world would be fundamentally different if the systems were given and we therefore had to adapt our problem to fit them; yet – tacitly – this is the view many philosophers take.

The arbitrariness in this flexibility is, however, only apparent. Given a problem, our choice is largely determined. If the problem is that the TV set does not work, we can certainly exclude the bathtub from the system we propose to investigate. But if some fiddling with the control knobs produces no response whatever, we may either have to contract our system so as to focus on the set's fuses, or else we must expand it to include the wiring and fuse installation in the house, possibly the transformer substation, or even the generating plant. In neither case is the change of system arbitrary; it is forced on us by the type of failure of the earlier system in solving the problem.

The basic *desideratum* in choosing our system is obvious enough: we must include in it all those things, objects, processes, factors, ... that are relevant to the problem to be studied, and exclude – as far as possible – all those that are not.

Relevance must be more closely circumscribed if it is not be vague to the point of uselessness. The starting point for determining relevance is the final aim; in our example, we want the TV to work before a certain programme comes on. But this statement is not adequate: we are outside the system to be considered, and – for reasons to be discussed below – must remain so; what we want are specifications that concern the system. Thus one might discover, say from the manufacturer's manual, that the TV set needs alternating current with a frequency of 50 ± 0.5Hz and a voltage of 220 ± 40 V. An electricity supply within these limits will give a satisfactory reception; outside them the image will be more and more spoilt, until it entirely disappears and the set conceivably is damaged. From the point of view we are presently concerned with, these limits constitute the tolerance limits of the electric current supplied to the set; anything whose effect remains within their bound is to be taken as irrelevant, while anything exceeding them must be taken into account and for this purpose included in the system.

However, the situation is not always as simple as this example suggests. Take the case of someone wishing to catch a train. He must not arrive at the station even as much as a second late; but he could arrive at any earlier time; two days early, or two years, or even before his birth. He will of course use further criteria to eliminate the absurd suggestions (absurd only within a wider context) and to avoid such undesirable consequences as uselessly wasting time; but these criteria linked to considerable extensions of the system originally considered, are not used again once they have served their purpose of setting a suitable tolerance, say five minutes, for his arriving early. We can therefore see them as auxiliary systems; epistemologically they will be dealt with by lateral epistemic cycles; and once they have played their role attention is again fixed on selecting as relevant those factors or aspects that could interfere with our achieving the tolerance thus determined, with arriving not after the train has left, but not too long before either.

In theoretical physics we rarely have this sort of problem of significant further considerations while attempting to fix tolerance limits; it does occur very commonly in experimental work and likewise in applied research. It occurs in a somewhat different guise when, because of preconceived ideas (which may be correct or not, but which are always present, of course, in the very structure of the higher-level epistemic cycles by means of which all of this work is done), we expect to find solutions of a certain type rather than another. But the notion we have introduced of auxiliary systems, used in lateral cycles, will turn out to be important in a number of further aspects. It should be clear that it is not possible to settle relevance problems and so determine the choice of system without explicit use of the corresponding image in the epistemic cycle.

The successive adjustments of the system to be considered parallel those of the model that describe it, and, just like these are carried out by the epistemic cycle. As one process converges, so does the other, and in fact convergence is achieved jointly. This is the optimal case, the one that occurs "ideally". There are evident exceptions, arising whenever there is need to recur to a higher-order cycle; we shall have occasion to examine several specific cases below.

We have so far discussed the systems that lowest-level epistemic cycles interact with, in order to modify their model according to the responses provoked by planned interference with them. But the process of system adjustment occurs also in higher-order cycles, in the form of modifications of the base that were discussed in the preceding chapter. What is relevant here is that modifying the base of a higher-order cycle frequently – though evidently not always – also requires an adjustment of one or more systems at the lowest levels. We shall encounter examples of this below.

(iii)

System adjustment (and the corresponding base adjustments in higher-order cycles) are easily observed in many activities of daily life. To the examples given above one more may be added: the course of a discussion between colleagues, let us say, of some problem. *Don't change the subject!* is a common exhortation in such a context; but closer examination will show that the 'subject' – the matter under examination, i.e. the system in the present terminology – is frequently changed by adducing further considerations that all participants feel to be relevant, and by forgetting other aspects that as the discussion advances are seen to be irrelevant. The exhortation not to change the subject signals disagreement about the proposed system adjustment.

In the field of artificial intelligence, system adjustment is still almost wholly unimplemented. Certain statistical techniques for the discovery and elimination of irrelevant aspects in a situation are known and occasionally applied; but the one technique that has been developed for picking out the relevant ones, factor analysis, is so beset by pitfalls and even conceptual difficulties that it has not, to the author's knowledge, found any use for augmenting the capabilities of the machine-intelligence programs. Moreover, the most important type of adjustment, system extension, is at present quite beyond our possibilities.

It is in the sciences that system adjustment has become a consciously applied part of experimental technique. We therefore discuss it as such below, after exploring two philosophical aspects of the non-constancy of the system during the process of knowledge acquisition.

(iv)

The first of those two aspects concerns precisely the possibility of the various system adjustments we have mentioned. It is a fact of practical experience that such changes of the system we contemplate (or rather: work on, since we do *not* merely contemplate) are always possible; indeed, we can all too easily attempt to construct quite unsuitable systems that do not facilitate the task of knowledge acquisition. But as we have seen, system modification and system adjustment are concrete activities, carried out by us on the real world. We therefore abstract from our experience to obtain an ontological principle; we include in it a further aspect which will, though implicit in what has been said, be discussed in detail below; we formulate what will be called the *principle of partial separability*: the systems we need for the epistemic process may be separated from the rest of the world both conceptually (i. e. by creating an image in an epistemic cycle) and physically (i. e. by taking steps to isolate them from outside influences); but the isolation achieved will in no case be complete or permanent.

This principle, like the concept of epistemic cycle introduced in the preceding chapter, is a distillation from the actual experience of knowledge acquisition. Unlike this concept, however, it is essentially ontological. Its precise meaning will become clearer as we discuss the actual process; before proceeding to do this, two general points concerning it will be noted.

Firstly, its basic importance must be stressed. It is a remarkable fact that, in a world characterised by its multifarious and ubiquitous interactions which generate its extraordinary unity, we can nevertheless hive off for consideration the contents, for instance, of a test tube and forget, in the explanation we forge of what goes on inside it, all the manifold nearby objects, the laboratory, the people in it, and even the earth itself and even the huger sun. But at the same time the limitations of this separability must not be forgotten. To begin with, it cannot be absolute. An absolutely separate system is beyond our interaction with it; we cannot experiment upon it and so come to understand it; indeed, we cannot even become aware of its existence. We could, therefore, happily postulate any number of such totally inaccessible systems – ghosts, alternate worlds, what have you – around us, without this making the slightest difference to anything that could conceivably affect us. Indeed, we might as well forget about them; for us, they can only exist as fantasies.

Much more important are the real limitations on separability. Not every conceivable system is sufficiently separable to be useful. I sit at a four-legged table; if I wish to understand how it comes to remain steady on the floor, I cannot form a system of only the table's left half. Yet for other purposes the left half of the table might yield a perfectly adequate system. But even the system best adapted to our purpose is still only partially isolated from the world: it must admit at the very least all those connections that allow us to interact with it, and in actual fact it admits many more. Below we shall see that it is possible to estimate the degree of isolation and to improve it by appropriate experimental techniques; but it must always remain less than perfect.

(v)

The second philosophical aspect that arises from the non-constancy of the system while we study it and get to know it concerns the sorts of ontological elements that may appear in our philosophical basis. Traditionally, philosophers who accepted the existence of the *external*[1] world always saw this external world as composed of some particular basic constituents. Atoms, in

---

[1] The flexibility of the adjustable-system approach is apparent in the answer it offers to the now famous question "What is the external world external to?" Any system we adopt excludes the mind of the knowledge-seeker, one might be tempted to

classical or modern guise; vortices; matter, characterised as possessing extension and weight; fields, sometimes seen as material; energy; objects; processes;
Wittgenstein, not precisely a materialist, at one time saw the world as made
up of facts; nowadays physics offers us quarks in endless variety, gluons, and
leptons.

No such fundamental constituents are required in the present approach.
Systems certainly are not in this way fundamental, and even less are they
constituents. They are subject to change, both by us as needed in our task of
getting to know them and through their own evolution, whether from internal
or external causes; they are constituted and finally vanish again; they may be
simple, they may contain several of even a large amount of subsystems, we
may want to consider several overlapping systems at the same time, as in the
case of the auxiliary systems introduced above.

From the work carried out on many systems, in random fashion at first,
and then – in scientific research – more systematically, we may conclude that
descriptions in terms of one or another constituent are convenient for a wide
range of situations and may be judged to hold considerable truth. But we
seem very far as yet from being able to state that one or another candidate is
sufficient by itself; at the present stage, physics offers us a bewildering range
of elementary particles, apparently made up of quarks and gluons as far as
the hadrons are concerned, together with four basic fields of forces, partly
unified through highly sophisticated and plausible but by no means very fully
confirmed general theories. Above and beyond constituents of this sort, there
remains the fact that much of what seems significant in our everyday world
arises not so much from these elementary particles and fields but from the
complex patterns that configure larger systems. Are these patterns in some
sense independent constituents of our world? We do not know.

But in the present approach the difficult problems raised here do not
belong in the philosophical stratum. They are matters to be investigated by
one or another natural science. The results from these researches are important
both in themselves and to the philosophical enterprise; they will both enrich
the philosophical picture and shape it through criticism or the need to revise
simplistic assumptions. But the fact that constituent X is suddenly discovered
not to be elementary, after all, will not upset the basic approach outlined here.

(vi)

A final philosophical aspect (into which we go in greater detail below)
derives from the non-constancy of the images successively developed to correspond to the various bases. This non-constancy implies that the set of possible

---

answer. This is so in most cases, of course. But we noted above the need to possess
an image of ourselves as we act on the remainder of our world, and saw that push-
down stacks in computers model this reflexion use quite adequately; the resulting
self-consciousness provides us with a more complete answer to the externality
question: the mind of the knowledge-seeker can simultaneously be internal and
external to the systems he or she works with, by being represented at different
recursive levels. Thus the intuitive answer remains justified even when we are
aware of our thinking, and aware of our awareness.

assertions relevant to one image will not coincide with that for another one; in other words, the universe of discourse cannot remain invariant throughout the process of knowledge acquisition, and this process is then not fully describable in formal logical terms. It can, however, be wholly rational. This is one aspect of a distinction of some importance.

(vii)

We proceed to examine a number of aspects of experimentation in physics that are often forgotten. They are in one way consequences of what has been said in the preceding paragraphs; it is perhaps more correct, however, to say that the principles laid down there are generalisations obtained from them. We begin with the matter of system definition.

System definition, though a common designation, is somewhat misleading as a term, in that it suggests the merely verbal or at best conceptual activity of providing a statement of what constitutes the system. In fact, of course, it is the first step in the actual experiment and therefore covers both the decision on what *should* constitute the system and the various experimental techniques that ensure that the system actually will be of the required description. In order to underline once again the need for active intervention in the system (the part of nature we are studying), we shall follow the physicists' usage and speak of system preparation rather than system definition.

In the physical sciences the experimental system to be dealt with may be prepared by designing and constructing an appropriate set-up, or through the selection and combination in suitable ways of the data as they become available. We describe these methods separately but note at once that in complex experiments they are often combined.

The experimental set-up offers four different ways in which a part of the world may be isolated from the influence of the rest of the world:

(a) Excluding a factor known to have an effect. This can be achieved by removing the system from the neighbourhood of an interfering source, by counter-acting the effect, or by shunting it away from the system. Thus, if the system is affected by the action of a magnetic field, we place it as far from any big magnets or electric machinery as we can; we counteract the earth's field by creating around the system another magnetic field, equal in strength but opposite in sign to the earth's field; or we surround the system with a mu-metal shield that will not let a magnetic field pass inside. We can of course combine such procedures.

(b) Uniformity. If a factor acting on the system does not mediate any exchange of energy, mass or momentum between the system and its surroundings, we can neglect it as regards the mechanical evolution of the system; this can be used with the gravitational interaction, for which none of the three methods of exclusion work. If the system is restricted to horizontal motion only, the gravitational potential remains constant and gravity does no work on the system; this is why we can describe the movement of billiard balls on a horizontal table without taking into account the earth's attraction, even though it is many times stronger than the other forces acting on the balls. Similar considerations apply to other types of interactions.

(c) Control. We can set up active compensatory measures, assuring non-interference by the factor controlled in this way. A typical case is the use of a thermostat to maintain a constant temperature during the changes effected on a heat-sensitive system. Controls of this type involve some kind of feedback, whether automatic or through human intervention; complex problems requiring a special theory arise when several feedback controls interact.

(d) Measurement. We may, lastly, measure the intensity of the interfering factor and, provided that a lateral epistemic cycle offers us the necessary information on how this intensity relates to the effect, correct for it. Doing this amounts to constructing an *ideal system* in which the factor never took effect: the process is that of interpolating an additional epistemic cycle between the physical system and the image being built; the image of the interpolating cycle has the properties of the corrected system and is used as the base instead of the physical system.

Case (d) really lies on the border line between the direct experimental ways of preparing systems and the indirect methods. Of the latter, two are particularly important.

(e) System selection. If alteration through active intervention on the chosen segment of nature is not possible, we can nevertheless often measure a set of relevant quantities and thus choose which among many alternative candidate systems we subject to further study, either more detailed measurement or some interactive procedure followed by appropriate measurement. This is the almost exclusive method of system preparation in sciences such as astronomy, where we cannot (yet?) take a star and change its state to the one we desire its evolution to start from, but where we can and do select from among the endless variety of stars in the galaxy precisely those that meet our criteria. It is worth noting that, because system selection is also a form of system preparation, the usual emphasis on the essential difference of astronomy as an observational but not experimental science is somewhat misplaced. Not only are various kinds of system selection widely used in other sciences that are recognised as suitably experimental (thus in the physics of nuclei and experimental particles, accelerating target beams of fast particles onto others, and coincidence set-ups select from among the outgoing reaction products those that meet some specified criteria) but, more than that, astronomy shares a common experimental basis with physics, since all its theoretical constructions start, quite explicitly, from one or another physical theory, which is then specialised by adding a great deal of detailed astronomical information and tailored to the particular circumstances by eliminating irrelevant aspects and introducing appropriate approximations.

(f) Aspect elimination. If system selection removes from consideration a subset of the systems available to us, aspect elimination removes from consideration those aspects we do not wish to deal with in the image that is being built. Thus in the problem of the billiard balls we ignore their colour, but a human being's colour may be very relevant in certain social problems. The aspects that are eliminated in this way go all the way from those whose irrelevance is trivially obvious, as in the case of the billiard balls' colour, to those whose elimination eventually turns out to be a mistake. In physics, temperature is sometimes of this latter kind. All of them, however, have one thing in common: the elimination is *not* done by physical intervention that alters the system (as in the first three methods of system preparation) but by operation on our image; the most adequate representation of this process is probably by seeing it as the introduction of an additional epistemic cycle that filters out the undesirable aspects and simply passes on the rest; but other mechanisms are feasible.

Aspect elimination is, in a sense, the most important of the six procedures, for it must precede any attempt at all to formulate a tentative image, since otherwise such an attempt would be swamped by a vast multiplicity of irrelevancies. It is one of the most remarkable achievements of the human brain that it can eliminate these irrelevances and concentrate on what are judged to be the significant aspects of even a very complex situation in an astonishingly short time, often only fractions of a second. At the same time it must be remembered that, like all activities involved in the cognition process, it is liable to error and must be subject to correction.

(viii)

In preparing a system for a specific experiment it is not enough, of course, to decide on a suitable strategy for defining and isolating it; the adequacy of this strategy must be submitted to tests. Such tests are, of necessity, experiments their own right, and in their turn require the development and testing of a defining strategy. An infinite regress is nevertheless not called for, because these tests may be broken up into partial tests that critically examine different aspects of the strategy and, being partial, are on the one hand less elaborate and on the other hand require much less theory; hence one arrives after very few steps at well-established results or at experiments that are trivial to do.

Three matters need taking care of: firstly, all the elements composing the defining strategy must be submitted to sufficiently exhaustive tests; secondly, it can happen that two different steps proposed in the strategy affect each other or even cancel each other out, and the testing procedure should reveal this; and thirdly, tests are needed to show that the strategy is complete, that no significant factors have been left out.

(ix)

The third kind of verification involves a set of experiments that also have a further purpose: checking the instrumentation. This is by no means a simple matter. In the first place, for each instrument we must verify (a) that it measures what it is supposed to measure (according to the image developed while defining the system) and not something else; no instrument is perfect in this respect, and we need detailed information concerning the influence of other factors on the instrument; (b) that it is correctly calibrated, i.e. that a reading of 3.66 volts means that a tension of precisely this amount exists between the terminals, plus or minus an admissible error which at this stage must be checked to be within the bounds stipulated; (c) that the instrument is sufficiently stable in its operation.

Once the instruments have been gone over one by one, the experimental set-up is put together; the first measurements will almost always be of two kinds; either so-called null experiments or measurements of already known results. If we can correctly reproduce results found by earlier workers, this verifies that everything is working as expected and in particular that the strategy developed for isolating the system is adequate.

It is very commonly the case that, though the individual instruments have been calibrated, the complete set-up cannot be: what we are attempting to measure is expressed as a combination of the instrument readings in which there remains a parameter (or even several) that we have not been able to determine either theoretically or experimentally. In such cases carrying out the experiment on known cases acquires even greater importance: it is in this way that the missing parameters are found.

A null experiment is rather different. In many situations we wish to generate and observe some effect against a certain background; typically, the line spectra of atomic physics appear as narrow high-intensity regions above a background of much lower but smoothly varying intensity; the null experiment now determines this background, and the intensity finally quoted will be the difference between the total intensity and that of the background.

Both calibrations and null experiments must generally be repeated; if possible, they should be interspersed with the "real" experiments, so that we may discover and correct any variations in the background and the instrumental calibrations – in the origins and scale factors of the quantities we measure, in other words. Much ingenuity and hard work is expended by experimentalists in realising these conditions in an effective way, and many strikingly original experimental designs owe their origin to this need. To a physicist the Michelson-Morley experiment is remarkable at least as much for the inventive and insightful way in which these problems were tackled as for the unexpected results it yielded. And the essential point here is that these two researchers created for the purpose an entirely new technique, which permitted so effective a cancellation of background variations as to increase the precision achieved in the measurements by a very large factor.

(x)

System preparation (in the ample sense we have so far employed) and measurement constitute, between them, the elements of any experiment; or rather, of that experimental element of which, as we have seen, several must be combined to form a complete experiment, some to yield calibrations, isolation checks, or (as well as experiments) basis values, others that yield the raw data that together with the previous ones form the physically meaningful results we are after. We consider further combinations of experimental elements below; here we analyse the individual elements in a different way.

System preparation may be split up into a more passive part, intended to keep experimental conditions constant, unaffected by "outside" influences, and a more active part, which brings the system we work on into the desired state, or even creates it in that state. Similarly measurement may be split into an interaction part, in which the situation to be measured is brought about, and the actual measurement procedure itself. Calling these four components of the experiment C, P, I and M, we obtain the scheme

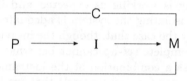

The division we have made is to some extent conventional and directed more by the combinations into which experimental elements enter than by general considerations. Thus the interactive I can just as easily be considered a part of the system preparation, and many experiments may conveniently be studied as if there were no I. P and M cannot, of course, be absent; but the separation between P and C, or between C and M is not sharp. In particular, the system preparation may fall into several parallel actions the results of which may come together in I; this is why we add an index m which numbers these individual actions. As a case in point, consider the typical experiment in which nuclear or subnuclear particles are accelerated to a chosen energy (preparing one part of the system) and shot at target particles (at rest, or prepared in some other appropriate state): only the two together constitute the experimental system. Similarly, measurements may be made simultaneously, a fact we indicate with the index n.

The scheme is open at both ends; the systems brought into the desired state by the procedure P must come from somewhere and cannot be created ex nihilo; after their final state has been determined by M they go somewhere, for even if they disappear as systems they leave their constituents. But it should be clear that P and M may have profound effects on them, often altering them beyond recognition. This is nevertheless of little concern to the experimenter; the state of the system before the action of P, or after that of M, yields no useful information.[2]

We need not belabour the point any further. In no actual experiment does the taking of an observation by itself suffice to produce a meaningful result; it must be accompanied by a large number of ancillary operations and experiments among which the suitably thought-out and properly executed preparation of the system one "observes" is the most significant. And it is this context that turns a unitless and incomprehensible number into a quantity with physical significance.

---

[2] One kind of experiment is to be excluded from this conclusion; those in which we follow up the time evolution of the state. Here each measurement is also the preparation of the next step, and for the whole process to be meaningful the measurements must not sensibly affect the state of the system.

That this is so is glaringly obvious to anyone who has participated in laboratory work; but it is fatal, nevertheless, to the still widely accepted views of philosophy of science where the basic element is the individual observation. Out of this, according to positivist thinking, the world – or at least what we take to such – has to be reconstructed. In fact, as we have seen, the basic element is a complex made up of a previously planned action on the world, together with the observation of the result. And if there must be a basic element, which may, I think, be doubted, it is the correlation of these two things.

We must extend the argument in order to understand why experimenters so often repeat their work. Two factors determine this, and give rise to two sorts of repetitions.

Firstly, as we have attempted to make clear earlier, the experimental system is never one hundred percent hived off from the rest of the universe, nor are the various arrangements with which we surround it, control it, prepare it, interact with it, and measure it. So if we repeated precisely the same sequence of operations, the result would be slightly different; we observe here what is called experimental error. *Error* not in the sense of a mistake, as when an air disaster is attributed to pilot error, but in the original Latin sense of random wandering, as in knight errant. Now such errors must be determined with some precision; in fact, experimental results without error indications are rather useless. But if we are informed that a certain length is 0.25 ± 0.01 mm, then we can be confident that it will lie between 0.24 or 0.26 mm (or perhaps a little bit outside these limits), and so we know in what circumstances we can use this result and when we must have a more precise measurement. Moreover, the errors themselves often prove a fertile source of knowledge (as they would not if they were mistakes). To observe them – since we have no direct access to the error-free system – we repeat exactly the same procedure a large number of times, perhaps 200 or 300 times, and do statistical studies in the collection of results: the mean estimates what we meant to measure rather well, and the deviations from it give us a good sample of errors.

Secondly, our system definition is only approximate. We have left out what we judged to be irrelevant, we have simplified and often distorted what we have kept in our images. Have we gone wrong there? Have we omitted something we should have included, or the other way around? We will not know until someone else does a similar experiment, but thought out independently, carried out with a different apparatus, or – for the best results – on a different basis. Such replications, unlike the sample repetitions that furnish us with information on statistical errors, will reveal systematic errors; and these could indeed be very much more like mistakes.

Thus to do all over again, but in a different way, the experiment that Mr. X already carried out is not because of our mutal distrust. On the contrary, Mr. X is likely to be very helpful about it, for this activity permits something that is essential to the scientific process: the critical examination of results, and because the results here were experimental, so must the critical examination be.

(xi)

We may sum up the situation by saying that actual experiments are more complex (in the sense of involving several essential concepts, not merely in that of being complicated) that the somewhat simplistic accounts commonly admitted by philosophers (with a few honourable exceptions, like Rom Harré) their role in the process of acquiring knowledge is active and central and not merely one of confirming (or otherwise) armchair theorisations; they are not carried out in isolation but form a part of sequences where the appropriate combination of theoretical notions and experimental work give both their meaning and their value – and incidentally, as we shall discuss in the following chapter, delimit the theory's scope. However, two points remain to be commmented upon.

The scientific experiment, as we have stressed, is a central element in any science, and in some more so than in physics. Nevertheless, experimentation is, and rightly so, regarded as an art. Firstly, it requires the acquisition of considerable skill which is difficult to teach in the classroom and which is best learnt by working with the outstanding craftsman in the art. Secondly, and more importantly, it has a creative component as significant as that in theoretical physics; but the latter has had many distinguished exponents, from Maxwell in the last century to Einstein and Dirac in this, while the creative nature of experimental work is not often acknowledged. Yet to think up a nice, clean way of showing the atom's internal structure – as Rutherford did – or of pushing the sensitivity of measuring relative speeds of light up by several orders of magnitude – Michelson's achievement – was no mean feat and required what in apparent contradiction one might call a well-disciplined imagination. A rich and fertile imagination, allowing them to open up the fields of thought that were wholly new; yet disciplined, so that their imaginings should remain within the bounds of the possible, even where they transcended what was merely known to be possible. A good analogy, perhaps, is a river. A small insignificant rivulet will never yield much; a large, strong one, undisciplined and unconfined, will flood a whole plain and merely wreak havoc; but a good strong flow confined to a properly shaped river bed, will drive dynamos and irrigate fields and feed numberless towns. But we can carry the "discipline" too far: we can dam the river; then it becomes dangerous, and if it ever breaks through the dam ...

(xii)

The other matter to be raised here concerns the distinction between theory and experiment. At first sight this might seem rather simple: what we do with our hands is experiment, what we do with our brains is theory. But on thinking it over the superficial simplicity evaporates. To begin with, as we have just stressed, the experimenter uses his brain; very much so; but in ways somewhat different from the theoretician's. Secondly, we have seeen that the raw data as given by the physicist's measuring instruments are in themselves insignificant. They have to receive their dimensions, be multiplied by their scaling factors to be expressed in suitable units, then they are corrected, collated, compared and connected with other data, and usually quite extensive calculations are carried out on them before the experimenter is satisfied that conclusions that are not merely misleading can be drawn from them. This process requires considerable use of various physical theories. Something similar, of course, goes on before the experiment is carried out, in order to design it and ensure its success. These processes are an essential part of the experimental method; but they are theoretical.

On the other side, the theoretician works quite happily with pencil and paper, or on the blackboard while he can; but often enough this process comes up against insoluble problems. Equations turn up that will not yield to our most sophisticated mathematical tools; still more commonly, we cannot even formulate any equations. Possibly we lack the concepts; sometimes, any set of equations we think up turns out to be inconsistent; and frequently, the data we have from experiment are too few or too many (this may happen) or too inconsistent to form a suggestive picture; or perhaps we have just not yet seen how to do it. In such cases nowadays we tend to have recourse to computer modeling – that is to say, we make numerical "experiments" which themselves are simulated in electronic "experiments". To talk of experiments is perhaps not entirely justified here, but the procedure certainly reproduces many of the features of the real experiment, by covering only a selection of cases, by requiring all sorts of special precautions rather analogous to those for isolating an experimental system, even by introducing error fluctuations of a rather uncontrollable sort; and certainly we are not doing straightforward theory any longer.

Thus the question where to draw the line is not trivial. Rather surprisingly, few philosphers of science seem to have realised so far that this is indeed a question worthy of their attention; and even fewer have been aware that the line is not fixed once and for all, but moves with the circumstances. In fact, in a very important sense, "experiment" is only the one experiment the physicist is at the moment engaged upon (whether in carrying it out or in interpreting its data); all else – including the results of other experiments – is theory. We can, of course, change our point of view; we may consider, simultaneously a group of experiments; but in order to do so, we must (at least in principle, and always in practice, if we are to do calculations) have access to the corresponding raw data, before any statistical treatment. Alternatively, though less satisfactory, the data may all have received identical treatment; in actual practive this may be very difficult to achieve. But – and here lies the essential point – any choice we make must be adhered to: it defines for the statistical techniques just what is "experiment" and what is "theory" and must not be altered before the calculations are concluded, else we generate a muddle. Once the statistical and numerical treatment of the data is complete, we can proceed to another experiment; and our theory-experiment separation will change correspondingly.

Note that though the line we thus draw between theory and experiment is quite different from the traditional rigid one in being relative and hence moveable, it is as little arbitrary as say, the relative speed of an object. This, likewise, depends on the reference object relative to which we determine the speed, and changes when we change the reference object; but given the reference object, it is perfectly well defined and objective, independent of the quirks of any observer. As likewise is the line between theory and experiment.

The relative character of the theory-experiment distinction has many philosophical consequences. Perhaps the most obvious is to render hopeless the many attempts at formalising physical theories: while their boundaries are thus made uncertain, such efforts are not likely to be successful.

(xiii)

The complexities of modern physical experimentation have, of course, a considerable history behind them that has issued in intricate and highly significant organisational structures.

Early experiments were done by isolated scientists, with perhaps an assistant. Eratosthenes had such an assistant when he determined the size of the earth, with quite astonishing precision, considering the simplicity of his instrumentation. Throughout the Middle Ages, and well into modern times, this was how science was done: one man developed theoretical structures, did the experiments, made the deductions and published the results – commonly not in organised form, as a book, say, but in letters to colleagues. The letters, were, of course, written in Latin, the common language of all scholarship. Communication was not yet felt to be so vital a need; even in Newton's days Robert Hooke published his famous law of force for springs [3] in the form of an anagram so that no one should steal his secret. Even so characteristic a thing as the lab coat is recent, going back only to the end of last century.

The great teaching and research laboratories developed around the middle of last century. The first was Justus Liebig's great chemical laboratory in the University of Essen, where he established a tradition of research in agricultural and general chemistry that had far-reaching consequence; but other institutions, first in Germany, then in England and elsewhere followed suit.

In our century, partly under the impulse of two world wars, partly because of internal changes in physics, partly on account of great advances in the relevant technologies and partly under the pressure of the steadily increasing involvement first of the huge industrial corporations and then, when these were no longer willing to bear the expense or the long-range risk, of governments, physics moved into immense laboratories organised by several universities, or by a government department, or nowadays even by several governments jointly. Here groups of as many as a hundred physicists, helped by dozens of technicians, and hundreds of computer people, engineers, vacuum specialists, machinists, and able organisers work on experiments that have needed two to five years' preparation, take three or four months of intensive work to carry out and then a year or so to get the final results.

Such experiments, because of the time scale, the cost and the many people involved, are very carefully planned. That this has proved feasible, while real

---

[3] The force is proportional to the spring's extension beyond its equilibrium length; in Latin: *ut tenso sic vis*.

physics is being done, and with surprise changes of plan occasionally occurring, seems to disprove a thesis that was current some years ago that planning was incompatible with scientific research. But by no means all experimentation in physics is of this vast, almost industrialised kind; there is a great deal of extremely important work being done in small groups of two or three people, in sometimes inadequately equipped laboratories. Yet the results appear, and show that neither equipment nor people by themselves make good science; it is the combination of the two that is needed, together with a certain minimal tranquility and stability and some basic services which gives them a chance to work.

4. The Structure of Theories

(i)

In the preceding chapter we examined the determination and isolation of physical systems as an essential part of understanding them well enough to do experiments on them. In this process, we saw that it was indispensable to build and constantly adjust an appropriate model for the physical system – i.e. a dynamical image of it, in the language we developed in Chap. 2. But as we observed in that chapter, such an image (or rather, set of images; for we adjust and develop it systematically, yet return to earlier versions when it suits us) presupposes higher-order epistemic cycles. And the images of such higher-order cycles, when they are generated by the social institutions we call the scientific community, are what are termed scientific theories.

This is not, of course, a definition; it is, rather, a description of their significance: they allow us to understand the structure and behaviour of selected segments of nature sufficiently well for us to build specific models by means of which we can control them, make use of them, adapt them to our purposes, or at least to predict their evolution. In fact, theory construction is in itself a constantly evolving activity, and attempting to define it is about as useful as attempting to define anything that lives and therefore changes. We shall here make some attempt to describe the rich fauna of theories in physics as it exists in the second half of the twentieth century and it will become clear that no simple definition could possibly cover all the variety unless it were almost wholly vacuous; but this is only physics, and that only at this present moment, while its past has been very different and even now the seeds of great future changes can be descried.

We discuss several aspects in which physical theories differ from one another, and also others through which they interact, in this chapter; we precede these matters by a discussion of some more general points.

(ii)

Physical theories are generated by the well-disciplined and hence creative imagination. But they are not fictions; it is their purpose to illuminate reality. The impulse to build a new theory therefore starts either from unexplained (or inadequately explained) experimental data, or from dissatisfaction with existing theoretical structures. There was no theory at all at the turn of the century that would explain the shape of the black-body spectrum as observed in the laboratory, and it was this gap that set Planck working to create his radiation law. Five years later, it was dissatisfaction with certain discrepancies in the transformation symmetries of electromagnetic theory, as then formulated, and not any experimental problem, that led Einstein on to the path of special relativity. Sometimes both factors are at work, as with Copernicus, for ptolemaic theory not only resulted in an ugly and awkward heaping of epicycle on epicycle but gave predictions that no longer agreed well with the slowly improving measurements of the astronomers.

But whatever may be the origin of the theory, it must apply to reality if it is to be part of physics; as we saw in Chap. 2, this means that it must possess an experimentally determinable scope. Hence it must be possible to derive models from the theory – possibly together with other theories – and these models then serve as the basis for experimental work in different areas of possible application. If the models fit the results to the required precision (and supposing that all the other theories and experimental techniques have already been shown to be within their respective scopes and that the approximations made are justified) then we have established at least part of the scope of the theory at the stipulated precision. But if this is to be at all possible, the theory must possess a minimal degree of generality. It sometimes happens that only one model is derived from it; the usual practice is then to treat theory and model as a unified whole and apply the term "model" to it as well[1], for such cases the model often may be completed with a fairly wide range of different experimental values, yet the physicist refuses it the title of theory because it is not general enough where generality is called for. A very typical case in point are the various "models" – the shell model, the optical model, the cluster model, and others – used in nuclear theory. Each is in a sense a full-blown theory: it possesses a complete armoury of concepts, a well-developed mathematical formalism, and a wide range of successful applications; in some cases the model's creators have even received Nobel prizes for their work. Yet they remain models: for the shell model solves problems the optical model cannot tackle, and the optical model explains many effects where the cluster model or the shell model are lost. And as a result, the nuclear physicist is often obliged to find an explanation of some peculiarity through a hodge-podge of two or more of these models – even though their basic assumptions quite thoroughly contradict each other.

A theory, then, is expected to have a considerable degree of generality. We shall see below that this fact has several philosophically significant consequences.

---

[1] Thus even within theoretical physics the term "model" is ambigous: it refers to the model in the strict sense, i.e. a conceptual structure extracted from one or more theories and completed by means of specific data concerning system structures, initial values, approximations to be used, and so on, so as to be an adequate description of one actual situation for a particular purpose; it also refers to the theory which only possesses one model.

(iii)

Many studies in the philosophy of science consider individual theories in isolation, as if they constituted the whole of science by themselves. The interrelations of the various theories in physics, for instance, are, however, sufficiently rich and complex to make such a point of view untenable. We here note briefly their chief forms and comment in more detail when discussing the relevant aspect of theory structure.

(a) Theories are linked, most fundamentally, because they describe different aspects of the same universe. But because they do so from different directions and with different aims, this underlying unity is rarely obvious; it is generally revealed only when a more fundamental theory has been developed, within the frame of which it is now possible to understand the very different phenomena described by the earlier theories. The lodestone's ability to find the north, the mutual attraction and repulsion of various bodies when rubbed with fur, and the old phenomena of so-called animal electricity formed the provinces of three distinct physical theories; these were gradually – not without a struggle – united under the umbrella of the electromagnetic theory developed by James Clerk Maxwell and experimentally pushed ahead by Michael Faraday.

(b) By far the most frequent connection between theories is their joint use in building specific models. Thus the nuclear models, mentioned in the preceding paragraph, generally combined quantum mechanics with certain relativistic elements. For molecular models, where we have to account for the properties of systems containing several nuclei and many electrons, we commonly treat the nuclei – which are much heavier – as classical Newtonian particles, while the electrons are considered as quantum mechanical and relativistic; we thus have three different theories combining to create this type of model. It is this sort of model building that is perhaps the physicist's most difficult but most characteristic and most satisfying task. The theories he or she combines are as a rule logically inconsistent; yet they can to a considerable extent round out each other's weaknesses, and so a judicious combination will furnish us a rationally sound if logically weak[2] physical model.

(c) In the process of model building another facet of theory interrelation is found to be significant: the existence of many physical concepts used by several theories. Typical of these are space and time, plus their derived concepts like velocity and acceleration, and mass. However, the roles played by such a concept generally differ in different theories; and the concept will, as a result, change somewhat. Thus in special relativity, space and time establish certain links (they do not simply fuse, as certain popularisations maintain, since causal relationships are relativistic invariants) they do not possess in Newtonian mechanics, and in general relativity they acquire new geometrical structures; likewise, mass is no longer constant in these theories. In quantum

---

[2] Concerning the distinction implied here between the rational and the logical, see below.

mechanics, angular momentum and energy generally satisfy certain quantification conditions, instead of being continuously variable.

Quite a different problem arises when concepts like temperature appear in nuclear physics, for instance. For here we have a case of a deliberate generalisation of a concept on the basis of an analogy which, as often, is useful and suggestive, but not complete.

(d) Physical theories tend to share certain mathematical techniques. This does not refer to the common background of standard mathematical principles, such as the differential and integral calculus, but to those procedures that are not quite standard, such as approximation methods and the like. Some difficult problems arise here, and the matter is discussed below, in connection with the formal elements in physical theories.

(e) Some theories are derived from others, by specialisation through the introduction of additional assumptions, or by combination with other theories. Only a very few theories in physics are wholly independent; perhaps only the three 'mechanical' theories, Newtonian mechanics and the two relativity theories, can be said to be independent. Quantum mechanics depends strongly on the underlying *mechanical* theory; electromagnetism depends weakly. Geometrical optics, that is, ray optics, can be seen as an independent theory; but it was found, well after its creation, that it could be derived from so-called physical or wave optics, which in its turn is a derivative of electromagnetic theory. As a result of such connections, whole hierarchies have sprung up. To give but one example: one particular application of quantum mechanics is the theory of the solid state; a specialisation of this again is the theory of metals, in which the theory of electric conduction is of particular importance. This theory, combined with statistical mechanics and some remarkable new concepts, gave birth to the theory of superconductivity. Now superconductivity – the fact that below a rather low critical temperature certain metals offer absolutely no resistance whatever to the passage of an electric current – is itself a fascinating phenomenon; but the theory developed to account for it, however derived (*super-derived*) and dependent it may be, is extraordinarily interesting and could well be the first of next century's physical theories.

(f) Finally, physical theories share the intellectual background which brought them forth. This is a much discussed and highly confusing matter. Before entering upon it, it seems important to discriminate between this background of common ideas among the origination of theories and the social influences at work during the process of theory formation. The social influences can, of course, act through such background ideas; but they also act in many other ways. Thus the extent and directions of the previously listed forms of theory interaction (except, evidently, the first) depend very largely on the social conditions under which they can manifest themselves; consider merely the influence of the various means of communications of scientists and of their steadily increasing degree of specialisation. Yet, when all these influences have been taken into account, it is still possible to perceive the effect of the intellectual climate within which scientific research is carried out. This is particularly noticeable because over the last twenty years this climate has changed. Where earlier generations still largely held to the idea of science as the most vital element in twentieth-century culture and therefore of our understanding of nature as its central aim, the twin technological applications in money-spinning and in making bigger and better bombs have take over; fundamental research in physics has diminished both in importance and in quality, to the point where the American Physical Society no longer lists it in its subject classification. Here then we have a very obvious common factor that acts on all theoretical work in physics at the present time; how many other, more subtly acting yet no less important factors are there that we simply are not yet aware of?

(iv)

Another question has generated an enormous amount of philosophical literature but on closer examination turns out to be largely a pseudo-problem. This is the validation of scientific theories in general, and of physical ones in particular. The difficulty that philosophy of science has attempted to resolve arises from two somewhat bizarre misreadings of scientific practice. The first concerns the sort of evidence provided by the experimenter; rather arbitrarily, this is reduced to a simple yes/no answer to the question whether the theoretical interpretation fits the experimental results. In reality, no theoretician would be content with such a meagre pickings, nor would an experimentalist even think of offering it: genuine data come with extensive details about the circumstances of their measurement, of the values of the relevant parameters, and above all of estimates of the experimental precision attained. It is this much more complete information that the theoretician exploits in evaluating the performance of the theory. The second misreading of scientific practice is found in the idea that the purpose of experimenting is to provide the data on the basis of which some theories are established as true, or tentatively true, while others are eliminated as false or rejected as outmoded. Examination of actual practice shows that this is not the case. To take but one example, classical Newtonian mechanics is generally considered to have been "disproved" by Einstein's relativity theory on the one hand, by quantum mechanics on the other. Yet this outmoded, disproved, and falsified theory continues to be taught in university courses of physics, indeed it is considered almost basic there. Is this only a piece of outrageous conservatism on the part of the scientific establishment?

Not exactly. The fact is that such experimental work as is concerned with evaluating theories has a different function than the one ascribed to it in the usual accounts; it serves to establish the scope of the theory. It is thus important to know just where the theory works well and for what values of the various parameters involved it provides results within a given precision; and to establish such scopes for different precisions we must clearly obtain experimental data both inside and outside the scope so as to know where to trace the borderline. In other words, the theory must be both verified and falsified before its scope is fully known. With theories of extensive scope this may not be achieved until a better theory arrives on the scene which gives us some indication where to search for the limitations of the earlier theory – and above all, how to do so. But this means that it is only when we have a new theory that we can fully appreciate the old theory; and not at all that then we throw it out. Just that is what happened to classical mechanics. We now know just where the limits to its validity lie; we also know that there are certain regions where neither relativistic nor quantum theories but only classical mechanis will work; we are also aware that quantum mechanics, though it replaces and improves upon classical mechanics in mechanics, at the same time requires it as a basis in a highly intricate and by no means fully understood relationship. And finally, as an ironic last shot for those who had declared its definite demise, classical mechanics has in recent years flowered in two new fields of tremendous excitement and great promise for the future, non-linear mechanics and chaotic mechanics.

Thus the real situation in a developed science such as physics differs very much from the sort of free-for-all battle between adherents of alternative theories commonly depicted by writers on the philosophy of science. Theories (or, perhaps better, incipient theories; what used to be called hypothesis) with no or very minute scope are indeed rejected mostly by those who proposed them; theories of more extensive scope maintain their validity and feritility even when other theories that in certain directions can better them have been developed. This does not at all mean that the sciences – physics included – have not had their revolutions; but they have not been of the simplistic kind alone envisaged by Thomas Kuhn, and instead of throwing out as worthless junk all past work they have generally resulted in freeing it of various hampering and in the end anti-scientific fetters.

(v)

The factor in the situation which Kuhn has forgotten is that even between his paradigmatic revolutions physical (and, in general, scientific) theories evolve and change. Indeed, every author of a research paper, every textbook writer, restates the theory in the way that seems best to fit his purpose; sometimes he is conscious of introducing significant novelty, on other occasions he merely represents one more step in a combining trend. The resulting forms of the theory differ sometimes so fundamentally from their beginnings as to be almost unrecognizable; yet there has been no decisive break, and certain basic elements have remained the same; but their place in the theory has generally shifted greatly. Nor are the older forms of the theory invalidated: they are reformulated, the models that gave rise to them continue to be used but are now derived in newer, sometimes much more perspicuous ways. New theoretical techniques are found that avoid or at least improve upon some approximations. New concepts are introduced. The formal parts of the theory are recast in new moulds, unified or split up in new ways. Other theoretical structures, earlier considered to be independent, are incorporated in it; conversely, certain parts may be hived off and thought to have been wrongly included in the theory's framework. Correspondingly, the scope of the theory sometimes broadens, sometimes narrows in the course of its history.

Most of these changes can be observed in the long, complex and as yet only very inadequately studied history of classical mechanics, of which we can here give only the barest sketch. Its prehistory, so to say, goes back to Aristotle, of course; and while we must certainly reject his views on motion as a general theory of mechanics, it must not be forgotten that they remain perfectly valid for the motion of bodies under conditions of high friction, e.g. in viscous liquids. The Middle Ages raised formidable ideological barriers against direct experimental investigation, which only a few brave spirits broke through. The first major steps beyond the aristotelian picture were made by Galileo; it was he who carried out certain significant experiments, it was he who used them to conclude that a body remains at rest or in uniform motion when no forces act on it – so that (to put it in our language) friction, though almost ubiquitous, is a force. It was he, likewise, who first satisfactorily defined velocity and acceleration in their present meanings. These elements were built up into a complete theory – which now included gravitational forces – and provided with a formalism based on the entirely new mathematical theory of fluxions by Isaac Newton. His starting points were the three fundamental laws of mechanics[3], and the law of gravitational attraction.

Newton's theory was physics at its best; it was bold and imaginative theory; it explained an enormous range of phenomena, so much as to seem almost miraculous for a time; and it possessed quite a number of flaws. The theory of fluxions, though intuitively plausible, is contradictory and Berkeley's objections to it are prefectly valid; only Berkeley could not see beyond the warts and failed to envisage what Newton's inspired intuition had foreseen; and it is true that it took two centuries of hard work to bring that vision into being, in the hands of Cauchy. Many of Newton's deductions were no more than plausibility arguments and had to be rethought; much of his experimental basis was inadequate or even wrong. The place of gravitation in the theoretical structure was unclear, and by the middle of the nineteenth century it had become clear that it was better to treat it as a separate theory. At least one fundamental concept (the undulatory) was wholly lacking in Newton's formulation of the theory of light and was formally introduced by Thomas Young at the beginning of the nineteenth century. For a time it was also thought that optics (or at least what we now know as geometric optics) could be combined with mechanics into one single theoretical edifice.

But above all, it came to be doubted whether the famous three laws that Newton had started from were really able to bear the weight of the entire theoretical construction. Part of the problem was conceptual; Newton had correctly foreshadowed one present understanding of the role of force, but only intuitively, and so had not expressed himself very clearly on the subject;

---

[3] The first, given above, was known to Galileo; the second is that the force acting on a body is given by its mass times the resulting acceleration, and in a somewhat different form, was also known to Galileo; the third states that action and reaction are equal and opposite, i.e. that if body A acts on body B with a certain force, the body B will react on body A with an equal force in the opposite direction. All three laws have often been misinterpreted and will therefore be discussed below.

while his followers only increased the confusion by also muddling the very clear meaning of the other two laws. Part of the problem arose from certain advances in the formalism; by restricting themselves to forces that could be described as derivatives of potential-energy functions (so-called conservative forces, because in closed systems held together by such forces the total energy is conserved), the great mathematical investigations, first of Lagrange and Laplace at the end of the eighteenth and then of Hamilton, Jacobi and Sophus Lie round the middle of the nineteenth century achieved entirely new and exciting structures, much more general and abstract in their approach than Newton had been, but precisely because of that also much more powerful. It is only by means of these highly developed techniques that Poincaré and later physicists could at last shed some light on what had until then remained a mysterious and even upsetting question: why is the solar system so stable? Does God have to intervene every now and then to repair its minor errors and deviations? Or is it self-correcting, so to say? It turns out that it is stable only to small perturbations, so that the time to its eventual running-down had simply been enormously underestimated.

The mathematical investigations alluded to actually evolved in several not always harmonious steps. In the first of these, Lagrange and Laplace developed rather general sets of differential equations; later, when the energy concept had become familiar and its conservation laws understood and accepted, Hamilton reformulated mechanics on the basis of a variational principle: if a particle moves from one point to another, its path is such that the integral of a certain quantity along that path is less than along any neighbouring path. At the end of the century, Heinrich Hertz was to attack Hamilton's principle because of its teleological overtones; Jacobi, on the other hand, found a way of transforming the original mechanical problem into a new one, expressed in a different set of coordinates, which is so simple that it can be solved by inspection; and then you just transform back again ... Later work, by Sophus Lie, among others, reconciled these approaches and showed them to be essentially equivalent; but they had opened the door to an increasing tendency over this last century to turn this part of mechanics into a mathematical discipline divorced from physics – and hence subject to all sorts of conceptual confusions – while the mechanics of continuous media or of friction, separated from it, was looked down upon as secondary. Nevertheless, one further great step forward was taken just after the first world war, when Emmy Noether developed a remarkable theorem connecting the various conserved quantities such as energy or angular momentum with the continuous transformation groups, here the time-shift and the rotation groups, under which the motions of the system are invariant. Once again, an entirely new approach to mechanics was opened up: from the dynamical symmetries of a movement one finds the quantities that are conserved, that do not change with time; if one knows them all, it is possible to deduce the trajectories.

Thus classical mechanics has gone through very profound internal changes; and the latest of them, its new flowering in two entirely unexpected and entirely new fields, is still so recent that no conclusions can yet be drawn –

except that this will clearly be a development of major importance. Suffice it here to add only that this rich and complicated history is not at all untypical; it is only longer than that of most other physical theories because this is the oldest and best organized of them.

(vi)

Theories, then are fairly general conceptual structures that have links to each other at various levels, possess scopes whose extension is only slowly established, and constantly grow and evolve. Within the framework of these general ideas, we now proceed to examine in more detail some of their components.

The first and in many ways most basic is the interpretive element of the theory. Many modern philosophies of science treat this merely as a set of rules on how to link the theoretical concepts with measurable quantities; but this is not a very adequate representation of the facts. It ignores the fact that at least one level of modelling lies between a theory and its experimentally accessible consequences, it ignores the role of the interpretation in the origin and later development of the theory, and above all it ignores the fact that the interpretation is much more than a set of rules – it is what gives meaning to the theoretical concepts and so allows us to apply them. It should not be forgotten that much of what constitutes an interpretation is intuitive and not well formalisable; indeed it can commonly be communicated better by example, analogy and precept than by description or abstract rule. Hence the physicist's nickname of "handwaving" for it. At the inception of a theory the handwaving is still vague and inchoate; it grows in coherence ... as the theory develops, and for a mature theory it is possible to develop a sure feeling of how it works, how its constructs link into other theories, how they build up into models and how much and what sort of reality they can carry; only it is not always easy to put in words. One acquires knowledge of it by working with a theory, using it to build models that work adequately.

Three main components go to make up the interpretive element of a physical theory. The most basic is an ontological conception of the sort of objects to which the theory applies: solids, perhaps, or only crystalline solids, or electrons, or the electro-magnetic field; or, more generally, fermions (i.e. elementary particles with half-integral spin). This conception gradually, as the theory grows and matures, builds up into a complete picture of the properties and relationships of these objects – as far as the theory covers them – but it is much more than that: a mere list of properties and relations known at any one moment would not suffice for the job of model building. This is essentially creative and requires an intuitive feeling for the kind of new properties or relations that a theoretical framework, with some development and commonly joined to other theories, might point up. Hence a genuinely ontological understanding of the objects studied by the theory is needed.

The second component is an understanding of the various concepts available within the theoretical framework for describing the theory's objects and their dynamical behaviour. Of such concepts we must know, firstly, how they are linked to these objects and their surroundings, in what ways they characterise (or fail to characterise) them; and secondly, their interrelations and properties, their dimensions and units, the range of possible values and their usual sizes, and so on.

On the basis of these concepts is built up the third component, an understanding of the forms of the dynamic behaviour and the changes undergone by the things the theory describes. This is the component of most direct heuristic value; its positive aspects lead us to intuitions needed for building appropriate models, its negative aspects (i.e. the information on what sort of behaviour is unlikely or downright impossible under the theory) help us in weeding out mistakes committed while deriving theoretical predictions.

Without these three components of the interpretation, moreover, it would be difficult if not impossible to connect a theory with other theories, for we saw in points (b) and (c) of paragraph (iii) above that the relevant forms of theory interaction depend heavily on their interpretation and in the meaning of their concepts; yet the majority of models used in physics require more than one theory. And even given the model we still require the interpretation before we can link the model's details on the one hand to the details of the experiment's preparation and on the other to the corresponding results.

It should be clear now that the interpretation of a physical theory, by providing an ontological commitment on an only partly formalisable basis, goes well beyond what a set of correspondence rules could offer: it allows the creative use of the theory, while the rule set could hardly go farther than routine applications, and even there the difference between the sorcerer's apprentice, who only knows the rules, and his master, who understands them, is notorious.

It should also be clear that "interpretation" is something of a misnomer; the term carries with it a misleading implication of something added on, a rough translation into a more easily comprehended language, perhaps. On the other hand, names like "meaning" or "essence" of the theory do not represent the clarity and organised structure of the interpretation in a developed theory. Thus the physiscists' preference for talking about "handwaving" has some justification. It might be added that the present generation of textbooks in physics are gravely at fault in presenting little of the handwaving; and the student, suddenly confronted in the research laboratory by an entirely unsuspected and bewildering richness of significance and relationships behind the dry bones is puzzled and anxious, while the philosopher of science, unable to go through that salutary experience, is misled.

(vii)

The building block of theories and models, at all levels, are concepts. No attempt at a general definition of what a concept is will be made, in accord with the general tendency of this work to break away from too rigid moulds; in the present case, moreover, this seems well beyond the bounds of the possible. The term, in fact, covers a rather wide range of different sorts of notions employed in constructing theories and models. We mention the more important of these sorts and outline some general characteristics of concepts that are sometimes overlooked; further discussion is left for the chapters in which certain particularly important concepts are discussed.

The simplest concepts arise as descriptive properties of materials. At first perhaps no more than the results obtained from specific techniques of measurement – here the most primitive form of operationalism find its justification – they develop into quantitative estimates of some aspect of material structure when our theoretical understanding outgrows the experimental technique. New methods of measurement are invented, operationalism no longer works, and the concept can now enter into another theory.

Even with a sound theoretical basis, many such concepts remain purely descriptive; it is then highly unlikely that they will prove applicable outside the scope of their theoretical framework. Other such concepts turn out to have a much higher explicative capacity: they really allow us some understanding of the behaviour of the things – objects, fields, whatever – the theory is about. This kind of concept often transcends the limits of one theory and turns up in several. An example is the concept of electrical resistance; originally formulated only for the conduction of steady electric currents in metallic conductors, it has been extended to all kinds of other materials, solid, liquid, gaseous and even plasmas; it has, in the form of impedance, also found application to alternating currents and arbitrarily varying ones. This example also exhibits another common feature: such concepts seldom come alone; they tend to form groups which are associated into structures that in different theories bear considerable analogy. Here the concepts of resistivity and conductance are linked to resistance in an almost unform way; and the concepts of electric current and electric potential form a slightly looser group.

Many concepts that in this way cross the theoretical borders do so by changing and broadening. A typical case is that of inertial mass; a single positive constant in classical mechanics, characterising the quantity of matter that constitutes a given object, it divides into two in relativity theory, dynamic mass on the one hand, which enters into the relativistic equivalent of Newton's second law but is a function of velocity, and rest mass on the other hand, which remains constant and takes over the specification of the quantity of matter. But concepts change not only when they cross the boundary between two theories; they undergo an often considerable evolution as the theory they correspond to evolves. Thus in the complex growth of classical mechanics the energy concept also acquired new and richer forms. It originated, even before Newton's time, in the somewhat muddled notion of *vis viva,* and only the use of Newton's and Leibniz' calculus showed that this should be $\frac{1}{2}mv^2$ rather than mv or mv^2, as had long been thought. Lagrange's work showed that the sum of this plus a curious quantity related to the force was constant throughout certain types of motion; Young, a generation later, called this sum the energy and its two terms came to be known as the kinetic energy (the modern descendant of *vis viva*) and potential energy. Another generation went by, and royal battles were fought over the extension of this sum to include also what we now know as thermal energy, to formulate the law of conservation of energy. Since then the meaning of "energy" has steadily expanded as other forms of energy have been discovered. But a last and surprising twist came from the work of Emmy Noether and her successors: as we mentioned above, in paragraph (v), she showed that to each invariance group there corresponds a conserved quantity. If a system has the kind of structure in which the possible movements do not change if we shift the origin of the time measurements, then the corresponding quantity is conserved – and this quantity turns out to be the energy. Biological systems do change when we shift the origin of time; a five-year old boy does not behave like a two-year old child. Correspondingly, energy is not conserved in them, they are open systems that constantly exchange energy with their surroundings. But many mechanical systems are very nearly closed; and for such systems the energy concept has acquired a new significance through the Noether theorem.

Most of the concepts we have so far discussed have a quantitative expression. They can be measured and expressed in numbers of appropriate units. There are also very many abstract concepts that lack such an aspect. Some of them concern patterns, as for instance the crystallographic structures, for which precise mathematical descriptions are available; other patterns, e.g. the approximate equality of protons and neutrons in stable nuclei are not easily expressed in mathematical form. Other concepts are those of the various forces occurring in nature, e.g. gravitation, and though there are various theoretical and sometimes highly mathematical theories linked to these concepts, one cannot say that any one of these theories, or even all of them together, in any sense constitute the concept; for there is always more to be said than these theories hold, and it is just this "more to be said" that in the long run creates the dissatisfaction with the existing theories and then guides the search for better ones.

Lastly, there are abstract concepts of great generality, usually concerning the type of object that a theory is about. Electrons, for instance, or crystalline solids. And at a still higher level of generality, the concept of object itself.

All concepts, including these last ones, including perhaps even the concept of concept itself, have a definite and limited applicability. In some cases this finite range of applicability may, as for a theory, be expressed as a scope. Such a case is that of temperature, defined as the mean kinetical energy of the random linear motion of the component particles of a many-particle system in thermal equilibrium. This notion is difficult to apply if there are too few particles in the system, or if the temperature is so low that thermal equilibrium takes a long time to establish itself, or if the temperature (in certain systems) is so high that the nature of the particles changes with small changes of the temperature; and if the system is far from equilibrium, then it may be difficult to state just what is meant even by a local temperature. But there are many concepts for which the limits on their usefulness cannot be expressed by means of a scope; in particular those concepts that do not have quantitative expressions have no scope. Thus the concept of an object is limited by its rather evident restriction to solid bodies (whose lifetimes, moreover, are not infinite) or semi-solid bodies; liquid or gaseous structures, on the other hand, have individuality only under rather special circumstances (stars, for example), and elementary particles, which come under none of these headings, transform into each other with a readiness which makes the object notion of rather dubious utility in dealing with them. This does not, of course, deny their objective reality; water is not less real merely because we cannot describe it in terms of so and so many separate and identifiable drops which we can count.

(viii)

The formal component of physical theories requires a more extended discussion, in view of the many misconceptions concerning it which are current in the literature of philosophy of science.

We note, to begin with, that many physical theories possess more than one formalism; the different formalisms are usually equivalent over part of the theory's scope but differ in the remainder, sometimes even having different scopes themselves. On occasion they give rise to discrepant and contradictory results. A case in point is that of classical mechanis. For this theory we have Newton's original formalism, the Lagrangian one, the variational one based on Hamilton's principle, the Hamilton-Jacobi formalism of canonical transformations, the invariance-group formalism, as well as several less successful ones, such as that proposed by H. Hertz at the end of last century. Newton's formalism also applies to non-conservative forces, while several of the others do not; the variational one can be extended to cover geometrical optics as well as mechanics; the invariance-group formalism can be generalised for relativistic mechanics, while the Hamilton-Jacobi formalism may be taken over into quantum mechanics. Yet the invariance-group formalism, conceptually perhaps the most satisfactory one, runs into difficulties when one attempts to base statistical mechanics on it, and it is probably the canonical-transformation method that here is the most satisfactory; though it cannot be said that all problems in this area have yet been solved.

A similar situation is found in quantum mechanics. This theory was originally developed with three different formalisms, due to Schrödinger (the wave-equation formalism), Heisenberg, Born and Jordan (the matrix formalism) and Dirac (the dual vector-space formalism). Later work by von Neumann added the Hilbert-space formalism, Feynman produced his powerful path-integral method, Kastler and others developed the approach via C^* algebras, and recent years have seen attempts to formulate so-called quantum logics. The interrelations between these various formalisms are complicated and by no means fully classified, among other reasons because, particularly of the last two, many different versions exist which are by no means compatible. In at least one case, however, the situation seems fairly clear; this is the relation between the wave-equation and the matrix formalisms. It was shown by Schrödinger that these are indeed equivalent (in the sense of leading to the same predictions) provided the system remains confined to a finite volume: provided we have a closed system, in other words. But we can only observe anything when our system interacts with something outside itself; and for such cases the two descriptions do not coincide. Many modern textbooks of quantum mechanics ignore this fact; they tend to present the subject in a wonderful hotch-potch of notations, concepts and techniques drawn impartially from Schrödinger, Heisenberg and Dirac – with little regard for its consistency – and to offer the whole in crude positivistic fashion as *deduced* from some axioms; often the axiomatic basis is taken to be the Heisenberg inequality, perhaps as counter-intuitive a starting point as one could well think of. The unfortunate student who has swallowed all this is then pulled up short when the consequent problems of reaction theory bring him face to face with its shortcomings.

But it is important to realise that such intermixtures and combinations of different and not always fully compatible formalisms is dangerous only when it is not understood as such, when it is presented as a consistent and even axiomatic deduction structure. Correctly understood and made use of in full awareness of its limitations and possible pitfalls, it becomes a remarkably flexible tool that precisely helps to overcome the weaknesses of one formalism by means of the strengths of another one. For this to be achieved, however, an understanding of the underlying physics is evidently indispensable.

If thus one theory may possess several different formalisms, one and the same formalism may also occur in more than one theory or model. This may be observed at three distinct levels. Firstly, the same formalism may, in considerable detail, apply to two models of quite different physical systems, between the structures of whose dynamical behaviour there exists a close analogy. Thus a massive body suspended from a spring and moving under the joint influence of an external force and a viscous damping proportional to its velocity behaves very much like the electric charge circulating round a circuit containing a capacitor (which acts like the spring), a resistance (which provides a damping proportional to the "velocity", i.e. the time derivative of the charge) and an inductance (the inertia of the mass); an applied electric potential parallels the force. The intuitive analogy is brought about by the fact that the total energy in either system (the same concept, not merely analogous) is divided into two parts – kinetic and potential energies in one case, magnetic and electrostatic energies in the other – which behave analogously. As a result, the formalism for both models is that of a second-order differential equation with constant coefficients. A second level of common formalism is more abstract and usually has both common elements forming the basis and specific details varying from case to case. Thus the so-called theory of wave motion, in spite of its name, is an essentially mathematical formalism that in one form or another applies to such diverse physical systems as vibrating strings, elastic membranes, liquid surfaces, sound both in air and in solids, the propagation of light, quantum mechanics, and even surprisingly, perhaps, to certain problems in urban traffic control. Another very general formalism that in recent years has served in an astonishing range of applications is that of the stochastic process theory. Lastly, at the most abstract level, the various mathematical disiplines, as applied to the problems of physics, constitute so many formalisms of very nearly all-embracing scope. Some questions that arise from this third level are reviewed below.

(ix)

The formalisms of physical theories and models are normally expressed in mathematical form. They are not, however, straightforward mathematical structures. We examine this matter under three headings, the quantities that enter into these relations, the nature of the relations themselves, and the kind of conclusions to be drawn from them.

The quantities in physical formalisms – let us call them physical quantities for short – differ from their mathematical counterparts in two respects. Firstly, they possess properties quite alien to mathematical quantities: they have dimensions (a velocity is a distance divided by a time, and so on) and are expressed in multiples or submultiples of certain units (miles, perhaps, or centimetres per second for velocities). Secondly, they bear only a superficial resemblance to the mathematician's so-called real variables: the reason for this is simple: they are defined only to within a finite precision. Not only, that is to say, do we not know their values better than to within certain error limits; if we try to specify their meaning with infinite exactitude, our attempts break down. We can talk about the length of a table to within a centimetre, perhaps even to within a millimetre if the carpenter has done outstanding work; beyond that it rapidly becomes no longer length of the table but only the length at certain points; if we try to improve on that, we shall have to specify how much pressure we put on the table by our measuring instruments, we must keep the temperature and the load on the table steady, the humidity in the air is a vital factor ... yet even if we keep adapting our concept of "length of the table" we are finally baffled because it becomes impossible to state just which molecules belong to the table, which adhere to it permanently without belonging to it, which are temporary – and in any case, just how does one set about defining the edge of a molecule? Our example may be thought too trivial; but even simpler operations run up against such difficulties. It is easy to count the number of people in a reasonably small room; to count the population of a country can only be done at great expense and with an imprecision – largely due to various indefinitions – of several ten or hundred thousands. The mathematical variable is defined with infinite precision; for a physical quantity the distinction between rational and irrational numbers is meaningless.

Three types of relations can be established between physical quantities. The dimensional calculus provides one kind, the unit-consistency condition a second kind, the physical equations the third[4]. The dimensional has in recent years been much neglected; indeed it does not seem to have received any philosophical attention at all. It is true that most physicists meet it only as a useful tool for analysing the kind of relations physical quantities might have. It is certainly that; but it is more. It provides, in the first place, signifi-

---

[4] The unit-consistency conditions are commonly, for the purpose of numerical evaluation, incorporated by way of "conversion factors" in the physical equations; in other circumstances they are often ignored, giving rise to statements like "Taking $\hbar = c = e = 1$, we find that ... ".

cant restrictions on the physical equations; and in the second place, it yields considerable insight into the structure of the physical theories to which it is linked. This last is merely because it constitutes a second and independent way of exhibiting the relations among physical quantities; but this time without reference to their numerical magnitude, concentrating attention on their qualitative content. Conversely, one may view the physical equations as establishing constraints on the otherwise too general qualitative conclusions of the dimensional calculus. Jointly, then, they condition each other and provide a far richer formal structure for the theory than either by itself.

The physical equations relate physical quantities to each other. This fact imposes two kinds of general conditions on them. Their structure must be dimensionally consistent; that is to say that two terms cannot be added or subtracted unless they have the same dimensions, the two sides of the equations must likewise be of the same dimensionality, and the arguments of all transcendental functions, logarithms, exponentials, sines, and so on, must be dimensionless. And nothing in the properties of the equation or of its solution can depend on the algebraic character of the physical quantities, i.e. on whether they are rational or irrational, and sometimes even the assumption of their continuity may have to be irrelevant; as we saw, the rational-irrational distinction makes no sense for physical as opposed to mathematical quantities. Naturally the resulting equations, though written in mathematical notation, do not always behave like mathematical ones; if we treat them as such – as we must, because these are the only techniques that we know – we often get unwanted or meaningless results. The equation for an essentially positive quantity is quadratic, and of its two solutions one may be negative; or a differential equation describing the time evolution of a system has several solutions, some of which are quite *unphysical*, in the sense that they do not behave as the handwaving behind the theory leads us to expect. One of the most frequent occurrences is that some quantity becomes infinite – diverges, as the technical jargon has it. A divergence almost always indicates that we are outside the theory's scope; the remarkable thing here is that it should be the theory's formal component which here sets a limit to the scope, rather than experimental evidence.

Such discrepancies between the formal part and the intent pursued by the theory are sometimes dealt with in other ways. Certain divergences, e.g. the infinities arising in quantum electrodynamics, are removed by means of elaborate and elegant mathematical tricks; since we have not succeeded so far in reformulating the basic notions of the theory in such a way that the divergences do not appear and the need for such tricks does not arise, we have only papered over the difficulty instead of solving it. Yet at the same time this conceptually somewhat unsatisfactory theory is also one that has produced results of astonishing precision over an enormous scope: judged on that criterion it is one of the most successful theories of modern physics.

In at least one other case, the discoverer of a seemingly unphysical solution to a wave equation, in the form of wave functions corresponding to states with negative energy had the insight to trust the formalism and look for an appropriate modification of the handwaving. The discoverer in question was Dirac, and what he suggested was the introduction of the concept of antimatter, a remarkable and highly successful solution to the problem posed by the apparently superfluous solutions of his wave equation for the electron treated relativistically. To modify the basic theory on the ground that the formalism suggests it is not, of course, to be recommended as a suitable strategy except in very skilled hands. We mention the case, however, because it underlines the extent to which there is give and take between the formal and informal components of a theory. And from this there follows a vital point: we may – indeed we must – isolate these elements in order to analyse and understand them; but this sort of understanding will be one-sided and incomplete until we see the elements again in their interaction, against their background.

(x)

The formalism of physical theories and models is further distinguished from mathematical formalisms by the frequent use of various types of approximations. This is made possible because, as we have seen, the descriptions furnished by the physical formalism are always of finite precision; and it is necessary whenever we are up against otherwise insoluble problems. It is the nature of these problems, which an appropriate approximation allows us to circumvent, that offers us a classification of approximation types by origin:

(a) In conceptual approximations one concept (or group of concepts) is substituted for another concept or group of concepts. The most fundamental such approximation is made in every theory and in a sense lies outside the frame of the formalism: the substitution of either real continuous or discrete (and generally integer) variables for the physical magnitude, which is only more or less well represented in this way. Within the formalism this kind of approximation is common in models and relatively phenomenological theories, but less so as we go down to more fundamental theoretical structures. A good example is the concept of the non-radiating stable orbit in the "old" quantum theory developed by Niels Bohr in 1913 as a model for atomic structure. This example also exhibits clearly what is the normal characteristic of conceptual approximation, namely the intrusion of one conceptual scheme into another, with the resulting incoherence and even contradictions; these problems are further discussed below. It is to be noted that we commonly resort to this kind of approximation when we do not entirely know the concept being approximated. Had it been understood in 1913 that the atomic electron absorbs radiation from a random background field and radiates energy back into it, but that these processes balance each other on the average when the mean motion is the Bohr orbit, the concept of a non-radiating orbit would have been superfluous; as it is, this approximating notion was important, not only because it solved the problem of the moment but because it paved the way for the later and more complete understanding.

(b) We have recourse to functional approximation whenever a relation among a group of concepts cannot be handled. It may simply not be known sufficiently well, and we substitute an approximation we can work with; it may be too complex mathematically and we replace it by a more manageable one; or – a very common case – it may involve physical quantities that we cannot measure easily for one reason or another (perhaps only because it comes too expensive), and we use instead a simpler relation that does not contain these awkward quantities.

(c) Then we have the numerical approximation. This is rarely relevant in high-level theories; but specific models always require the handling of actual numbers, both when making predictions and in their application to interpret experimental results. Calculation with numbers, whether done by hand or in the electronic computer, cannot be other than approximate, except for the very few cases when everything can be reduced to (not excessively large) integers. We cannot compute the value of $\sqrt{2}$, nor could we represent it exactly in a finite number of decimals (or bits) if we could compute it; we can only find an approximation whose error limits may be estimated rather exactly. And the growing use of numerical simulation even in very abstract theoretical problems is steadily increasing the importance of this kind of approximation.

(d) Finally, we have the observational approximation. Many theories, and all their applications, contain quantities that can only be measured experimentally. The so-called fundamental constants, the velocity of light, the mass and the electric charge of the electron, and so on, are of this sort. As we saw in Chap. 3 such experimental data always have an experimental imprecision, expressed usually as an error limit, which from the very moment of their introduction into theoretical approximations renders them approximate.

The four approximation types we have mentioned do not, of course, exclude each other. A conceptual approximation requires a series of functional approximations in order to accommodate it into the theoretical framework; a numerical approximation may conveniently be represented as a functional one; and it involves a basic conceptual approximation, in that we represent a continuous variable as a discrete one which in the computer, moreover, has only a finite range. The observational approximation is an approximation only because we make the basic conceptual approximation of trying to express as an exact number what in fact is not an exact number, in the mathematical sense. Lastly, a conceptual approximation, particularly when it concerns an entire physical model, can be quantitatively evaluated only at the numerical level. There is thus a complex interplay between the types; little work has so far been done to elucidate it on a more general basis than that of the practitioner.

A different classification scheme for approximations appears when we consider their actual use. Three types may conveniently be distinguished, in terms of the behaviour of the approximation value as a function of the parameters of the theory or model, and upon repeating the approximating. To fix ideas, let us consider a physical quantity ξ which is a function of some parameter α (which could be several; but nothing except a more complicated notation is added by considering the more general case). We aproximate $\xi(\alpha)$ by $x_i(\alpha)$, where the index i numbers the successive attempts to obtain the approximation for a given value of α. Then the three types of behaviour are:

(I) The approximation depends very much on α, but not on i. The difference

$$\delta(\alpha) = \xi(\alpha) - x_i(\alpha)$$

is a (reasonably) well defined function of α.

(II) The approximation depends on i, but little or not at all on α. Then

$$\delta_i(\alpha) = \xi(\alpha) - x_i(\alpha)$$

and since the repeated evaluations of the approximation have nothing to do with each other, the best description of the whole set of different δ_i's is statistical: we find the average, the spread among that average, and ideally we determine the distributing function.

(III) In the mixed case, the approximation depends both on α and i,

$$\delta_i(\alpha) = \xi(\alpha) - x_i(\alpha) \, .$$

There exist mathematical tools for handling approximations. They are chiefly of two kinds: we have descriptive methods for giving with some precision the details of how an approximation behaves, and we have constructive methods for building a desired kind of approximation. Behaviour of type (I) is described by the O, o notation which provides comparison functions for $\delta(\alpha)$ and establishes comparison criteria; various techniques exist for constructing such approximations. Both the descriptive and the constructive problem of type (II) are dealt with by means of statistical techniques, sometimes highly sophisticated ones, in which the assumption is made that the particular case in question is a member of an ensemble of similar ones. When the approximation is of observational origin, as in (d) above, this is directly justified because almost always we have a series of slightly different measurement results, and the arguments given at the end of Chap. 3 apply. When the approximation is numerical, as in (c) above, exact repetition of the procedure used will give the same result; but since the physical significance should not be affected when we use a different but equivalent procedure, or work on a different computer, and so on, we may lump all these possibilities together and again treat the approximate result, viewed as a member of an ensemble, by statistical methods.

What these methods cannot do is deal with any question of conceptual approximation; because of this limitation, no method will establish what in a given case will be the most appropriate approximation; this choice has to be based on the physicist's experience and intuition and must be justified *a posteriori*, by its results. The limitation springs from the nature of conceptual approximation: the intrusion within one conceptual scheme of concepts belonging to a different framework. The structure generated in this way will usually be inconsistent; this is, among other reasons, the explanation of why the attempts at formal representation are so unsatisfactory.[5] These inconsistencies and the consequent possibility of actual contradiction are further analysed below.

---

[5] The attempts have been made within the structuralist formalisms of the Sneed-Stegmiller group. They suffer from the further defect of offering, in the positivist

(xi)

That experiments are, in the now customary formulation, theory-laden was discussed in Chap. 3. It might correspondingly be expected that theories and the models derived from them are experiment-laden, and this is indeed the case, though in varying degree, and in two essentially rather different ways.

One kind of experimental element in theoretical structures is the presence of characteristic constants whose value is needed before the structure can be used. Many fairly fundamental theories possess at least one such quantity: Planck's constant in quantum mechanics or the velocity of light in special relativity are good examples. Such constants cannot be determined by the theory but must be found experimentally. In less fundamental theories several such constants appear; and in the limit of a phenomenological generalisation the experimental data provide a very high proportion of the theory's structure. When models are constructed, two further places appear where experimentally determined values are needed: in the specifications for the system structure that the model describes, and as the initial or boundary values that fix the model predictions. The way in which such experimental data acquire a considerable degree of theoreticalness has already been discussed in connection with the relative mobility of the theory-experiment distinction.

The other kind of experimental element appears, in a much more subtle form, in the agreement between experimental fact and theoretical picture. The experimental element is present in the theoretical picture because one has been adjusted to fit the other. Yet this conception must be used with care: it is clearly valid for all those theories and theoretical models that we make use of while developing another one; but the one we target on, whose relation to experimental results is precisely what is under examination, cannot yet be said to contain these same experimental results without risking misrepresentation.

---

tradition, no more than a description of approximation, with no further analysis; and even this is restricted to a single rather restricted kind of approximation, the distinctions we have made above and the multiple mathematical techniques available being entirely ignored.

5. Induction and the Scope of Theories

5.1 The Unnecessary Induction

More than fifty years ago, Karl Popper (1936) gave a forceful argument to the effect that induction is not a tenable way to establish either that a scientific theory is acceptable or that it is more acceptable than rival theories. This chapter is motivated by two considerations: (1) as the steady flow of publications on the subject shows, the old confusions about induction remain alive, and (2) Popper's argument, while conclusive, is formal; there is more to be said. The weakness of inductivist conceptions lies in a mistaken view of the nature of scientific theories; they are seen as – ideally – universal propositions, while (as I shall argue) their abstract component is a set of propositions of finite scope, the determination of which is one basic purpose of experimentation.

(i)

The term 'induction' is used in several basic senses. The original sense, introduced by Bacon and used by many later writers, including Newton and, in our day, Herbert Simon (1977), is that of finding some appropriate theoretical account for a set of observations; whether this account is derived by using these observations to improve an earlier account, or is found by means of one or another heuristic, or is simply invented, is not relevant: it is the passage from the data to a theoretical model that is described as an induction. In this wide sense induction is not simply unobjectionable, it is a fundamental activity in scientific research: we do pass from experimental data to theoretical descriptions, and to considerable effect. The study of induction in *this* sense is meaningful.

A rather different sense is that of an extrapolation from a past series of similar occurrences to the prediction that another one will happen. The sun has come up in the east on numberless mornings, and therefore it will do so again tomorrow morning. The prediction is rarely so certain; more commonly we cannot say more that it is likely. Can we give a numerical value for the probability that the prediction will come true?

This question leads us directly to the most sophisticated sense of induction, namely the use of probability-theoretic arguments to estimate or increase the degree of confirmation of a scientific theory. If the probability of a theory A

being true is greater than that of theory B, then A is to be preferred; or, if only one theory is being considered, the question to be answered is whether the probability of its being true exceeds some threshold. It is generally recognised that it is not possible to establish with probability 1 that any given theory is true; the threshold is therefore taken to be less than 1.

Induction in this narrow sense is what forms the centre of interest in the remainder of this chapter; but because of its origin in the extrapolation version, and because the question of how induction is to be justified is generally formulated with reference to the latter, I consider this form in the next section.

(ii)

The difficulty with justifying induction in the sense of extrapolating from repeated past observations to a new one is that any such justification would itself be an inductive one. We cannot expect a formal, deductive proof of induction, since it is not always valid; but an inductive justification would be circular.

Though much has been written on this question, what is generally ignored is the implication of the uncertain nature of inductive extrapolation, namely that if we go on extrapolating, sooner or later our prediction will fail. Indeed, no sequence of similar events is known to have occurred that was not finite; and even for the sequences that have not yet ended there is not one for which we can safely predict that it will continue forever. Even the sun will rise next morning only as long as the earth continues to rotate; and astronomers tell us that tidal friction will bring it to a stop within a foreseeable time.

Thus the evidence of the past suggests the truth, not of any inductive principle, but rather of the anti-inductive one that every induction will, sooner or later, fail. But to arrive at anti-inductionism we have used an inductive argument; hence anti-induction is true only if induction holds. However, if anti-induction is true (in the sense just stated), then induction must alway fail, and must fail also in this case. Therefore we cannot validly conclude that anti-induction holds; it must therefore be possible to find inductions that work indefinitely.

We need not labour the point. Induction is evidently not a completely consistent concept. Nor is it difficult to see that in fact the question of whether induction is valid is ill posed. What we need to know in practice is *how long* any given induction will go on holding; and the answer to this question cannot be found from abstract considerations – such as could be advanced to provide a general justification of induction – but must come from a concrete understanding of the specific case in hand. Inductively, it should be possible to argue that the longer I live, the oftener I have woken up the next morning in good health; therefore the longer I live, the higher the probability becomes that I shall live another day. More realistically, I am aware that human beings do not generally live more than four-score and ten years, so that the probability of my staying alive in fact diminishes as I go on living, rather than increasing. It is thus to be expected, and is indeed the case, that while inductive extrapolation is not an argument to be found in scientific work, the determination of repetition lengths, life expectancies, halflives, and so on, is a frequent task.

This reformulation of the problem of inductive extrapolation has two significant aspects: firstly, it no longer involves any need to show that, in any general sense, induction will "usually" or "probably" or "often" work and can therefore be justified either inductively or on different grounds; and secondly, there are only specific questions to be answered, each of which poses a task for relevant scientific research. The *philosophical* problem of inductive extrapolation has evaporated, and we are left with a scientific one of very different nature. Of course in many cases we are not yet in a position to provide an adequate answer to this new problem; for instance, we cannot say why men rarely live beyond a hundred years; but the fact that they do not do so, as a matter of practical experience, is sufficient for determining what each of us may expect.

Is the argument just given an inductive one? If it were, I would have given a counterexample to my thesis. In fact, it is not. In other cases, a well founded scientific theory allows us to construct a theoretical model for the situation in hand, and from this model we can derive an evidently non-inductive prediction for the number of times an extrapolation will work. Here we cannot do this. But though we do not have a good theory of human aging, we have every reason to expect that such a theory is possible. We can therefore argue that *if* we had such a theory we *could* predict such lifetimes; these averages will then justify the conclusion arrived at above, simply because any future theory will account for them if it is sound. Such an argument may be considered suspect on quite other grounds, of course, e.g. because it is counterfactual; but it is clearly not inductive.

To sum up, the apparently inductive extrapolations from past experience are not in fact inductive; they are based either on already established theory or at least on the usually justified expectation that such a theory can be found. Both in scientific research and in everyday life the need for an extrapolation is a stimulus towards the formulation of a theoretical model; this tends to bear out what has been said. Note that where a theoretical model cannot well be considered possible, no extrapolation whatever can be done. A case in point might be the discussion whether miracles could happen in the future.

5.2 Probability and the Scope of a Theory

Just what is meant by the notion of *accepting a scientific theory* is in a sense basic to all of philosophy of science; for the interpretation put upon it colors one's conceptions of how to arrive at an acceptable theory, of how to decide whether it is acceptable or not, of what to do with it in either case, and finally of what meaning to attribute to it.

There appear to be several conflicting ways of interpreting the acceptance of a scientific theory. For some authors, acceptance means that the theory is

true (e.g. Braithwaite 1953, Caws 1968); the truth of a theory may be a dichotomic variable (true or false) or it may admit of degrees, almost universally
identified as the probability of the theory (see e.g. Barker 1957). For others
the acceptance is at best provisional until something better can be found.[1]
For still others a scientific theory has no relation to truth or untruth but is no
more than a convenient summary of experimental data, providing an easily
handled mechanism for interpolation and limited extrapolation (Mach 1897,
Wigner 1971).

Correspondingly one finds very different attitudes to the problem of induction. Acceptance of a theory as true (or partly true) commonly goes hand in
hand with the conviction that inductive procedures are required to establish
the degree of acceptability – interpreted as the probability – of the theory. If
theories are only provisionally accepted, this need not be incompatible with
the use of induction; but Popper adduces strong reasons against using inductive methods and yet maintains the idea that a theory can be more or less
probable. Finally, the extreme Mach-Duhem point of view has, of course, no
use for the notion of the truth of a theory, probable or not, but may still be
combined with induction.

All these approaches are thus united only insofar as they employ the concept of the probability of theory. A further bond might be said to arise from
the fact that they all run into puzzling problems for which no very satisfactory solutions are available. For the inductive procedures there is the question
whether their justification is itself inductive and therefore circular. But even
if we take induction to be well established, there are questions. How should
we choose the observations to be used with a given theory? Should they be
as diverse as possible, or should they form a closed group? And should they
exclude all repetitions, or admit some? And how do we eliminate the spurious
confirmation provided by the endless repetitions of one and the same observation? Then there are the two well-known paradoxes of the black raven (Hempel
1937, 1945) and the grue emerald (Goodman 1955). The first one leads us to
the disagreeable alternative of either saying that different but logically equivalent formulations of a theory can have different degrees of confirmation or
else accepting that seemingly irrelevant observations contribute to confirmation. The second one appears to show that whenever we add confirmation to
a theory we also tend to confirm other and quite distinct theories. Finally, an
irksome problem appears when the logic of the inductive procedure is formulated as an application of Bayes' law: the smaller the prior probability assigned
to the theory, the greater must be the accumulation of positive evidence for
the theory to achieve a given degree of confirmation, until in the limit of zero
prior probability no amount whatever of factual data could do so-yet this prior
probability seems to be quite arbitrary and a value of zero cannot be excluded;

---

[1] Thus Popper (1959a) accepts a theory "but only in the sense that we select it
as worthy to be subjected to further criticism, and to the severest test we can
design".

conversely, a theory of prior probability 1 could never be disqualified by any negative evidence.

I shall argue here that these difficulties confirm that the concept of probability cannot reasonably be applied to the validity of a scientific theory; that instead the scope or range of applicability of the theory should (and, moreover, does in actual practice) serve as the basic criterion for its use; that this substitution makes both impossible and unnecessary the use of inductive methods; and that on this basis the puzzles mentioned above either evaporate or become useful common sense.

(iii)

That attributing a probability to a scientific theory is not meaningful becomes clear when one attempts to spell out in detail its possible significance.

For any objective interpretation of probability, a theory as such cannot of course have a probability; at best one can speak of the probability of success in applying the theory to specific cases: "theory T has probability p" will then mean the prediction that if one acts on the basis of T, the results will correspond in a proportion p of cases to what T envisages. This is ill defined, as may be seen from two considerations:

(a) If one and the same action is repeated, success (or failure) is also repeated;[2] thus by a suitable choice of the sequence of actions we can achieve any desired probability for the theory. The source of the trouble may be seen in measure-theoretic terms: in order to determine quantitatively a probability we require a sample space over which Borel sets can be defined, together with a measure for these Borel sets which can be normalized to 1. Now the sample space is presumably the "space of possible actions" to which T is relevant; even if it is not evident that we can talk here of a space in the mathematical sense, at least the analogy seems to be licit and even useful. But any definition of a measure over this "space" is clearly quite arbitrary. Its purpose would be to specify the proportions in which different kinds of actions among those in our sample space should appear in any sufficiently large sample; but any possible criterion would have to depend on the future applications of the theory and thus be both unforeseeable and dependent on factors which are irrelevant to the validity of T. To put the matter crudely, scientist A chooses his actions in a field f_A where T works very well, and so drives a high probability for T's being true; while scientist B works in a field f_B where T tends to fail so that he adjudges it bad. Clearly what was intended as an objective and generally acceptable estimation of the values of theories turns out to be at best an estimation of our highly subjective personal prejudices.

(b) A very important use for the probability of a theory would be the comparison of different theories; but even if theories T_1 and T_2 had well-defined probabilities p_1 and p_2 in their respective spaces of possible actions,

---

[2] This does not exclude statistical theories such as quantum mechanics: since they predict only probabilities, a single action based on them must, in the present context, be taken as an experimental series long enough to establish a relative frequency for comparison with these probabilities.

we cannot say that T_1 is better than T_2 when $p_1 > p_2$ unless these two spaces coincide. Otherwise a theory which is applicable to only a single case but there works perfectly would be superior or at least equal in value to any other theory whatever. One might try to form a common space by extending each theory's space of possible actions to their set-theoretical union; but that this alteration of our original conception is inadequate is clear when we observe that now the probability p_1 of T_1, say, would depend on what other theory T_2 it is compared to.

Such arguments apply straightforwardly to the frequency interpretation of probability when we attempt to use it on scientific theories. We merely have to substitute 'relative frequency' wherever 'probability' occurs, and then go to the limit of very large numbers. The case is a little different for a subjective interpretation, since clearly a phrase like "rational degree of confidence" (Jeffreys 1961) applies to a scientific theory if it is applicable at all – a point we shall not discuss here; on the face of it such an interpretation would therefore permit us to speak of the probability of a theory's being true. However, as soon as we attempt to obtain numerical values for such a probability, the difficulties we have discussed will appear. The probability of a specialization of a more general theory can presumably be derived from that of the general theory; one might use the same value for both, since the more special theory, as a logical deduction, is based on the same experimental evidence as the general one. But in this case there is no validation problem for the special theory, and the question need not occupy us further. When, however, a theory is derived from several logically independent ones with probabilities, or when it is an original theory not deducible from other theoretical structures, its validation must be based on experimental evidence. This is of course the very basis of scientific research; but any attempt to do so via a probability for the theory will run into difficulties entirely analogous to those mentioned.

A further problem for any subjective view arises because the subjective probabilities of any two theories whatever form a basis for comparing them. We saw above that the objective probabilities of two theories are not strictly comparable even if they cover the same ground, because each theory will also cover ground the other does not; here we have the opposite problem, for theories with completely disjoint "spaces of possible actions" should not be comparable as regards their degree of confirmation. Such a comparison finds its raison d'être in establishing grounds for preferring one to the other; but it is not meaningful to prefer, say, the theory of nuclear reactions to the chromosome theory of heredity; they could not conceivably be alternative explanations for a given phenomenon.

Other problems appear with other interpretations of probability. It is not clear, for instance, what meaning could be attributed to the propensity of a theory to be true; apart from the difficulty of applying such a notion to anything but material objects, the propensity for a certain event to happened depends, as Popper (1959b) very rightly points out, upon the experimental set-up as a whole. But when considering the probability of a theory we do not have in mind a specific experiment, for which indeed the validity of the theory is determined without any probability or propensity considerations entering into it; but to consider the general situation of the theory under test as the appropriate analogue to the set-up of hand, air resistance, table and so on, when tossing a coin merely leads again to similar difficulties as previously discussed.

In another interpretation (Brody 1975) the probability of an event is identified with its measure in an ensemble (i.e. an indefinitely large set) of theoretical models in which all experimentally uncontrolled factors are allowed to vary; this conception is theoretical and so to some extent subjective, but achieves an adequate representation of objective reality precisely insofar as the theory on which the models are based fits that reality. Here the impossibility of considering the probability of a theory is evident because there is no way of formulating a suitable ensemble; and of course the problems of actually determining a numerical value for such a probability appear in the same shape as before.

(iv)

The conclusion is clear: on no interpretation can one speak meaningfully of the probability of a scientific theory. This would be a catastrophic conclusion if in actual fact such a concept were ever used; but fortunately this is not the case: I have not been able to find a single case in actual research where a numerical evaluation of the probability of a theoretical structure has been attempted, and though phrases such as "it is probable that theory T holds" are common enough, it is always in a loose and metaphorical way which denotes simple uncertainty.[3]

To see what this implies for inductive methods, it is necessary to make some distinctions among them which are often neglected. As discussed above, there are three rather different types of arguments which are usually classed as inductive: they have to do respectively with making specific predictions, with the discovery of theoretical formulations, and with their justification. Let us look at them in turn.

(a) *Making Predictions for comparison with experiment.* Two slightly different cases arise under this heading. The first one involves a prediction of the type 'the next observation will be x' from an already established theory which yields that x has probability p. In practical applications the predictions are often much more complex ('among the next n observations the frequency of x will be between the limits $n(p \pm \delta p)$ with probability $1 - \varepsilon'$, and so on), but the given type sufficiently shows what is involved. On a subjective and

---

[3] This must be carefully distinguished from inverse probability statements (which are equivalent to 'on the basis of theory T such-and-such an event has probability p' where the space of possible events is well-defined and in general does not coincide with the "space of possible actions" for T, so that nothing concerning the probability of T in the sense used above can be derived from them) and from the likelihood of hypotheses in the statistical sense, which is a distinct concept and does not obey the probability axioms – for instance, it is not additive over hypotheses (Fisher 1922).

mentalistic interpretation of probability such a prediction is problematic, for to suppose that a mere idea in our heads can make the prediction come true in a proportion p of cases (in the long run) either attributes extraordinary mental powers to those conversant with probability theory or is consistent with our accepting that not only this but all the rest of the world is the creation of our ideas. Moreover, how can we explain that probabilities seem to have exactly the same effect whether we actually work them out or not? And even if we get them wrong, nothing happens differently – unless we act on the basis of our calculations. It is such considerations (which we need not elaborate or formulate more precisely) that form the most substantial stumbling block to the acceptance of any subjective account of probability.

For an objective interpretation the type prediction should be verified in a proportion p of cases in the long run; this is precisely the import of the frequency view, and in other views the required argument is well known. That we have derived the probability p of observing x's from an already established theory means that it has been possible to define (though perhaps not explicitly) an event space in which the observation x has measure p; and insofar as the theory is satisfactory, the making of observations will be precisely a procedure of sampling from this event space. Yet it must be noted that this agreement between prediction and observation is not automatic: for one thing, the theory may prove inapplicable to our case, as even well-established theories sometimes do; or the sampling procedure may not correspond to the stipulations of the theory; and for another, as is well known from theoretical statistics, if the theory yields a probability p for x's to occur, then in a series of n observations we will not normally have np x's, but some number that probably lies in the range $np \pm \sqrt{np(1-p)}$, and the probability of this being so may in turn be evaluated.

The other case of deriving a prediction starts, not from a theory, but from a sequence of past observations. "The sun has risen every day for many thousands of years; therefore one may predict that it will rise tomorrow, with extremely high probability". Only when we note that this argument is extremely elliptic can we discover if – and under what circumstances – it is valid. If the missing portions are not supplied, it is possible to fall into the trap of Laplace's so-called secession rule that after n repetitions of an observation another one has a chance of $n + 1$ to 1 of occurring: hence the chance of my waking up alive tomorrow grows larger the longer I live .... The missing portion of the argument is the formulation of a theory which the given sequence of observations serves in part to confirm; if this proves possible we may obtain the required prediction from the theory as in the first case considered; but if the theory is not viable, the argument collapses. Hence this case reduces to the first one, but only after we have found a theory – to be discussed below in (b) – and confirmed it – as in (c).

(b) *The discovery of theoretical formulations.* It is commonly thought that rules may be formulated for inductively obtaining a theoretical description from experimental evidence; the notion goes back at least to J.S Mill[4] (1843), and in our days H.E. Kyburg (1968), for instance, appears to consider it so obvious that it does not need stating; but it does not square with scientific practice. Even in the last century Whewell (1858) pointed out that "a supply of appropriate hypotheses (among which the right one is to be chosen by inductive methods) cannot be constructed by rules" and called them "felicitous and inexplicable strokes of inventive talent"; and since then almost every distinguished scientist has insisted on the essentially creative nature of theory formulation, in which the researcher's intuition plays a central part. And the reasons are clear enough. Firstly, any hypothesis which goes beyond the extreme phenonemological limit of merely providing a convenient summary of the experimental evidence involves concepts which no process of abstraction from that evidence can generate and which must therefore be invented. Secondly, we expect to gain from a scientific theory not only predictions that go beyond (and sometimes considerably beyond) the data used to justify it, but also an insight into the structure and behaviour of the particular segment of the universe it concerns; indeed, without that insight we could hardly achieve the predictions.[5] Thirdly – and perhaps surprisingly for those who have never taken part in theoretical research – a new theory is often proposed in direct contravention to some part of the available experimental data which the theoretician feels to be somehow irrelevant.[6] The roots of theory creation lie in fact both in the scientific background of the problem and in the social and cultural conditions of the time, as well as in the individual quirks of the scientist's psychology. The techniques of theory building, insofar as they are not rooted in the problems of the particular scientific field, thus belong more to the psychology and history of science than to its philosophy; here we need only remark that the term 'induction' applied to them is evidently a misnomer: it would be better to speak of invention.

One aspect of this problem requires some further comment. Once the general structure of a theoretical formulation has been found (or sometimes in the process of finding it) it may be necessary to determine numerical values for certain parameters; if they can be formulated as parameters of a popula-

---

[4] The term 'induction' was of course used much earlier, but usually in a somewhat different sense; thus Newton used it to mean a process of generalization to an as yet undetermined scope.

[5] Some writers (e.g. Paulette Fevrier and J.L. Destouches (1959) still treat physical theories as nothing more than prediction machines. This is to ignore rather arbitrarily the considerable and largely unanalysable skill required for actually deriving the predictions; moreover the view is fully consistent only with an idealist or at least positivist ontology.

[6] Wegener's theory of continental drift, for instance, contradicted most of the facts known fifty years ago, only to be triumphantly vindicated by discoveries made after his death. Or, as Dirac (1963) puts it: "... it is more important to have beauty in one's equations than to have them fit experiment. ... If there is not complete agreement between the results of one's work and experiment, one should not allow oneself to be too discouraged, because the discrepancy may well be due to minor features that are not properly taken into account and that will get cleared up with further development of the theory."

tion distribution which is sampled experimentally, this is the problem known technically as statistical estimation; it may also be a problem in curve fitting. There is much misunderstanding of how such problems are solved. The point to be stressed here is that we always require some theoretical notions before even a suitable statistical technique can be chosen, let alone applied: thus if we take the parent population to have an approximately Gaussian distribution, we may use the arithmetic mean over the available data as an estimator of the population mean; but with a Cauchy distribution $(dP(x) \approx (l + x^2)^{-1}dx$, where x is the variate) the sample mean is known to be inadequate. Similarly, the function fitted to a curve should ideally be derived from the corresponding theory; polynomial fitting is only done if no such theory is available (or for simplicity in secondary questions). When such statistical methods are used actually to develop a theoretical notion, they often go astray; thus if the number and general nature of the factors to be hunted for in the popular technique of factor analysis is not established beforehand, the results can easily be quite meaningless, however impressive the computer output.[7]

(c) *The justification of theoretical formulations.* The problem here, how to use experimental or observational material for judging the validity of a theory, is the central problem of confirmation theory. It has two aspects, the acceptance criterion for one theory considered in isolation, and the criterion for preferring one among several contending theories. Both involve the concept of a probability for a theory in an essential way; in fact the main idea is usually formulated in terms of Bayes' law:[8] the new probability q say, of a theory T in the light of the experimental evidence e is given in terms of its old (or prior) probability p by

$$q \sim p\Pr(e \mid T)$$

where $\Pr(e \mid T)$ is the probability of finding e if T is assumed, and the proportionality constant depends on what other theories are compared to T. This tells us what to do once e is found; but where and how are we to look for it? The strategy to be followed used to be stated in terms of J.S. Mill's celebrated methods of agreement, difference and concomitant variation; today their much more sophisticated descendants, derived from statistical theory, hold sway whenever the factors are too numerous or not easily controlled. But though these strategies raise a good many questions themselves, here we are interested more in how to deduce consequences from their results. If these consequences are found by stipulating how a theory's probability is raised or lowered by experiment, then we shall use the term 'induction'.

---

[7] They may also be misleading if the essentially linear nature of the underlying mathematical theory if forgotten – an ever present danger in the social sciences.

[8] But not always: Popper's confirmation theory makes a theory the more acceptable the less probable it is; Bayes' law cannot then apply. And his 'degree of corroboration' has not even the formal properties of a probability (Popper 1959a).

(v)

The conclusion drawn from what was said in Sect. (iii) is that inductive methods (restricting, then, the term to the probabilistic justification of theories) are not viable. Of course the strategies outlined by J.S. Mill and since grown to maturity in the analysis of variance, for instance, remain valid; the gap that opens is in the interpretation of their results. None of the alternative criteria sometimes adduced can fill it. Compatibility with all known facts (Giedymin 1960) is an impossible criterion to satisfy except for the ultimate, all-embracing theory to end all scientific research – and (let us hope) we are as far from it as ever;[9] indeed, as was mentioned, deliberate incompatibility with certain facts is sometimes useful. Compatibility with all previously established theories is a self-defeating requirement, for it would result in the sterilisation of research, condemned to remain within the limited framework of present-day theorization; it is also impossible, for at no time have the well-established theories been at all logically consistent. Simplicity has been much discussed, notably by Bunge (1963), who has shown in great detail how confusing and even misleading such a notion can be; in fact, if it were used as the sole or even the main criterion, it reduces research to absurdity: the limit of a simple (and even irrefutable!) theory is "something happens somewhere". Yet if we look for the simplest theory which is also consistent with facts,[10] we have resurrected the ghost we had hoped to expel: this is no more than a restatement of the problem we started from.

Fortunately an examination of what is actually done in the course of scientific research offers a clear and reasonably simple way of filling up this gaping hole left by the disappearance of the probability concept as a support for inductive methods. Scientific theories are not in fact either accepted or rejected *en bloc*; nor does the mere number of confirmatory instances play any role in judging their usefulness.

Instead of merely counting instances, the theoretician tends to examine their circumstances and to draw from them conclusions of an entirely different nature: he tries to delimit the sort of cases where his theory "works", and those where it does not. In other words, he tries to determine the range of values of the theory's variables for which it is suitable – which here is called the scope of the theory. This procedure thus established simultaneously regions of acceptance and regions of rejection for the theory. The question is not whether there are such regions – this is taken for granted; it is rather where the limits of these regions lie. The value of a theory is not really known until these limits are known.

It must be observed that the scope is not a simple notion which depends only on the theory in question: whether any particular application falls within the scope of a theory is decided on the basis of the precision of the predictions

---

[9] There is no need here to enter into the hornet's nest of questions involved in establishing "all known facts".

[10] Wittgenstein (1961): "The process of induction is the process of assuming the simplest law that can be made to harmonize with our experience."

required. If the approximation needed is rough, then the scope will be much wider than it can be for very precise predictions. Alternatively (and as we shall comment below, the alternative is of practical importance), a much simpler theory might do instead of a very good and hence often very elaborate theory.[11] Thus classical Newtonian mechanics yields quite good predictions for satellite orbits – good enough to pick up their images in a telescope; but if we want to use the satellite orbit to deduce the shape of the irregularities in the earth's gravitational field, then the orbit must be calculated from the more precise theory of relativity.

That the scope is the basic criterion used in research for evaluating a theory is easily documented; but to avoid prolixity we will cite only one example, taken from a well-known text in the theory of nuclear reactions:[12]

"... consider the interaction of a single nucleon with a nucleus ... Such a nucleon, if it enters the nuclear field of the target nucleus, will interact with the target nucleons in its path and will either be absorbed or emerge on the other side after considerable buffeting. ... At lower energies, the wavelength of the particle becomes comparable with nuclear dimensions and it is not so useful to think in terms of a series of nucleon-nucleon collisions.

The simplest model of the nucleon-nucleus interaction ... [was] studied by Bethe as early as 1935, and it was found that some of the [consequences] ... were not in accord with experiment ... It was therefore superseded by the compound-nucleus model (Bohr 1936), which was ... in turn shown to be inadequate by the new results on direct interactions at higher energies ... It was then found that the old potential model, with the addition of an imaginary component of the potential to produce absorption, is able to give a good account of the data above about 10 MeV and also of the data at lower energies when the cross-sections are averaged over many resonances. This new model occupies an intermediate position between the shell model with its weak absorption and the compound-nucleus model with its strong absorption ...

It is thus not possible to treat elastic scattering on its own without considering the accompanying absorption processes. To a good approximation they can, however, be all lumped together and treated just as a process that removes particles from the incident beam. Provided no single absorption process dominates the interaction, it is not necessary to differentiate between the various absorption processes and consider each of them in detail."

(pp. 1–3)

[Discussing a semi-classical approach to the optical model]. "Such a model cannot account for the features of nuclear reactions that depend on the characteristic structure of the compound nucleus; ... Nevertheless, it was found

---

[11] It might be added that if we require absolute precision, then no theory whatever will do ...

[12] P.E. Hodgson (1963). I am grateful to Dr. Hodgson for permission to quote from his book. Note that 'nucleon' is a generic term for either proton or neutron, and that 'imaginary' refers to the second component of complex numbers and does not have its everyday meaning; a 'potential' is a mathematical way to describe the interaction between two particles.

to account for the main variations with atomic weight and energy of the total and differential cross-sections for neutron scattering, and to give qualitative agreement with the measured neutron strength functions and cross-sections for compound-nucleus formation."

(p. 9)

"The [optical] model is not equally successful in all these interactions, and it is useful to examine in a little more detail the approximations made, to see if the range of applicability of the model can be more accurately defined.

At low energies the compound nucleus can usually exist in a series of fairly sharp states, well separated in energy. If the energy spread of the incident beam is less than the mean level spacing, the scattering cross-sections will show characteristic variations that cannot be accounted for by an optical potential."

(p. 16)

"The optical model gives σ_{CE} and σ_A, while experimental measurements of the elastic scattering and absorption give $(\sigma_{SE} + \sigma_{CE})$ and $\sigma_R [= \sigma_A - \sigma_{SE}]$. This leads to serious difficulty in applying the model to low-energy scattering ... At energies greater than about 10 MeV this difficulty gets progressively less serious ... The higher energies make available many more channels for the decay of the final nucleus, so that the compound elastic scattering falls to zero and the absorption and reaction cross-sections become equal. For nucleons interacting with the medium-weight nuclei this happens in the region of 15 MeV.

From this energy region up to several hundred MeV the optical model gives a good account of the experimental situation. It becomes less useful above about 300 MeV because the number of partial waves making a significant contribution to the interaction becomes so large that numerical computations become rather lengthy. Furthermore, relativistic effects become significant, and there are ambiguities in the formulation of the optical potential for incident particles of spin greater than zero.

Generally speaking, unless explicit account is taken of resonance and compound-nucleus effects, the optical model is most useful in the energy range 10–300 MeV.

The optical model is usually more successful for heavy and medium-weight nuclei than for light nuclei, as the more nearly approach the limit of uniform nuclear matter."

(p. 17)

These quotations from the introductory chapter, where the main aims of the book are set out, might almost constitute a textbook example for the thesis here propounded. The insistence on defining accurately the range of applicability – which is here called the scope, for brief; the numerically evaluated energy range and numerically not specifiable range of phenomena well represented by the model; the dependence of the validity of the model on the experimental parameters (which define the application and do not, of course, depend on the model); the limitations of the model due to certain factors being ignored, and those due to computational problems; the simplifications which can for certain purposes be introduced: all these and other points are directly relevant to an understanding of what 'the scope of a theory' means, how it is determined and how it is used. Similar though perhaps not always so clear statements may be found wherever an overall picture of a scientific theory is presented.

One further point is worth mentioning: the very generally accepted requirement of giving experimental error limits to make any measurement result meaningful makes sense if it is to be used either for circumscribing more closely the scope of a theory or to decide if a given theory will serve for the purpose at hand; but it seems difficult to fit into a picture of determining a probability for the theory. The scope concept, then, is seen to be both in actual use and very relevant to the epistemology of theory construction. On the other hand, a relatively superficial examination of actual research practice also reveals that the inductive methods (which we outlawed above) cannot in fact be exemplified; they are not used in research. It might be thought that a least one inductive argument is almost always needed, though never spelt out in detail: since the search for the scope cannot cover the ground in continuous form but must restrict itself to a finite number of sample points, we must extrapolate from the results inductively and so justify the conclusion; otherwise, how does one arrive at the justification of the continuity of the results? The answer is, of course, that one does not: the continuity or otherwise of the scope's frontier is part of the theory and established just as the rest is. In any cases the theory predicts the there should, indeed, be a discontinuity. If it is found where predicted, this adds another confirmation; but if a continuity appears unexpectedly, or a discontinuity is found where there should be none, the scope of the theory is again limited in an appropriate way.

We must thus conclude that what dominates in actual research is the effort to establish the scope of the theory under study; essentially no other tools are used for this purpose.

The implications of this view are rather far-reaching.

A first point is that the adoption of scope implies a change in the demarcation criterion for what can be held to be a scientific theory. The most widely accepted answer so far has been Popper's (1959a), who proposes that a candidate theory must in principle be falsifiable. It has been pointed out (Hempel 1965) that this is unsatisfactory because if a theory T is scientific according to this criterion then the conjunction $T \cap s$ also satisfies it, where s may be non-falsifiable. The scope concept can avoid this sort of problem; this is seen if we push the analysis a little further. An experimental measurement is compared, not to the whole theory under study, but to a specific prediction which generally requires only some part of the theory for its derivation. In other words, we must speak, rather, of the scopes of the various laws derivable from the theory; from these scopes we can establish the scopes of the basic concepts and relations of the theory – in fact, if it has been formalized, of its axioms. The scope of the theory as a whole must then be taken as their intersection: the range, that is to say, where any part of the theory can safely be applied.[13] If such an analysis were carried out on the combined theory $T \cap s$, it would reveal at once that s has no scope associated with it and thus has no significance. A more serious difficulty with Popper's demarcation criterion (Archibald 1966) is that it does not work very well for theories which do not decisively exclude certain events but only assign them very low probabilities – a common situation. But the concept of scope has a "fuzzy edge" because of the uncertainties in experimental data and the resulting finite precision required for any application. Thus a demarcation criterion which calls an abstract structure scientific when we can determine a scope for it is well adapted to such situations; at the same time it does not suffer from the peculiar asymmetry of Popper's criterion as regards verification and falsification.

Secondly, as was mentioned above, we cannot any longer speak simply of accepting a theory: it will be acceptable within its scope and unsatsifactory outside it. The question of where to draw that line of separation is a very practical one, to be answered in each case by appropriate research, and not a matter for abstract discussion. This sidesteps several of the oddities in the probability-and-induction formulations. A theory is not automatically refuted by a single negative instance, as Popper (1959a) would require, nor on the other hand does it have but little influence in the face of very many confirmatory instances, as for Carnap (1950): its importance now depends on *where* in the hoped-for scope it is located. At the edge it may indeed be welcomed, while it must be taken as almost certainly fatal if it falls in the very middle. One could even say that to establish a theory properly (that is to say, to determine its scope) we must, experimentally, both verify and falsify it. But the apparent paradox only underlines the fact that the probabilistic and inductive formulations ignore the significant element which actual research requires.

But since the scope concept implies that a theory is accepted for some purposes and rejected for others, one may ask whether we have not made the concept of truth irrelevant to any scientific theory, in the fashion of the pragmatic philosophies. This problem can be only briefly discussed here. While no scientific theory can be unrestrictedly and wholly true, any theory of more than negligible scope must contain some truth (i.e. form to some extent an adequate model of reality) in order to make that scope possible. At the same time, within the framework of that theory it is quite impossible to isolate what is true and what is false (if only because we would then automatically have the makings of an improved theory). Indeed, we can only begin to sift what truth it may hold once we have a much better theory to cover the same ground.[14]

---

[13] As an example, it may be observed that the conservation of linear or angular momentum, a central element of classical Newtonian mechanics, is valid well beyond it in quantum mechanics and relativity theory; other elements, such as the vector additivity of velocities, are not.

[14] Even a thoroughly discredited theory such as the phlogiston theory of combustion contains some valuable elements: from the standpoint of our much better understanding we can see it as a sort of mirror image of what "really" happens in combustion. The example underlines that the partial truth in any theory cannot

And of course anything that might be called an absolute truth can at most be achieved asymptotically. Within its scope, however, we are well justified in accepting a theory as a good approximation to the truth and not merely as a convenient prediction tool; and it is neither possible nor necessary to demand more.

Related to this question of what truth there is in scientific theories there is another matter. If the scope of a theory delimits a region of high "truth content" (to speak somewhat loosely) then another theory whose scope includes the same region may be substituted for it and nothing should be lost: within that scope their *truth contents* should coincide, but outside it they can differ widely, of course. It will be observed that this is no more than a restatement of Goodman's (1955) paradox that finding a green emerald confirms not only the theory that all emeralds are green but also the one that they are "grue" – that is to say, green before the year 2000 (for instance) but blue afterwards. Yet this is paradoxical only for the inductivist trying to confirm one theory probabilistically as against the other; there is nothing paradoxical or even exceptional in two or more different theories being applicable to a certain range of phenomena. Indeed, it is of common occurrence in actual experience, and extremely useful in research: it is the basis of all successful approximations. A good example will be found in the quotations given above concerning the optical model: here (under certain limiting conditions which are spelt out in some detail in the original text) we need not treat the various absorption processes individually but can substitute for the ordinary real-valued potential a complex one whose imaginary component describes a single non-specific absorption which approximates rather well the joint effect of the different absorption processes. It is possible, at the expense of considerable labour, to calculate these processes one by one; but not all of them are amenable to a more or less simple treatment, and we lose valuable insights. In other circumstances the approximation is indispensable: a computer, for instance, is unable to calculate an exact square root, but it can give us a sufficient approximation. If now we can show that the approximation has no significant effect on the scope of the end result of the computations, then we have justified its use; yet there seems to be no way of justifying it by means of probability considerations. In view of its importance, what appears in the literature as Goodman's paradox should really be called Goodman's theorem; and those who attempt to argue it away (e.g. Barker and Achinstein (1960)) are inappropriately curtailing the scientists' armoury of tools. Goodman's new concepts of 'projectibility' and 'entrenchment', however, are not needed, because logically conflicting theories can have overlapping scopes.

It should be pointed out that this sort of "scope equivalence" cannot as yet be handled by the methods of formal logic, for the two theoretical structures which are equivalent in this sense are united only by the requirement that inside the scope where the equivalence holds their predictions must not

---

be analyzed out by the manipulations of formal logic: it will usually be analogical, so that it is understood only in the light of new and profounder notions.

differ by more than the margin of tolerable error which is a parameter of the equivalence relations; there will, in fact, often be decisive discrepancies between certain key concepts in the two theories which make them logically incompatible. Possibly because of the peculiar nature of this relation there has been such confusion over the replacement of scientific theories by newer and (presumably) better ones. Some would say that the new theories invalidate the older by proving them false; others hold the process to be a purely sociological one or even one of mere fashion. Deaf to the unending discussions, scientists sometimes use the new theories and at others continue with the older ones; eyebrows are raised at this and the scientific establishment is accused of ultra-cautious conservatism. Now scientists are no doubt quite as liable to human failings as other groups; but in the light of our discussion of the scope concept it is clear that no charge of systematic timidity is justified: for all problems where an older theory offers predictions within the scope (as found from the required precision) there is no reason to prefer a newer theory, except for convenience.[15] As time goes on, of course, extrascientific reasons arising from the development of technology or from economic and social changes will create new problems whose greater need of high precision will preclude the older theories; thus the sociological factor is by no means irrelevant, but it acts not quite as often supposed. Hence there is nothing reprehensible about the use of Newtonian mechanics in much of engineering and scientific research, provided we have recourse to relativity theory where necessary. On the contrary: doing so is only congruent with the idea we mentioned above that within its scope a theory does in fact contain a good deal of truth; if we had to abandon it altogether when a new theory was found we could not really say that it had ever held any truth whatever – for what happened to that truth when the better theory came along?

It remains to see whether the black raven (Hempel 1937, 1945) is not indeed a red herring. The paradox is this: Since "all S are P" is (logically) equivalent to "all non-P are non-S", the two statements must – if the notion has any meaning – have the same probability; but then the discovery of a non-P which is also non-S not only confirms the second statement but also increases the probability of the first one – to which the discovery is quite irrelevant. Mackie's careful survey (1962) makes it rather evident that none of the many ingenious arguments about the paradox really solves the puzzle, his conclusion notwithstanding. Without going into detail, it seems clear that the reason must be sought in their using the probability concept for theories. With the scope conception the matter is much simpler. The scopes of the two statements do not coincide; in fact if the statements hold they are disjoint, since that of the first must lie within the set of S's, that of the second within the set of non-P's. But then any experimental evidence which only makes one scope contract or expand is irrelevant to the other, since (unlike probabilities)

---

[15] As sometimes happens: the author has had occasion to use a newer and more complex theory for which a computer program was available instead of an older one which would have required laborious calculations by hand.

there is no upper limit for them and even less for their union. Note, however, that this applies only to certain types of logically equivalent statement pairs: in many cases the scopes coincide, or overlap, or are contained one in the other. For such cases confirming evidence concerning one scope is relevant to the other, and no paradox of the Hempel type can appear. Another possibility arises in a universe where all scopes are restricted to a finite number of cases (a much stronger restriction than that to a merely finite universe): then there might exist a limit on the set-theoretical union of the two scopes through which the expansion of one could influence the other. We have reason to suppose the case of practical importance, but it must be considered in connection with certain counter intuitive arguments among those listed by Mackie.

The other puzzles mentioned above disappear together with the notion of induction that gave rise to them.

(vi)

To sum up, it seems quite clear that it is not any notion linked to probability which provides the criterion for the usability of scientific theories but rather their scope, a concept of entirely different nature. Most important, perhaps, is that the applicability of a theory cannot be decided in general but depends on what the proposed application is, together with the precision it requires; it is thus no longer an abstract epistemological problem but a concrete and practical one, to be decided by the investigation of each case. This account is, as we have seen, borne out by scientific practice. (I am not arguing that what is actually done is the overriding norm: the philosophy of science cannot be reduced to a mere description of usage. It is to be desired, rather, that it should develop to the point where it can signal deficiencies or new possibilities in scientific work. I am only saying that in the present case scientific practice bears out the conclusions of philosophic argument.)

Why then has the notion of a probability for scientific theories persisted so stubbornly? Part of the reason, I suggest, is that we have here no more than a somewhat ossified metaphor. Whenever we deal in genuine probabilities, we cannot make an absolutely certain prediction; it is then a small step to extrapolate the same terminology to circumstances where our uncertainty has a different origin, namely that the theoretical structure we work with has an as yet unknown scope. Another reason lies in the fact that a scope can often (though not always) be expressed in numerical terms; thus in our example above the optical model works well in the region from 10 to 300 MeV, approximately. If we compare such a scope with what one would like to achieve, the fraction we obtain has at first sight many of the properties of a probability (thus both are numbers between 0 and 1); and a probabilistic terminology seems not inappropriate. But this does not justify an elaborate philosophical structure built on the metaphor, such as Carnap's (1950) theory of probability. It must not be forgotten that two theories whose scopes may have the same measure need not therefore be scope equivalent (in the sense suggested above): a nuclear-reaction theory that works between 310 and 600 MeV is by no means a satisfactory substitute for the optical model!

Finally, it should be stressed that even if the basic ideas outlined here are accepted, there remain many open questions, apart from those already mentioned. There are in the first place a number of technical questions: the relationship between scope and precision must be made more specific seeing that in some cases the scope can change drastically with changes in precision and in others be almost insensitive; I have spoken of the experimental determination of the scope, but frequently there are theoretical considerations that help to delimit it; the relation of scope equivalence requires more detailed study, as do the interrelations among the scopes of the various elements that go to make up a theory; and the question arises whether a scope can meaningfully be assigned to experimental techniques. Secondly, one somewhat exceptional case should be examined: if the values of some relevant parameters cannot be determined (whether because of experimental difficulties or other reasons does not much matter) then one cannot decide if a given situation falls inside or outside the scope, and even before that it may be impossible to delimit the scope. Such a state of affairs could justify a rather induction-*like* procedure of accumulating as much experimental materials as possible and using statistical arguments; but then the corresponding theoretical predictions would have to be of the probabilities to which the experimental runs are compared. Thirdly, the nature of the scope itself calls for further investigation; I have treated it here as essentially an uncountable set of satisfactory applications of the theory defined by parameters (numerical and otherwise) that generate a partly metrical space, but this may need revision. Fourthly, as mentioned above, it is usually taken for granted that the scope of a theory exists (i.e. that it makes satisfactory predictions in some cases but not in others); but though experience has so far borne out this assumptions, its ontological basis should be made clear and requires adequate underpinning. It may well be, after all, that there could be scientific theories to which the notion cannot be applied – in which case the present theory would itself acquire a definite scope; this does not however mean that such theories have an indefinitely large scope, for that would turn them into dogma. Lastly, there remains a question of considerable relevance for scientific practice: that of the *provisional* assessment of a theory, before its scope has been determined.

References

Archibald, G.C. (1966): Brit. Philos. Sci., **17**, 279

Barker, S.F. (1957): *Induction and Hypothesis* (Cornell University Press, Ithaca, N.Y.)

Barker, S.F., Achinstein, P. (1960): Philos. Rev., **69**, 511

Braithwaite, R.B. (1953): *Scientific Explanation* (Cambridge University Press, Cambridge)

Brody, T.A. (1975): Revista Mexicana de Física **24**, 25

Bunge, M. (1963): *The Myth of Simplicity* (Prentice-Hall, Englewood Cliffs, N.J.)

Carnap, R. (1950): *Logical Foundations of Probability* (University of Chicago Press, Chicago)

Caws, P. (1965): *The Philosophy of Science* (van Nostrand, Princeton, N.J.)

Cooley, J.C. (1957): J. Philosophy **57**, 293

Dirac, P.A.M. (1963): Scientific American **205**, No. 5, 47

Fevrier, P., Destouches, J.L. (1959): in Henkin, Suppes, Tarski (eds.) *The Axiomatic Method* (North-Holland, Amsterdam) p. 390

Fisher, R.A. (1922): Trans. Roy. Soc. **A222**, 327

Giedymin, J. (1960): Studia Logica **10**, 97

Goodman, N. (1955): *Fact, Fiction and Forecast* (Harvard University Press, Cambridge, Mass.)

Hempel, C.G. (1937): Theoria (Goteborg) **3**, 222

Hempel, C.G. (1945): Mind **54**, 1, 97

Hempel, C.G. (1965): "Postscript (1964) on Cognitive Significance", in: *Aspects of Scientific Explanation* (The Free Press, New York) p. 121

Herbert, S. (1977): *Models of Discovery.* (Reidel, Dordrecht)

Hodgson, P.E. (1963): *The Optical Model of Elastic Scattering* (Clarendon Press, Oxford)

Jeffreys, H. (1961): *Theory of Probability*, 3rd ed. (Clarendon Press, Oxford)

Kyburg, H.E. (1968): in I. Lakatos (ed.) *The Problem of Inductive Logic* (North Holland, Amsterdam) p. 98

Mach, E. (1897): *Contributions to the Analysis of Sensations* (Open Court, Chicago)

Mackie, J.L. (1962): Brit. J. Philos. Sci. **13**, 265

Mill, J.S. (1843): *A System of Logic, Ratiocinative and Inductive* (Longman, London)

Popper, K.R. (1959a): *The Logic of Scientific Discovery* (Hutchinson, London)

Popper, K.R. (1959b): Brit. J. Philos. Sci. **10**, 25

Whewell, W. (1858): *Novum Organum Renovatum*, London. Many relevant passages are quoted in R.E. Butts, *Willeam Whewell's Theory of Scientific Method* (University of Pittsburgh Press, 1968), pp. 133 ff.

Wigner, E. (1971): in B. d'Espagnat (ed.) *Foundations of Quantum Mechanics* (Academic, New York) p. 1

Wittgenstein, L. (1961): *Tractatus Logico-Philosophicus* (tr. Pears and McGuinness) (Routledge Kegan Paul, London) par. 6.363

6. The Incommensurability of Theories *

In recent years, a certain conception about the relationship between scientific theory and experiment has come into fashion, according to which the terms that describe experimental results are to such an extent "theory-laden" that their meaning is inescapably linked to the theory in question; therefore, it is argued, experimental results cannot be said to constitute decisive proof of a theory, and furthermore, they are irrelevant within the framework of any other theoretical formulation. The conclusion appears to be that different theories are incommensurable, and that we cannot establish the validity of one over the other (see Note 1 at end of chapter).

This concept has been severely criticized, mostly because it leads to an unacceptable relativism, where the selection of one scientific theory rather than the other is determined by extrascientific factors, from the "aesthetics" of a theory to sociocultural pressure and personal taste. This relativism has its source in the notion that we cannot refer to one theory as being closer to the truth – or as being a better representation of reality – than another. Such criticism, therefore, attempts to refute the thesis that criteria used to judge a theory, i.e. experimental data, are intimately tied to the theory and change with it.

To criticize the thesis of incommensurability for its relativism appears to me to be entirely justified; however, analysis must be pushed further, because we cannot retreat with impunity into a view of experimental data as being totally independent and rigorously objective; this view is in fact incompatible with the critical activity which scientific investigation requires in practice.

It should not be forgotten, in other words, that the thesis that different theories are not commensurable was erected in counterposition to the ideas of logical positivism, according to which science, statically conceived as a series of propositions, was divided into observational propositions (uniquely gifted as direct bearers of meaning), and theoretical propositions, which have meaning only to the extent that they can be reinterpreted as consequences of observational propositions. It should not be necessary to mention the inadequacies and fallacies of these discredited notions, which falsify and distort reality absurdly (see Note 2).

---

* "Sobre La Inconmensurabilidad de las Teorias", Revista Mexicana de Física **35**, Suplemento S103 (1989). (Translated by J. Brody.)

The situation is to some extent reminiscent of the mood which reigned when Descartes put forward his notion of radical doubt, in opposition to the official conception, which refused to accept any doubt. It is understandable that there should have been an almost violent reaction to the domination of extremely rigid philosophical structures, which were not up to the task of depicting a shifting and complex reality; but the Cartesian reaction missed its mark. The method of radical doubt could only, in its turn, lead to a rather sterile idealism; and, in fact, Descartes himself did not use it in his scientific work. Analogously, we can sympathize today with those who are enraged by the sterile rigidity of positivist ideas; but we cannot accept a notion which is purely and simply its opposite.

There is another aspect of Cartesianism which is relevant here. The two opposed points of view took "all or nothing" extremes: either there could be no doubt, or everything must be doubted at once. The scientific method which finally indicated the way out can be conceived as the method of doubting at a precise point, chosen through the highly practical criteria that doubt must be fruitful, and not doubting (for the moment) everything else, whose turn will come in time. Correspondingly, we are dealing here with a different manner of doubt, which is experimental and theoretical, and does not take the route of intellectual speculation.

Analogously, in the discussion about the distinction between theory and observation, both sides have erred: those who seek an absolute and universal distinction, and those who deny its existence. There is a distinction; and, recalling the Cartesian lesson, we might suspect that it is a practical distinction, which varies from case to case. And this is corroborated by study of what actually happens in research. What is, properly speaking, experimental is what concerns the experiment at hand; any extraneous information – concepts, equations, data – is "theoretical", even if it comes from another experiment. No problem is caused, either, by the "theory-laden" character of the terms in the experimental description (see Note 3).

Firstly, a large part of this theoretical load comes from other theories, and is unaffected by our acceptance or rejection of the theory under scrutiny. Secondly, what is relevant in our judgement of the experimental criteria's objectivity is not so much the meaning of concepts, but the numerical values cast up for them by the experiment. Having carried out the experiment and obtained a specific set of numbers is a fact, unalterable by theory. Any relevant theory can put forward an *interpretation* of this data; the meaning of the concepts related with numerical data will vary according to the theory; but only one of two options is possible: either a theory yields an account of experimental data which is better than that of its competitors, in which case it is better, or all theories explain data in a more or less equally acceptable way, in which case the experiment was indecisive. Neither of the two sides seem to have grasped these points, presumably because it did not occur to them to verify their ideas in a scientific laboratory: there they would have found that this practical means for distinguishing theory from experiment is well known, and in fact has its foundations in certain important parts of statistical theory, which has treated and solved this problem long before it gained the attention of philosophers (see Notes 4 and 5).

To conclude, I would like to point out another problem created by the concept of incommensurability. In science, the relationships between different theories are multiple. They exist basically because the theories describe different aspects of a same reality. They manifest themselves in many concepts shared by the theories, which often extend over vast regions of theory – not without suffering some transformations in the process. Dynamic relationships also exist: these use one another, complement each other, and occasionally come together. Finally, there are experimental relationships, because *one* experiment usually requires several theories, which are sometimes contradictory. These relationships amongst theories constitute an important problem for philosophy; but, should we accept the thesis of incommensurability, this highly interesting field would appear to be off limits for us.

Notes

1. The notion of incommensurability is due to Kuhn, T.S. (1962): *The Structure of Scientific Revolutions*, Chicago University Press. Analogous ideas are expressed by Feyerabend, P.K.: *Problems of Empiricism*, in Colodny, R.G.(1965): *Beyond the Edge of Certainty*, Prentice Hall, and other later essays. The notion that the terms we use to describe experimental results are "theory-laden" seems to be due to Hanson, N.R. (1958): *Patterns of Discovery*, Cambridge.

2. Literature dealing with Logical Positivism is extensive, ranging from Schlick, M. (1918): *Allgemeine Erkenntnislehre*, Kreisverlag, Wien, to Ayer, A.J. (1959): *Logical Positivism*, Macmillan.

3. The theoretical load upon the experimental terms contains a great number of elements which originate in theories other than the one which is being examined. This is because we require multiple experimental techniques, based upon as many theories, in order to gain control over experimental variables which are not susceptible to measurement, and to measure the variables we have selected. This load, therefore, often includes a theory which is opposed to the one under scrutiny. A typical example can be found in the 1919 experiment designed to prove the prediction, made by general relativity, that the sun's gravitational field should deflect light. This experiment called for elements from optics, both geometric and physical, in order to construct and operate telescopes; from thermodynamics, required to correct the thermal expansion of instruments and photographic plates; from chronometry, essential to the coordination of measurements carried out on different locations; from positional astronomy; and from Newtonian mechanics, the foundation both for the requisite astronomy and for the techniques necessary to the construction of instruments. If all these positions were taken seriously, general relativity would be proved inasmuch as Newtonian Mechanics are valid ...

4. For brevity's sake, I have discussed the meaning that the results of an experiment may have regarding a theory as if a) no other theoretical suppositions intervened, and b) the issue were definitely the theory's acceptance or rejection. In fact, none of these conditions obtains. The requisite additional suppositions are precisely those which generate the "theoretical load" in the terms used to describe the experiment; the experimental result is therefore relevant to this group of suppositions, and the theory under scrutiny; if there is a discrepancy, we can only conclude that some element within the group is incorrect, whilst if there is agreement, we cannot exclude the possibility of mutually compensating theoretical defects. The point is not to accept or reject the theory, but rather to define its scope; therefore, a failure does not render it necessarily false or useless, just as a success does not guarantee a definitive validity. Moreover, shifting our focus onto a search for the theory's scope will allow us to understand how it is possible for us to use two theories which are apparently contradictory in order to conform "mutually"; for example, positional astronomy and the construction of instruments fall within the scope of Newtonian Mechanics, whilst the behaviour of light as it passes through the solar limbo, which is studied through these theories, fall without it. For more detail, see Chap. 5.

5. In statistical mathematics, the distinction between theory and experiment takes the form of a distinction between variables which characterize the population and those which characterize the sample; the latter are defined as the numerical values obtained in the experiment at hand. These obtained conclusions can then serve as the population parameters (i.e. the theoretical parameters) for another experiment. A good example is the χ^2 distribution, which is often used as a test of good fit. This distribution depends on a quantity known as numbers of degrees of freedom, whose determination directly involves the number of parameters we have calculated from the experiment under specific scrutiny; but it does not involve the number of parameters we obtain from other sources, be they theoretical or experimental. For more detail, see, for example, Kendall, M.G, Stuart, A. (1958 – 1966): *The Advanced Theory of Statistics*, 3 vols., Griffin, London.

7. A Minimal Ontology for Scientific Research

(i)

Traditionally, the various philosophical disciplines, and epistemology in particular, have been felt to be quite irrelevant to scientific research; or at least that has been the view of most scientists. Philosophers (and even many philosophers of science, one is sad to note) have reciprocated by feeling that science has nothing to do with *their* business. This entails a definite risk of sterility on both sides; and at least one philosophy – the Oxford linguistic school – seems deliberately to have embraced that risk. But as far as science is concerned, it will not do; there are too many open problems, troubling ones, too, that require philosophical answers. This chapter will outline a way out of the impasse.

In part, the problem is due to the peculiarities of our academic structure. Where, in order to advance, one is expected to write more and more papers that closely circumscribe the narrow specialisation one has chosen among those that are in fashion; where it is anathema to sit back, to contemplate what one has done and is proposing still to do, to attempt an overall view that does not respect the boundaries laid down by departmental organization; where it is bad manners, to say the least, to dabble in another person's intellectual domain – there the philosophical study of scientists' work and the scientific study of philosophy cannot flourish; and the loss is on both sides. The social roots of this compartmentalization have not been elucidated adequately so far; but my present purpose is rather to examine the philosophical side of the problem.

Here there is a fundamental obstacle, in the long-standing dichotomy between the materialist and the idealist. They have in common their desire to provide a solid foundation for their philosophical positions, a foundation moreover that will render their position relevant to extraphilosophical activities and to science in particular. But traditionally both points of view have attempted to approach this aim with inappropriate means and starting from a mistaken viewpoint.

The materialist argues that if we have knowledge, then we must have knowledge of something, of something, moreover, that must logically precede the knowledge, for that word to have any meaning. And he has knowledge, even if it is only the everyday knowledge that enables him to catch his bus in

* First published in: Revista Mexicana de Física **35** (1989) 107–116

time. Wherefore he concludes that the bus is real and would exist even if he had been too late to catch it. The idealist, noting the family resemblance of this argument to St. Anselm's ontological proof of the existence of God, replies, that to justify our notion that what we call knowledge actually is *knowledge,* we must put it inside for the moment and examine its foundations; but then one can hardly argue about what something is like before it is known; so we must abandon all idea of showing in this way that our knowledge is firmly based, and since no other way offers, we must abandon the concept of "something behind our knowledge" as mistaken.

The two sides in this hypothetical discussion share the idea that what is required is a *logical* reconstruction, starting from unassailable axioms, of the basis of our knowledge, and hence also of our world, an idea that goes back to Descartes. But this idea of how to proceed would be appropriate only to a wholly deductive process of knowledge acquisition. And here is where the mistaken starting point makes its appearance: acquiring knowledge is seen in just such a way, as a passive registering of what our senses report, later to be logically interpreted. But in fact we do not obtain our knowledge in this curious armchair fashion; instead, we go out and *work* for it: that is to say, we do things to various bits and pieces of the world, and from the effects of our actions, from the correlation between what we did and what came of it, we derive knowledge of what they are, how they behave, and so on. It is this active aspect in finding out things that is usually forgotten; it is this also which make the logical-reconstruction program altogether unworkable as far as searching for indubitable first principles and linearly deducing everything else from them is concerned. The actions of knowledge gathering themselves are as subject to examination, interpretation, comparison, revision and even rejection as any other kind of action, and to study their background and justification is certainly licit. Must we then accept the idealist's stricture that to use knowledge in order to argue about the basis of knowledge is circular? Yes, indeed: it is precisely this circularity – or rather, a very significant point, quasi-circularity – which turns the whole thing from an exercise in academic ingenuity into an affair of the utmost practical importance.

The "quasi-circularity" I mention arises from the fact that knowledge acquisition is an *activity*: going around the circle thus alters the state of our knowledge, and so the circle never quite closes. We must get the ends near enough to make sure that the whole procedure makes sense; yet they must remain far enough from each other to ensure that in going round the circle we really have acquired new knowledge. If the ends meet, we have logical consistency: but at the the price of no new knowledge, now or ever, for once the circle is closed we either go round it, gaining nothing, or break it completely. If the ends are too far away, we will certainly have obtained vast stores of new results: but it may be quite impossible to work them up into actually usable knowledge.

Thus the circular argument advanced by our idealist turns out to be a pale, static image of the very real process of acquiring knowledge, what I call the epistemic cycle. (See Chaps. 1 and 2.) Instead of being an argument against materialism, the cycle provides its mainstay. There is a price to be paid: materialist philosophies can no longer rest on a fixed and unassailable basis, but must share the two essential features of scientific theories, namely corrigibility and continual development. There is, on the other hand, a very considerable gain: as we shall see below, epistemic cycles form a dynamic structure of some complexity; but this structure and the increasing philosophical content of the cycles in the upper levels require constant critical intervention from a philosophical viewpoint. In the past, this has often been ignored and so done haphazardly, half unconsciously, by the scientists themselves; not always with happy results. It is high time that philosophers take over their rightful job in this field; nor is this a matter of setting things once and for all, since each time the cycles turn, they change, and each time the philosophical problems have to be studied anew, from a different perspective.

Finding ways of incorporating the philosophical effort into the business of scientific research is vital for the latter: for without this collaboration it is apt to lose its direction. There are ominous signs on the horizon here; and no one will deny the immense significance of putting science's house in order. That in the process academic philosophy will change out of recognition seems a small price to pay; and perhaps there will be many who will welcome the change.

(ii)

From the long tradition of empiricism we have inherited the hypothetico-deductive model of scientific research: faced with a problem domain, the scientist formulates an axiomatic basis for a solution and builds a theory by the deduction, following strict logical methods, of all required theorems. These are then applied to specific cases, yielding observational predictions subject to experimental confirmation. Since the latter cannot be unambiguous and exhaustive, inductive logic is called into to being to resolve the question of how well the theory is supported by the facts. Should this support turn out to be inadequate or even negative, the whole structure is scrapped and the search for a new axiomatic basis is initiated. The different theories so developed hang together in one consistent whole, sharing many concepts and a fundamental world picture.

A first attack on this pretty picture was delivered by Popper's insistence on falsifiability rather than the verificationist doctrine taken over from logical positivism. But his rigid rationalism (a better word would be logicalism) left the hypothetico-deductive conception largely intact. A more serious attack came from the historicist views of Hanson, Kuhn, Feyerabend and others, who say the theoretical structure as built up on a foundation introduced (as a paradigm, in Kuhn's terminology) into science from outside; between different theories there is little in common and they may even be incommensurable, and it is therefore an illusion to think that one theory is in any sense better than another, that there is scientific progress. And experimental method is so heavily laden with the theory the experiments are expected to confirm that the naive observer might wonder why the scientists bother to carry them out
. . .

Thus we have, on the one side, a picture of stately progress as theory after theory unfolds, but unfortunately the hypothetico-deductive epistemology (either in its original form or in its Popperian variant) fails to provide us with a mechanism to explain the creation of theories, or for that matter for the many controversies among which actual scientific research evolves. On the other side, we focus on those controversies and their social origins, but unhappily they are never really settled, only superseded, and so there is never any genuine scientific progress.

Progress but no mechanism to bring it about, or a mechanism for change but no progress; these are, rather absurdly, the possibilities we are offered. In actual fact, scientific research is not like either of these images, which are no more than linear extrapolations of originally quite justified insights; they approximate certain specific cases quite well, but the reality is far richer and more complex than either of them suggests.

In the first place, scientific theories are not, except in the initial stages of development, thrown out because better theories appear on the scene. A not untypical case is that of classical mechanics, which so far from disappearing under the twin onslaught of relativity theory and quantum mechanics has not only remained as the backbone of physics teaching and as the underpinning of endless applications but has recently come to flower again with astonishing richness in the fields of non-linear and of ergodic mechanics. The mistaken conception arose because theories were assigned truth values as a whole, thus giving rise to the extraordinary gyrations and amusing paradoxes of inductive logic and so-called confirmation theory; in reality, a theory has a range of validity (which elsewhere I have called a scope) within which it holds true to a given approximation, and outside of which it fails increasingly badly. One of the main aims of experimental work is to determine the scope of the theory it is concerned with, in the sense of determining its limits; this evidently requires detecting both where the theory fails ("falsification") and where it works ("verification"), neither is sufficient by itself to conclude anything about the acceptability of the theory. In combination, however, they trace out the theory's scope, and if this covers a useful terrain, it will continue to be used and upheld, even though according to Popper's criteria it should be eliminated. It may fall out of use if a new theory has a scope that is everywhere wider; but this is not generally the case, and is not so in particular for classical mechanics. Thus accounts of scientific change which seek to explain the replacement of one theory by another one will be wide of the mark: their explanations will not have many real applications.[1]

---

[1] Some theories are, of course, discarded: those that are found to have too restricted a scope. There is one important exception, namely those theories which, though they have essentially negligible scopes in themselves, provide a basis for (usually approximate) extensions which do yield a significant scope. Such theories are what below I call indirectly applicable. Note that the outright elimination of theories (or attempts at theories) is inevitably very frequent in the initial stages of the development of a science, but the situation changes as it reaches maturity. The case I have quoted of classical mechanics is typical and could be paralleled in many

Secondly, theories are not static. Mechanics – since we used that example – begins with a formulation starting from the familiar three Newtonian laws and lacking various concepts that to us seem vital, *e.g.* energy; the eighteenth century mathematicians reviewed and revised its structure, starting a process that led to the Hamilton-Jacobi formalism; from this the study of canonical transformations and invariants led to the insights embodied in Noether's theorem, while applying the ideas to special types of systems opened up new fields like those of continuum mechanics or ergodic theory; it is chiefly the latter that has provided such startling new results in the last few years. A recent textbook on theoretical dynamics might not easily be comprehensible to Newton; yet it describes something that has evolved in continuous fashion from his ideas and has its roots in them. Neither of the two epistemological viewpoints that dispute the field takes much notice of such evolution of theories, though it needs but little acquaintance with the history of science to be aware of the significance of this phenomenon.

Thirdly, theories, are by no means all of one kind. Even in a relatively unified and mathematically well-structured science such as physics, theories come in all shapes and sizes. Physical theories are conveniently classified according to at least six different criteria: profundity, quantitativeness, independence, structural exactness, generality, and directness of application. A theory may be more or less profound, as opposed to being phenomenological, in the sense of having more or less explanatory power; a phenomenological theory does not do very much more than summarize the experimental evidence.[2] A theory is quantitative in the degree in which it involves concepts with which numerical values can be associated (which usually can also be measured experimentally). Most theories in physics are highly quantitative, which has misled some observers into neglecting or believing non-existent their highly significant qualitative aspects. An associated error is to believe that the quantitative and qualitative aspects (the formal part and the contents) of a theory can be neatly separated, to yield an uninterpreted calculus for which a deductive structure descending from a suitable axiom set can be derived, plus an interpretation based on an *observation language*[3] or the like. The independence of a theory is a characteristic that has received very little serious study as yet; its meaning

---

other fields; but too many philosophers still argue as if classical mechanics had been refuted.

[2] This distinction, vital to the working scientist, is obscured by the empiricist philosophies and by logical positivism in particular, which treats all theories as phenomenological. An almost opposite error is committed by those (and they are often the same people) who ignore the phenomenological or semi-phenomenological theories, concentrating their analytical efforts on just those theories that are currently viewed as the most profound; they are misled, perhaps, by the terminology, since 'profound' could also be taken as the antonym of 'superficial' or 'trivial'. But it is not: physics thrives, in fact, on the constant interplay of theories along the whole scale. More than that, a phenomenological theory, far from being valueless, can be extremely "profound" in the other sense: for instance thermodynamics.

[3] Such conceptions have had so widespread an echo that they have been dubbed the received view of theories (Putnam: What Theories Are Not, in Nagel, E., Suppes, P., Tarski, A. (eds.) (1962): *Logic Methodology and Philosophy of Science*, Stan-

is therefore best conveyed by contrasting examples: fluid dynamics is a highly dependent theory in that many of its main concepts and their interrelations are taken from mechanics (classical or relativistic), while relativity theory is independent.[4] If a theory requires approximations to be made in building it up, for instance in order to eliminate what are judged to be inessential factors, then it is structurally inexact. A very typical example is quantum electrodynamics with its need for renormalization to get rid of inessential infinities; this example also shows clearly that structural inexactness has nothing to do with inexactness in the theory's numerical predictions, which here are amongst the most accurate in physics. A theory is general when its scope is not simply wide but covers a range of very diverse phenomena. And finally, a theory is of indirect applicability when it describes ideal cases that do not occur in nature but provide convenient conceptual simplifications for building the theory; thus the mechanics of Newtonian fluids, non-viscous, incompressible and non-dissipative, is very elegant but of limited usefulness since all real fluids deviate more or less markedly from this idealization. Quantum mechanics is another case in point, the fundamental theory having essentially no applications but providing the basis for a large number of approximations and extensions which do have many and significant applications. (This fact and its implications make otiose a good many philosophical disquisitions about the foundations of quantum theory, but this matter cannot be further explained here.)

A fourth point, which is ignored by many philosophers of science but is vital to the scientist's professional life, is the coherence of different theories and indeed of all the different sciences. We must not here think of the illusory logical unity that logical positivism at one time set up programmatically; rather it is a tough but elastic network of interconnections that allows for contradictions and inconsistency to creep in (and even to be built in deliberately, as we shall see below) almost without limit and yet does not break. What conditions this sort of fishnet coherence is the underlying fact that all these theories and sciences examine various aspects of one and the same world; but they examine different aspects from different points of view and for different purposes, and these differences constitute one of the main factors (not the only one) for the

---

ford University Press, Stanford Cal., p. 2409. Whatever justification the attacks of Kuhn (1962): [*The Structure of Scientific Revolutions*, Univ. Chicago Press, Chicago Ill] and Feyerabend (1970): [*Against Method*, Minnesota Studies in the Philosophy of Science Vol. IV, Univ. Minnesota Press, Minneapolis, Minn.] may have arise largely from the mechanical rigidity and inadequacy of these views. Some writers even consider that completeness and consistency questions can be posed and solved, as if the only difference between mathematical and physical theories were the existence of an interpretation for the latter. A more balanced view is found in Suppes (1968): Journal of Philosophy **65**, p. 651.

[4] Newtonian mechanics is perhaps the most nearly independent theory in physics. The dependence of a theory is clearly seen where there exist alternative theories for it to depend on. Thus quantum mechanics theory is a decisively dependent theory, for the two versions of it, using classical mechanics and special relativity, respectively, as their basis, differ markedly. Less significant but by no means negligible differences also arise in the analogous case of statistical mechanics.

appearance of logical discrepancies. Nonetheless, the theories in different fields of a science do not become incommensurable, for this would preclude the sort of interaction and joint use which is a commonplace – and a very necessary one – in research. Such interaction occurs in various ways. There is the employment of common concepts, such as mass, length or time.[5] Then theories borrow conceptual structures from each other; thus nuclear physics has taken the idea of the level gap from the theory of superconductivity, that of temperature from thermodynamics, and that of an ensemble theory for energy levels from statistical mechanics, adapting them in each case to its own needs and sometimes returning new ideas to the theories it borrowed from. Mathematical techniques such as group-theoretical methods of numerical analysis also cross theoretical boundaries with great mutual benefits, and so do approximation methods. And a special question is that of using one theory to describe subsidiary effects. (e.g. boundary conditions) for problems treated within another theory.

Fifthly, we have the whole vast chapter of approximations in research, again a matter that philosophy of science has olympically ignored; presumably this is a hangover from last century, when scientists considered this whole question, together with the fluctuations lumped together as "experimental errors", as only a matter of practical technique and not worthy of fundamental consideration. But an example shows how little such disdain is justified today. Consider the problem of phase change in a solid. In its full complexity, the problem is completely intractable. A first approximation is to neglect the presence of chemical impurities and lattice imperfections; since the faces and edges of a real crystal also involve distortions of the lattice, we consider instead an ideal, perfect crystal which extends on all sides to infinity. But even this holds too much complexity: we substitute a regular lattice of simple spins, capable only of pointing 'up' or 'down', for the real ideal perfect infinite crystal; this is the so-called Ising lattice. Even this proves too difficult; first, we treat only the two-dimensional lattice rather than the three-dimensional one; and then, instead of an analytic attack, we model it on a computer. Since here

---

[5] These are common, without a doubt, between theories based on the same kind of mechanics; but even microscopic and macroscopic masses, in spite of much argument to the contrary, and in spite of their occurring, one in quantum mechanics, and the other in classical mechanics, are shared between fundamentally incompatible theories. Two examples: Some years ago, there appeared a discrepancy in atomic masses as determined by mass spectrographs (these are, through their calibration, essentially macroscopic) and from nuclear reactions (quantum-mechanical in nature), and physicists insisted instead of accepting this as evidence of incommensurability, in further investigations that finally brought certain deficiencies in the calibration to light whose removal brought the two mass scales to agree very well. The mass of a macroscopic body (obtained by weighing – a classical procedure if ever there was one) is expected to agree with the sum of the masses of the elementary particles constituting it; there is a discrepancy which is explained as due to the binding energy in forming the nuclei, atoms molecules and crystals that make up the body; this binding energy is determined quite separately, and the resulting number check excellently: we have here $(E = mc^2)$ an agreement found from the contribution of three distinct and supposedly incommensurable theories.

we are faced with the finiteness of the computer's memory, we approximate the infinite Ising lattice by a finite square one (note that now we have used two approximations in opposite directions which however by no means cancel each other out), and finally use harmonic boundary conditions to reduce the edge effects, i.e. we consider the lattice as if it covered a two-dimensional torus, so that the last spin in one direction is immediately followed by the first one along the same line. To crown this edifice of approximations, we now approximate the continuous time evolution of the system by a succession of discrete if small intervals and so follow the dynamic behaviour of our pseudo-pseudo-... pseudo-crystal in the computer. The astonishing thing is perhaps that the results still bear sufficient resemblance to what the real crystal does. This case is perhaps an extreme example of how different kinds of approximations accumulate in attempting the description of a physical system; but the individual approximations are by no means unusual, either in type or in frequency of employment. It is in fact not possible to develop a theoretical picture without using the most basic of these approximations, that of inventing an idealized system for study, rather than the real one whose complexity exceeds all practicable bounds and contains far too much which is irrelevant or too particular.[6] But further approximations are almost inevitable: thus classical mechanics contains the concept of the free particle[7] which also appears in quantum mechanics, together with other such idealizational approximations. Quantum mechanics gives rise to its own peculiar approximations, due to the extreme mathematical difficulty of the differential-equation system found for all but the most trivial applications. What is noteworthy is the strong development of successive orders of approximation, both as a mathematical tool and as a physical conception (good examples can be be found in the Feynman-diagram method). So far the interaction of the various kinds of approximations are not well understood; the scientist mostly handles them on a purely intuitive basis or with trial-and-error methods.

Lastly, we must mention a point that is well known to all scientists and theoretically known but apparently not digested by philosophers: the existence of the so-called experimental error, i.e. the fact that successive measurements of the same phenomenon under identical circumstances do not lead to exactly identical results but fluctuate more or less widely around a mean which is normally taken as the result. For far too long, the scientists almost universally saw experimental error as merely a problem of laboratory techniques, for dealing with Gauss and the statisticians had developed ingenious methods,

---

[6] It might be objected that here we *abstract* from irrelevances rather than approximate. It is true that irrelevant elements in the global description on the system are removed: but most often something else has to be substituted in their place. In the example of the crystal, chemical impurities are merely left out of the account; but the finite size of the crystal, leading to edge effects of various sorts, is eliminated by the positive assumption or infinite size (in a universe tacitly supposed Euclidean, moreover). The questions of irrelevancy and insufficient generality are further discussed below.

[7] In the Newtonian formulation. Equivalent formulations are also found in other formalisms.

but which could be ignored by the theoretician once he had left the laboratory doorstep where the experimenters had handed him the final result. It is this entirely mistaken view among the physicists, above all, which explains the philosophers' attitude; but it does not justify it: had they been doing their job properly, they should have pointed out the erroneous conception and perhaps even shown the way towards its solutions. Of late, there has been a growing awareness among scientists that the question of experimental errors is of great theoretical significance. It is, of course, the ultimate justification for the pervasive use of approximation methods that we have mentioned; but it will occupy us also for much more general reasons. It must be noted here that philosophy has quite generally paid too little attention to the experimental side of scientific research; and where it has done so, it has not usually understood what it saw. Witness the unhappy extrapolation, as operationalism, of what were quite reasonable insights, or the more recent attempts to define an 'observation language' – a matter quite unrelated to the problem theoreticians and experimenters might have in understanding each others' jargons. The real issue here is double: on the one hand the failure to see that experimental work (even in the observational sciences such as astronomy or human genetics) is much more a matter of 'doing' than of 'observing' and passively describing what one has observed, and therefore all talk of 'language' is out of place until the activity involved has been understood and taken into account; on the other the mistaken attempt to draw a hard-and-fast line between theory and experimental fact, when in reality the distinction is relative and changes from case to case: for each experiment, its design, carrying out and interpretation require a vast amount of background and a great deal of detail specific to it; the latter is properly called experimental (it includes the actual observations, but also all the work that went into system preparation, preliminary measurements, calibration, control experiments, and so on), while the former may be qualified as theoretical, even though it normally includes much information derived from *other* experiments. When a different experiment is now contemplated, the presently experimental facts become theory, not because we talk about them in a different way but because we use them in a different connection. In other words, the theory-experiment distinction is a matter of their place in the constantly changing structure of the epistemic process; once again the expression 'observation language' is thus merely absurd. This second aspect requires, evidently, a considerable reorientation of our epistemological structures; it is the first aspect that will help us in doing so.

(iii)

What may we conclude from all this? Scientific theories are intellectual constructions based on the knowledge we gain from our planned interference with nature and the accompanying observations; they have definite, limited scopes and form at best an approximate representation of the nature they study; they evolve ceaselessly, sometimes very gradually, sometimes very drastically; and what makes this constant evolution and piecemeal elaboration is that they are only partly consistent (in the strict logical sense) either internally or among each other. Quite analogous comments apply to the concepts that form the raw material of theorization.

A first and obvious deduction is that the hypothetico-deductive model of how scientific research goes about its business is, if not downright wrong, at least seriously inadequate, and this on three counts:

a) it does not take into account the essentially active nature of the process of knowledge acquisition, a fact which is altogether alien to it. Indeed it plays down the experimental side or research, reducing it to an accessory procedure ancillary to the central business of the logical reconstruction of theories;

b) Though it leaves room for accepting that scientific hypotheses arise in an inventive and creative phase in research, it gives no account of this phase and offers us no tools for analyzing it and understanding it;

c) Finally, it misrepresents the one phase that it deals with explicitly, seeing the deductive process as one that can and should be formulated in purely logical terms.[8]

A second, not so obvious, point springs from a study of how new conceptualizations are incorporated into the epistemic cycle. These new ideas arise from a creative process concerning which we have the testimony of many outstanding scientists (a very clear one will be found in Einstein) but not very satisfactory theoretical understanding. But since these ideas must fit into the epistemic cycle it is easy to see that these ideas must be both free and disciplined: free in the sense that our past knowledge cannot determine and often not even condition new conceptions; but disciplined and unfree in the sense that a new hypothesis becomes a contribution to knowledge through the epistemic cycle, by being subjected to the searchlight of actual use and of confrontation with the results of that use, by being criticised and adjusted until it works satisfactorily. In that process it may of course be eliminated altogether.[9]

A third point has already been touched upon: if the *scientific* process of knowledge acquisition has the characteristics I have described, then the process of philosophic study must share them in order to be relevant. In particular

---

[8] Here, it may be said, the hypothetico-deductive model confuses what is rational with what is merely logical; but we cannot enter here into this rather intricate point.

[9] Note, incidentally, that these epistemological considerations shed light on a longstanding problem of academic politics, namely the need for and limits of academic freedom. Academic freedom, the liberty of propounding new ideas and of discussing and criticising them, is necessary because without it the epistemic cycle is distorted or even halted; this much is fairly generally recognized. But it cannot be unrestricted without falling into opposite dangers; the point being made here is that it must not be extended (i) to those people who are either not willing or not able to participate in the social process I have called the epistemic cycle, nor (ii) to those ideas which cannot be adjusted, tested and improved in the epistemic cycles: dogmas, for instance, which by definition are unalterable. Such people and such ideas are obstacles to knowledge.

we cannot expect to be able to postulate certain first principles and then to deduce everything else in an orderly and logical fashion. This much the recent history of philosophical discussion seems to have made rather clear; but on the other hand we must not throw out the baby with the bath water and conclude that we cannot arrive at all at any general ideas. On the contrary, once we have understood the epistemic cycle to some extent, we can apply its methods to the philosophical field also. In the remainder of this paper, this will be my main purpose.

8. The Determinisms of Physics *

Debate concerning determinism or indeterminism in physics has persisted now for a number of years. The main question is often whether physical theories are deterministic or merely probabilistic; it has been said, for example, that classical physics – that of the 19th century – was deterministic, whilst quantum mechanics has forced us to abandon this position. Certain writers have gone further, attempting to distill an ontological juice from these positions which, as we shall try to show here, is illicit.

A typical case can be found in quantum mechanics, which is often described as being essentially indeterministic; the most orthodox line of thought perceives the source of such indeterminism as being the central role played in quantum mechanics by the wave function whose square only provides us with the probability of observing things, rather than granting us the kind of firm prediction which results from the mechanics of point particles. From this, it has been concluded that the world is "essentially indeterministic", which entails – we are told – the crumbling of one of the bastions of materialist philosophy. Moreover, as chance and probability are regarded as being due to our ignorance, and not as an expression of an objective aspect of the studied system, an even more overwhelming conclusion is reached: in quantum mechanics, chance, i.e. our ignorance, is of the essence, which means that we are on the limits of what can be achieved through scientific investigation; quantum mechanics, in other terms, constitutes the definitive general theory of physics, its final frontier.

Such arguments are based on confusions and oversimplifications that must be drawn out of the darkness which has allowed them to take root amongst generally admitted notions. With this aim in mind, four theses will be put forward here which, though unable to end the problem completely, will – hopefully – allow us to place it in a terrain more akin to reality.

The *first thesis* is that determinism is not an ontological characteristic, but rather an epistemological notion. Which is to say that it is not the world we inhabit which can be characterized as deterministic, but only this or that description which we provide for some fraction of it. An example will make this distinction obvious: the behaviour of a certain volume of gas can be described (in principle, with the existing methods of calculation we are restricted to

* *Los Determinismos de la Física*, Revista de la Universidad (UNAM, Agosto 1989) (Translated by J. and C. Brody)

cases of roughly 1000 molecules or less) in terms of the trajectories described by its molecules, calculated through classical mechanics. This description – or model, as phyisicists often say – is deterministic, in the sense that, given all initial positions and velocities, positions and velocities at any subsequent moment are perfectly determined. An alternative model, deterministic though in a different sense, is that of classical thermodynamics; without referring to molecules, without even dealing with the fundamental variable which is time, it establishes relationships between variables of a different kind: pressure, volume, temperature, and entropy; here, the knowledge of some of these variables allows us to deduce the rest. The connection between these two models is provided by a third model, that of statistical mechanics. In this model, molecules are maintained, but instead of following their individual trajectories, we establish averages over the initial conditions we require; here is, therefore, a probabilistic model which does not make exact or unconditional predictions, but only provides us with the probability of one or another event taking place. Nonetheless, this model is more powerful than the other two.

The three are models for a single physical system, and each has its own advantages and peculiar limitations. It is now evident that we cannot attribute the property of determinism to the system itself, when we have two descriptions which are deterministic (though in different senses) and one which is indeterministic. To suggest that the molecules of gas "are" deterministic or not is senseless; such a description can only be applied to a theoretical model.

This point appears less surprising when we recall that it is precisely through the model that we can make predictions about the system's future behaviour. It is only when we understand (or feel we understand) a system in depth, when our model is extremely adequate, that we can sometimes grant ourselves the luxury of confusing the model with the modelled system. This is precisely what we require from a good theory, that it allows us to replace the system with the model, so that we can theorize rather than experiment. But when we examine this same theory philosophically (or when we study its validity) we cannot confuse two things: the description and the described system. An analogy: it would be inadmissible to confuse "Mexico has 15 million inhabitants" with "Mexico is a six-letter word" in order to obtain "Mexico has one letter for every 2.5 million inhabitants", or to ask whether Mexico is composed of letters, rather than inhabitants.

It is worth pointing out here that recognizing that the epithet of determinism does not apply to the world, accepting that – in today's philosophical jargon – using it this way constitutes a categorical error, invalidates a large part of the traditional discussions regarding our topic. In much the same way, becoming aware of the distinction between a thing and its name invalidates much magical thinking. The models we construct of reality have no effect upon it, except inasmuch as we use them to guide our actions upon this reality.

The *second thesis* to be put forward here is that the term "determinism" in fact covers several distinct notions which should not be confused. We have already seen how two different models may entail different kinds of determinism: one linked to the evolution of trajectories in time, and a second to the fact that variables, independently of time, are conditioned by one another. There can also be spatial determinisms, to be found, for example, in a tapestry, where we infer the pattern of flowers in one corner from what we see in another. Analogously, in physics the crystallographic structure of a macroscopic crystal can be inferred from observations of minute fragments of it. We can go further: from experiments made upon a few samples, we may draw conclusions of the type "copper (i.e. any sample of copper) is a reddish metal, a good conductor of heat and electricity, with a green oxide". This inference is not temporal, but it has the structure of a determinism: the observation of certain data in some samples, a theory that these properties originate in the material constitution of these samples, and the conclusions that if other samples have some of these properties, they will also have the rest.

Even temporal determinism, in its most traditional form, does not lack for relevant distinctions. As an example, let us take Maxwell's theory of electrodynamics, which predicted the existence of radio waves. The Maxwellian equations admit solutions expressed in the terms of two types of Lienard-Wiechert potentials, retarded and advanced, which respectively allow us to calculate the state of the electromagnetic field in the past and in the future. There have been extensive and muddled discussions about which of the two potentials should be employed in the applications; some have argued that using the advanced potential, written as a function of time to come, violates the principle of causality. But by 1909 Einstein had already provided a very simple answer to the problem, which has since been forgotten: as long as we have a system which is confined to a finite region, the two types of potential are perfectly valid, and correspond to two different situations. If we take as our starting point the field's condition on the boundary of the finite region at an initial moment, then the retarded potential can be used to calculate the system's future evolution; if, on the other hand, we use the final conditions, then the advanced potential can tell us how the field evolved in order to reach this final state. In one case we have a prior (or forward-directed) determinism, in the other a latter (or backward-directed) determinism. This is decided, neither by the system nor by the model, but simply by the type of conditions to be found on the boundary from which we initiate our calculations. In fact, as Einstein demonstrated, there is no inconsistency: from initial conditions we obtain the field's final conformation, whilst from the latter, going backwards, we can recover the initial conditions. We can even use mixed conditions, both final and initial ones, if this is what is required by the problem at hand.

In this latter case, we calculate both the future of the system and its past from the same border data. It is clear that determinism here is bidirectional; and in fact the distinction between this kind of determinism and unidirectional determinism – which allows only predictions or retrodictions – turns out to be of greater philosophical significance than the latter-prior contrast. A process described by the bi-directional model is reversible: if we invert the sign for time in every equation (if we "run the film backwards") we obtain the description of another, equally feasible, process. A process which is unidirectional cannot be inverted in this way. On the other hand, it may display phenomena governed by chance: for example, non-linear processes may exhibit an extraordinary range of different forms of chance-governed behaviour, whose complexities and rich structures are now the topic of many investigations. (The appendix contains more detail.)

This electromagnetic example underlines a further aspect. As we have mentioned, the compatibility of prior and latter determinisms is obtained only within finite and limited systems; on the other hand, the determinism of the equations of classical mechanics has no such restrictions. This distinction between what we could call confined determinism and global determinism has philosophical implications: the global and bi-directional form can be extended until it becomes total, engulfing each and every property which could conceivably enter into a description of the world. This form, associated with the name of Laplace, acquires an ontological cast when it implies that behind models with other, less "perfect", determinisms there must be the possibility of a model which exhibits it perfectly, totally, globally and bi-directionally. Laplacian determinism, to the extent that it becomes ontological, takes on a fatalistic character: everything, absolutely every event is rigorously determined, and, in its turn, enters into the determination of every future event. No free will, then, can be admitted, Unfortunately physics – which supposedly originates and supports this brand of inexorable fatalism – presents no examples of Laplacian determinism. Those we find are in their majority confined, unidirectional, or both. Even the determinism of Newtonian mechanics, global and bi-directional, is not total: it refers only to variables, such as position and velocity, which describe the movement of particles. The extrapolation that all the properties of the physical world can be reduced to the movements of its constitutive particles is a metaphysical appendage which is strange to physics itself. On the other hand, the exact determinism of classical mechanics is the determinisim of the theoretical model: its application to reality is limited by the inevitable imprecision with which we must establish boundary conditions, due to the impossibility of taking into account all the forces between particles, or even those which act amongst particles and bodies which are external to the system. We are faced with a fruitless confusion between what is solidly founded in physics and what is usually propagated under the protective cover of academic philosophy.

A similar confusion can be observed in the widespread practice of referring to causality and determinism as though they were synonymous. It would be perhaps less absurd to present them as incompatible. In order that we may speak about causality, it must make sense to distinguish between those factors in our model which are causes and those which are not – which, should they turn out not to be effects, either, can be eliminated from our description; but many determinisms, and in particular Laplacian determinism, are total, and refer to all the factors involved in the model. If causality is related to some determinism, it must be to a confined and partial form. Moreover, the concept of causality transcends the framework provided by determinism. A deterministic model establishes a relationship amongst certain variables, which is expressed in physics through equations. These equations allow us to write some of these variables in terms of others, but they can be transformed in many ways, and the selection of variables which figure as results appears arbitrary. Which of these variables, then, represent causes, and which effects?

The answer cannot come from within the system. This surprising conclusion can be cleared up through an example. A volume of gas is often described (with enough precision for our purpose) by the equation

$$pV = nRT \ ,$$

where p is the pressure of the gas, V its volume, T the temperature, n its quantity whilst R is the universal constant of gases. This equation can also be written as

$$p = \frac{nRT}{V} \ ,$$

if what we wish is to determine the increase of pressure which would result if we heated the gas to a certain temperature T'; or we can write it as

$$T = \frac{Vp}{nR} \ ,$$

in order to ascertain the temperature reached in a gas which has been compressed to a pressure p'. In the first case, the increase of temperature must be considered as the cause of the change in pressure; in the second, the inverse holds true. To what can we attribute this difference? In the first case, we heat up the container of gas, placing it over a heat source; in the second, we move a piston, using mechanical energy originated from without the system. And it is our decision to do either one or the other which transforms change in one of the two variables into a cause, change in the other into an effect. Causality has, therefore, an important subjective aspect; but, against all traditional ideas, this aspect pivots on our action upon the system, rather than around the ideas which turn in our head (except, of course, all those concepts which we utilize in order to take the decision of how to act).

But action does not have to be ours. Although this is the case which most interests us, and which undoubtedly originated the concept of causality, there are many other situations in the universe to which the term applies. Generally, what normally establishes the direction of a causal connection – what discriminates between cause and effect – is the interaction with another system not comprised within the model which we have been using. But if this further system is external to our model, it must also be external to any of the forms of determinism which occur in physics. Except, of course for Laplacian determinism, where there is no causality anyway as the distinction between causes and non-causes makes no sense: everything is a cause if it precedes the present instant; everything is an effect if it succeeds it.

Causal description goes beyond the simple deterministic framework, taking its place on the ontological plane of the relationship between different systems. It is an inherent property of our world which, in spite of its essential unity, allows itself to be divided up into systems which are in good measure independent of one another, but which nonetheless interact in such a fashion that each system's influence can be expressed in terms of cause and effect. To become aware of this fact, to realize that we constantly move from one division of systems to another, sometimes forming these systems theoretically, sometimes through our experimental interference (and even a combination of both), to observe all this is not only to understand the ontological implications of causality; it is a first awareness of how it is that we can firstly come to understand nature, and then take advantage of this knowledge to act upon nature in our benefit.

This discussion can be summed up as our *third thesis:* causality is compatible with certain determinisms but it goes beyond them, as it contains substantial ontological elements; those forms of determinism which are compatible with causality are also compatible with the occurrence of chance phenomena. Causality and chance, therefore, are not counterposed; they are complementary. The only opposition is between chance and those determinisms which, like Laplacian determinism, can have no place in physics.

This thesis evinces the lack of solidity that affects traditional arguments regarding the causal implications of quantum mechanics. From the fact that quantum mechanics utilizes the concept of probability in a form which is not susceptible to reduction we can neither conclude that it does not admit causality, nor that we can go no further in physics; rather, it indicates that our understanding of the links between a quantum system and the rest of the universe is incomplete: which is precisely what Einstein was telling us half a century ago. This idea is confirmed by the efforts to "complete" (or, more correctly, to reformulate) the theory, which in the last ten or fifteen years appears to have finally taken the right direction. The key lies in considering as open those systems which display quantum behaviour, as in constant interaction with the rest of the world. But these interactions cannot be described explicitly and in detail, because then we would find that we have amplified the system which our model contemplates to the point were it becomes unmanageable; the description must be statistical, it must establish averages about how the system behaves in all possible interactions. The resulting temporal evolution is a stochastic process which displays quasi-stationary states of equilibrium: quantum states. Much ground must still be covered in order to flesh out this theory (I shall not go into any further technical detail), but some success has been achieved. Most of all, this theory is characterized by a conceptual clarity which will not allow for aberrations such as those we have mentioned already, as well as others which figure in the available literature on the topic. Regarding the issue discussed here, the theory displays both deterministic and indeterministic aspects whose relationship is clear, and it can originate causal models and probabilistic ones, as well as models which combine both elements. In this, at least, it resembles the classical theories of physics more than quantum mechanics in its traditional form; in other aspects, of course, it differs enormously ...

The discussion which preceded the third thesis introduced the notion of a system as something which is in good measure, if not totally, isolated from the rest of the universe. It is this notion which is at the root of our *fourth thesis:*

those quantities which are of interest in a scientific investigation (with scarce exceptions) cannot be determined with absolute precision; they all display what could be called the limits of experimental error, limits which are really quantitative estimations of the importance of fluctuations or perturbations caused in these quantities by many factors which are external to the system. This must be made fully clear: the fact that limits of error must be established has been a part of experimental practice for a long time; our fourth thesis only explains the fact through the concept of a partially isolated system. The fact in itself – which, along with its implications, has often been ignored by philosophy – has an important consequence for us, in the sense that trajectories predicted by deterministic models (mathematical curves of zero thickness) are no more than an approximation. Sometimes a good approximation, other times a grossly inadequate one: but always an approximation.

In those cases where approximation is not enough, we can consider tubes instead of curves, tubes whose dimensions at their initial point are given by the limits of error at this point. Temporal evolution can make the tubes contract, but it is far more common for them to widen: uncertainty in the initial values causes increasing uncertainty as time goes by. Relatively recent studies have shown that growth can be of two kinds, polynomial or exponential. In the polynomial kind, the tube increases far less violently than in the exponential kind; the precision of predictions diminishes gradually, but they are useful for fair lengths of time. Often, in this case, the approximation of mathematical curves is enough. On the other hand, a tube which grows exponentially increases so rapidly that in a short time it fills all the accessible space, with which predictions are reduced to stating that the particle must be "somewhere". The distinction we have outlined is different from the distinction between determinisms in that it depends on the system's inherent characteristics and not (in general terms) on the initial conditions; it is therefore more fundamental than the difference between "determinism" and "indeterminism" (in a rather unclear sense) which so preoccupies certain philosophers.

The essential distinction between processes which diverge in a polynomial fashion and those which do so exponentially is that the former are (within not too long a period) predictable within the bounds of acceptable error, whilst the latter are unpredictable. However, "unpredictable" here is not intended to mean that such process are not susceptible to scientific investigation; we have simply chosen our methods incorrectly, and possibly we must use a different model for this type of system. On a different level, such unpredictability can provide excellent and even deterministic models. This can be illustrated with the ideal gas example. The model on a molecular level, the first discussed above, turns out to be of the exponential kind, and as such it will not yield very interesting predictions; but this same fact will allow us to construct a statistical model which is extraordinarily useful, although its predictions are only probabilistic; and as we have seen, it will serve as the basis to go further, constructing models which are perfectly deterministic and also timeless, thus negating the problem of exponential divergence.

These four theses reveal that the prevalent fashion of formulating problems about determinism does not truly correspond to the practice of physics. Determinisms or indeterminisms appertain, not to the real world, but to the theoretical models which we use in understanding it. The very term "determinism" covers a great variety of different relationships amongst physical variables, and many of these variations are amongst models used by physics; some forms of determinism admit chance, whilst others are incompatible with it; determinism and causality, far from being synonymous, differ on a fundamental level, some forms of determinism being compatible with causality whilst others are not; and we finally come upon a concept whose implications have yet to be explored: predictability.

There is a great deal of work left to do for philosophers.

Appendix

Based on the description of a system at a given instant one may, in a bi-directional model, calculate both past and future system states. That is, if a variable x (or a set of variables, a case indispensable in physics, but to which there is an immediate generalisation, so we can restrict ourselves to studying the simplest situation) represents the state of the system at time t, so x is therefore a function of the form $x(t)$, then there is a transformation $T(t, t')$ with $t' > t$ such that:

$$T(t, t')x(t) = x(t') \,,$$

where $x(t')$, the state of x at time t' is the unique result of applying T to $x(t)$; and there also exists $T'(t', t)$ such that:

$$T'(t', t)x(t') = x(t) \,,$$

and the original state $x(t)$ is reobtained as a unique result. T' is therefore the inverse of T, and the mapping represented by T is bijective.

In a one-directional model there exists a transformation $T(t, t')$ with $t' > t$ if the future can be predicted; or there exists $T'(t', t)$ if the past can be uniquely determined; but whichever of these two exists, it has no unique inverse. There will instead be more than one possible value; if T exists, then T' will take us to several different possibilities, all of which are predecessors of the state from which we started going backwards. (At least one must exist, since otherwise the system would have been mysteriously created the instant we observed it; but this state can be anything.) That this situation gives rise to possible random phenomena will be shown here through an example rather than rigorous deduction, which would require advanced mathematical techniques.

Let us take the case in which instead of time t (a continous variable) we merely have a variable $i = 1, 2, \cdots$ which *numbers* all instants. Then a simple example of possible transformations could be

$$x_{i+1} = T(i, i+1)x_i = 2x_i^2 - 1 \,,$$

(with x_i lying between -1 and 1). This transformation is one-directional: given a value for x_{i+1} we can't tell which of two possible values for x_i preceded it in the series of "times" i. For example, $x_i = 1/4$ and $x_i = -1/4$ both lead to $x_{i+1} = -7/8$. Now, if we choose a value for x_0 and successively generate $x_1, x_2, \cdots$, then one of two situations may result: (a) after a certain period of time all the values are equal; for example, if $x_0 = -1/2$ or if $x_0 = 1$, then all numbers in the series also have that value; if $x_0 = 1/2$ or if $x_0 = -1$, then x_1 is different but all following values are equal to x_1; in technical jargon one says that the system has become stationary, from the i at which all values started being equal. (b) On the other hand, all x_i may be different, without any repetitions.

Case (b) strictly obeys the mathematical definition of a series of random numbers (a "collective" in von Mises' sense); among their properties we may note that:

(i) Successive members of the series are independent in statistical terms, which means that if we calculate the nth value and $n + 1$ value starting from many different x_0, there will be no correlation between them. Therefore products such as $x_i x_{i+k}$ summed over all values of i give 0 for any k not equal to 0.

(ii) If we take two different series, one with $x_0 = a$ and another with $x_0 = a+e$, then no matter how small e is, the difference between the two values of x_n will grow exponentially as n grows (the difference is approximately $(-4)^n e$), and the two series rapidly lose any similarities or correlation. For large n, the difference can no longer grow, since the x_i are limited to lying within -1 and 1; but their correlation disappears completely, and the difference between the two series, initially small and regular, becomes itself a random number.

(iii) If we take many initial values $x_0, x_0', x_0'', \cdots$, the statistical distribution of $x_n, x_n', x_n'', \cdots$, for large n is the same as that observed along one of the series; in technical terms, the transformation $T(i, i + 1)$ is ergodic.

(iv) If we select a subseries, according to any criterion which does not depend on the value to be selected, this subseries will have the same statistical properties as the original series. This is the randomness criterion given by von Mises.

Despite all these characteristics, which we would only expect to see in a chaotic series, in a random one, the transformation that generates it is one which corresponds perfectly to what one would intuitively call determinism $\cdots$

Note that there is an infinite number of possible values for x_0 which lead to case (a): two when the system becomes stationary with $n = 0$, another two for $n = 1$, then 3 for $n = 2$, 6 for $n = 3$, etc. But the number of x_0 values which lead to case (b) – that is, produce series which never end – is still greater; the first is "merely" the number of integer numbers, while the second is the number of points on a straight line. And even if through bad luck we had a series which terminated, as long as its length is large it is impossible to differentiate statistically from other, non-terminating series. Inasmuch as the concept is definable for finite series, these too are chaotic.

There are of course transformations with these (and other even more curious) properties for the case when we use a continous variable to truly represent time. But the simple case we have seen already helps to understand why the sequences of numbers that we generate in computers (through so-called random number generator routines) have all the properties of random series. With them we can simulate, as precisely as necessary, stochastic processes and all random phenomena; this type of simulation, known as Monte Carlo calculations, has become an extraordinarily useful tool in such varied applications as nuclear reactor design and the study of social processes.

An electronic computer is the archetype of a deterministic machine, one could say, of precisely predictable behaviour. That it can generate random numbers and with them imitate perfectly random phenomena is surprising for traditional philosophical conceptual frameworks; we see their impotence in von Neumann's famous saying: "he who speaks of generating casual numbers through a deterministic algorithm is in a state of Mortal Sin". On the other hand, the thesis presented here on the multiplicity of different determinisms offers an immediate explanation. Given that the computer wipes out the previous number when writing a new result into its memory, the determinism which it obeys, rigid though it might be, is one-directional and therefore compatible with random phenomena. The routines which generate random numbers, and the Monte Carlo calculations that use them, do nothing more than take advantage of this opportunity in a systematic way.

Part II

The Theory of Probability

9. The Nature of Probability

The discussion of what probability is about has over the last few decades become a special topic in philosophy, linked more to questions about induction and the confirmation of theories than to the problem of how physical theories are structured and how they interrelate. Yet statistical theories are fast becoming the larger part of physics; indeed, one of them, quantum theory, underlies most of modern physics; and even in originally non-probabilistic theories the use of probability has increased. We shall therefore attempt to formulate a concept of probability that is suitable for understanding its use in physics; in view of the position here adopted that scientific research is the continuation on another level of common-sense ways of getting to know things, we expect that a concept of probability that works in physics should also work in everyday situations. The criterion to be used in judging conceptions of probability will be that if it has meaning as a concept within a science it must, as discussed in the preceding chapters, possess both a theoretical and an experimental aspect, and these two must be firmly yet flexibly linked through appropriate approximations.

The many views of probability to be found in the philosophical literature may be grouped for present purposes into three large categories: (i) The propositional views, both objective and subjective; for these a probability is a property of statements or propositions, expressing how certain they are (in the objective varieties) or how strongly we believe in them (for subjective ones). (ii) The frequentist views; here the variations are less extreme, and we shall consider almost exclusively the most explicitly formulated version, due to von Mises. (iii) A group characterised only by differing from either of the preceding ones; here belong both Popper's propensity concepts in parallel. We shall depart from the usage of the other chapters and discuss at least the first two groups in some detail, since this will ease the understanding of the probability conception here proposed.

What I have called the propositional views are often termed subjective, because according to them a probability is assigned to an element of our description of the world rather than to an element of that world itself. They can, however, be objective in the sense that the probability that is assigned does not depend on who does the assigning. The result is then at least intersubjective; however, whether it is more than that is debatable. The most 'objective' version of these views is due to Keynes, who defined probability as the *rational degree of belief* in a proposition, given a set of propositions that constitute the evidence for its validity. At the other extreme there is de Finetti's conception, according to which the probability to be assigned by anyone is to be determined by what he is willing to bet on it; since two different people need not be willing to bet in the same way, a probability à la de Finetti is purely subjective. A considerable variety of probability definitions which lie somewhere along this axis has been proposed; certain versions of them have become relevant for physics since they have been made the basis of physical theorising by Jaynes and others. In what follows, I shall not distinguish between these variants; in the present context only what they have in common will prove to be relevant.

Let us then take a probability to be the probability of a proposition's being true, in any of the various meanings given to this phrase. We also ignore the (philosophically very interesting) question whether it is a proposition or a statement, i.e. an instance of a proposition, or the judgment expressed by the proposition that the probability refers to. We are concerned only with the fact that the probability statement does not refer to an actual state of affairs in the real world but to some element in our image or description of it. It is clear that such a probability is purely theoretical; indeed, when we make a probability statement in connection with a physical theory, it must be about some proposition or other suitable element of the theory, and is therefore not itself part of theory but belongs to a metatheory. In this exclusively theoretical and even supertheoretical nature of the propositional views lies the root of their weakness: it is difficult to link these probabilities to reality in a convincing way.

For other concepts in physics there exist a measuring instrument – a voltmenter for electrical potentials, a thermometer for temperatures, and so on. But there is no *pitanometer* for measuring probabilities, and therefore it is by no means obvious how a probability we have calculated is to be connected to anything observable. The phenomenon normally taken to correspond to a probability is the *relative frequency* with which some property occurs in a set of measured values, all obtained in like fashion. While a property is not itself a proposition, that one member in such a set possesses it may be framed as a proposition; but what is the connection between our belief, rational or otherwise, in the truth of this proposition and the relative frequency of the cases for which it holds in a given data set? In all variants of the propositional conception of probability, a further postulate is required in order to establish this connection.

Keynes uses, under the name of *principle of indifference*, an idea that goes back to James Bernoulli: where nothing else is known, the different possible outcomes are assigned equal probabilities. Once such a set of basic probabilities is given, all further ones may be deduced by means of the well understood

calculus of probability. Unfortunately the principle of indifference is very unsatisfactory. It works where the number of possible outcomes is finite (and
even small), and there are good reasons for assigning equal probabilities, as
in the case of a six-faced die, manufactured to high standards of symmetry.
In most other cases, however, it is either inapplicable (as for unsymmetric
dice) or ambiguous (for an infinite number of discrete outcomes, or a continuous range, where it gives rise to well known paradoxes). Keynes' attempts to
remedy these defects have not withstood later criticisms; and the additional
postulates other writers have proposed have all turned out to be unsatisfactory in one way or another. I will here mention only de Finetti's requirement
of *exchangeability*, i.e. the requirement that the *a priori* probability assigned
to a sequence $(e_1, e_2, \cdots, e_n)$ of possible outcomes should not depend on the
order in which the events occur in the sequence. At first blush this sounds
much more reasonable than the principle of indifference; but it has two fatal
weaknesses: (i) its application requires all relevant probabilities to be worked
out beforehand, so that *a priori* probabilities are fixed for all times, and only
the *a posteriori* probabilities, found as conditional probabilities using Bayes'
law, change according to the observed results; this is very far from the common man's way of thinking when working out his chances at cards or roulette,
and even farther from the way a physicist derives probability estimates from
a theory and then compares them with actually observed relative frequencies
– either judging the theoretical model to give satisfactory predictions or else
adapting the model and starting over again. And (ii) exchangeability implies in
effect that the successive events in the sequence are statistically independent;
it can therefore not be applied when this is not the case, as in stochastic processes, where the probability of a certain event depends on what has happened
before; yet (in a sense we shall spell out below) stochastic processes belong
to the very foundations of physics and indeed of almost all science. What is
wrong with this principle is simply that it has been introduced in order to
make de Finetti's definition of probability work, without any consideration of
the extent to which it in fact applies to the real world.

It is perhaps an oversimplification to say that for the defenders of the
propositional point of view, probabilities are about ideas in our heads, while
physics, for instance, is about the real world; yet this oversimplification expresses and essential common point which accounts for the weaknesses of this
point of view. On the one hand, their probability concept often does not satisfactorily cover probability uses in physics; for instance, difficulties crop up
when time-dependent probabilities are considered: the physicist would have
to be pictured as changing his mind – and that perhaps several thousands
of times every second). On the other hand, this probability concept extends
beyond what is reasonably to be seen as belonging to probability; as we shall
discuss below, it furnishes a basis for the dubious notion of induction as well
as certain uses of the term *probability* in common parlance, where in fact
uncertainty, in the sense of lack of relevant knowledge, is meant.

(iii)

Two years before John Maynard Keynes gave the first complete and detailed description of the propositional view (which he called 'logical', since he conceived it as a generalisation of logical entailment to the case when this was not either true or false), the mathematician and neopositivist Richard von Mises published the first full account of the frequentist view. Since then much further detail work has been published, but no major reformulation has appeared; we shall therefore base our discussion on von Mises' ideas.

Von Mises saw that earlier interpretations of the probability concept were unable to make contact with the reality of relative frequencies. He therefore took these as the starting point, and defined the probability of some event to be the limit of its relative frequency in a series of observations when the series becomes infinitely long:

$$P(e) = \lim_{n \to \infty} m(e)/n \,, \tag{9.1}$$

where n is the number of data in the sequence and $m(e)$ is the number of times the event e is observed in it. But for this definition to be unambiguous, the sequence of data cannot be arbitrary. It must be what he called a *collective*, that is to say it must have the property that any infinite subsequence in it must have the same limit of Eq. (9.1) as the full sequence. Of course there must be a restriction on how we pick the subsequence. If there are only 0's and 1's in the full sequence, the subsequence of the 1's could not possibly have the same limit as that of all the 0's. von Mises introduced his *principle of place selection* for this purpose: the rule that determines whether a datum is picked from the sequence, but not on its value (it may however depend on the values of earlier data).

This conception of the nature of probability is radically different. Most important, it refers probability, not to propositions or sentences or any abstract concepts that could in their turn refer to reality, but directly to the events we are concerned with. It also requires these events to form a sequence with rather remarkable properties. The first point is clearly in its favour; it explains why so many physicists have felt that a frequentist view is the only tenable one in that it fits the requirements of physics. But the second point raises numerous difficulties.

First and foremost, a probability is defined as the limit of a relative frequency, $m(e)/n$, when n tends to infinity; here $n(e)$ is the number of times the event e is observed among a total set of n measurements. This frequency is that of events e that actually occur: in physics the typical event is that a measurement found in an experiment lies within a certain interval. No such infinite series of experimental data is possible; at least not in a finite time; and whether it is conceivable even in infinite time is highly debatable. The notion of a collective has therefore generally been seen as an abstract concept, as a type of infinite sequence of mathematical objects. Even then it has turned out to be extremely difficult to determine whether they possess a limit such as

(9.1), and whether this limit is unique in the sense of von Mises' requirement. So far no specific example has been constructed.

For the present purpose, these technical problems, though in themselves extremely interesting, are irrelevant beside the point that if the sequence is not one of actual events, then the construction of a collective has not yet established any contact with reality: how can any sequence of experimental data (which will be finite) possess the properties of a collective? There is no limit for the relative frequencies in it, and the relative frequencies for subsequences will almost always differ from that in the main sequence. True, as the sequence grows longer, the differences will usually become smaller; but can we conclude that *if* we could reach the infinite it *would* be a collective? von Mises assumed that the differences would always tend to zero, and called this the 'Urphänomen' of probability. But as we know now, the most one can say is that the probability of their going to zero tends to 1; using this in order to define probability would then be circular. Moreover, it is by no means always the case; even the probability limit we have just alluded to does not always exist, for instance when the probability depends on time.

Another difficulty arises when there can be no sequence of data, when in other words we have a single case. For the frequentist, the probability that it rains tomorrow morning is not definable. This rules out the major part of statistical techniques, where very generally one calculates, on the basis of hypothesis to be evaluated, some appropriate combination of all the data (calling it a 'statistic') and determines the probability of it occurring. If this probability is too low, the hypothesis is rejected (the discrepancy between it and the data is 'statistically significant'). But in each case, however long the data sequence, the statistic has only one value; for the frequentist this should be meaningless. It is a curious sidelight that in von Mises' major textbook he first gives a detailed and carefully argued exposition of the frequentist view and then, seemingly on this basis, develops standard statistical theory; he remains unaware of the contradiction.

Thus the frequentist view of probability suffers from a weakness that is precisely opposite to that of the propositionalist. Instead of there being no natural link between the probability concept and actually observed relative frequencies (hole having to be plugged by *ad hoc* postulates), we now have too rigid a link: probability is actually identified with the (limit of the) relative frequency. The link breaks even as it is set up; there can be no infinite limit of experimental relative frequencies, and we have fallen into a confusion of the experimental and the theoretical that characterises logical positivism. It is one thing to grasp the close connection between the two; it is another to mix them up.

(iv)

Karl Popper was led essentially by epistemological considerations concerning the status of quantum mechanics when he developed an idea about probability that does not clearly fall into either of the two chief categories we have discussed. He observed that a penny when tossed and allowed to fall has the ability to exhibit two types of behaviour: it may fall heads up, or it may fall tails up; following a venerable tradition, we ignore the remaining possibilities, namely that it should stay on edge, that it drops into a crack in the table and vanishes from sight, that a crow carries it off in flight, that it falls into a beaker of acid and is dissolved before settling on one side or the other, ... (In this we simply idealise somewhat the system we consider, along lines discussed in preceding chapters.) For Popper this is the Urphänomen, and consequently he sees probability as the property the entire set-up, coin + tossing fingers + floor it falls on, has of behaving in different ways; to state this more precisely, the probabilities for the different outcomes represent the tendencies of the system to evolve in these directions – the *propensities*, to use Popper's term. Popper expressly denies that a propensity to exhibit this or that feature is not in any way an Aristotelian *potentia*, unfortunately his conception remains incomplete, so that it seems impossible to determine what a propensity is exactly if it is not a *potentia*, and therefore also impossible to clarify the relation between a probability as propensity and the relative frequencies of different behaviour options on the part of the system. Popper's conception has obvious virtues in being firmly wedded to the nature of the actual system his probability concept is intended to describe. In view of its incompleteness, however, we shall not further examine it.

There have also been authors (notably R. Carnap and J.L. Mackie) who, seeing virtues in various conceptions of probability, adopt them all in parallel. Carnap proposed a Keynesian probabilty$_1$ together with a frequentist probability$_2$; Mackie envisages no less than five different notions of probability at the same time. That this is no way out of the perplexities surrounding the probability issue should be evident. Such combinations not only accumulate the difficulties each component conception raises – for they do not cancel out – but also have to face the awkward question of when to use the different parts and, above, how to operate any necessary transitions between them. And since none of these 'polyglot' views are at all satisfactory as far as the needs of physics and other sciences go, we shall not discuss them any further.

It could be objected to much of the preceding discussion that the non-scientific uses of probability have not been taken into account. "It will probably rain today", "he is most unlikely to get it wrong", or even "it really isn't on the cards" can – just – fit into some of the conceptions we have mentioned; but "the length of Cleopatra's nose was probably not a factor in the then Egyptian political scene"? Two points must be made.

Firstly, there is here some confusion to be analysed in the common usage of *probability* and similar terms. "Will it rain today?" "Probably." The second speaker – unless he is a meteorologist – is generally not estimating, however roughly, the probability of its raining; rather, he is expressing the fact that he does not really know but prefers to take his raincoat along. It is uncertainty, and not probability, that we are dealing with. And uncertainty, unlike probability, concerns our state of mind; I or you may be uncertain about something, while the other knows. "It is uncertain" means either that nobody (at least in our circle) knows for sure, or else that the people responsible for deciding the matter have not yet done so. (Uncertainty in an objective sense is quite another thing: indeterminacy; in Chap. 8 I argue that indeterminacy, like its opposite, does not make sense in any absolute, metaphysical sense.) No satisfactory way of assigning a numerical value to uncertainty has yet been found, nor is there any reason to suppose that such a degree (or its opposite, the degree of certainty) has the formal properties of a probability. If we are uncertain to degree p about one statement and uncertain to degree q about another, unrelated one, are we uncertain to degree pq about their conjuction?

Secondly, a related notion is that the need to use probabilites rather than conclusive logical argument is due to our greater or lesser ignorance of relevant fact. This notion not only is tacit in many ordinary uses of probability terms, but also appears explicitly in Keynes' principle of indifference. Our ignorance is, of course, no possible source of knowledge in physics or elsewhere; yet this "ignorance interpretation" of probability has numerous followers. We shall examine it below, once we have laid bare the roots of the misunderstanding that gives rise to it.

(v)

A different idea about probability belongs to the folklore of physicists. It is that Kolmogorov has specified what probability is in great detail, so we no longer need to worry about its interpretation. This is a confusion to which Kolmogorov himself would be the last to subscribe. What he did was to provide a formal statement of the mathematical properties we must expect any probability concept to have; while his work is of great importance and has borne fruit in many directions, it is so far from being an interpretation, i.e. a link to experimental and other properties of the probabilities we deal with in applications, that it, quite on the contrary, points up the need for one. To make this clear, and because other matters hang on it, we briefly state the axioms Kolmogorov took as his starting point:

(i) The objects to which we attribute a probability will be called *events* and are subsets of a basic set Ω, called the event space, sample space or phase space (because of its use in physical applications, we shall mostly use the latter term). If an event e belongs to Ω, then so does its complement $\Omega - e$; hence Ω contains the null event $\oslash$. If a (finite or infinite) sequence of events all belong to Ω, then so does their union. The events containing only one member of Ω are the *elementary events* of Ω.

(ii) To each event e in Ω a function $P(e)$, called the *probability* of e, is assigned which has the following properties:

(a) $P(e) \geq 0$ for all e in Ω;

(b) $P(\Omega) = 1$

(c) If $e_1, e_2, \ldots$ are in Ω and pairwise disjoint, i.e.

$e_i - e_j = \emptyset$ when $i \neq j$, then $P(e_1 \cup e_2 \cup \ldots) = P(e_1) + P(e_2) + \ldots$

We note that though the subsets of the phase space Ω are called events, in a terminology which conflicts with any propositional interpretation, this is merely a convenient name. If Ω is interpreted as the universe of discourse for a Boolean logic, the axioms apply, the elementary events now being atomic propositions. It suffices actually to assign probabilities only to the elementary events; this must of course be done in such a way that axiom (ii$b$) is satisfied; then the probability of any event is given by expressing it as a union of elementary events and using (ii$c$).

If the number of events considered in (ii$c$) is finite, the axioms suffice for describing the structure of the probabilities for discrete events; but if infinite sequences are also admitted (Kolmogorov stated this part separately), continuous phase spaces – indispensable in physics and many other applications – become possible, probability *densities* (derivatives of probabilities with respect to the continuous variable describing the event) can be defined, and so on. But for this extension to be consistent, not all subsets of any phase space can be admitted, only those that together form a Borel field (or ω-algebra). This is relevant below and is not a mere technicality, but we need not enter into detail.

It is sometimes stated that the Kolmogorov axioms are incomplete, in the sense that they do not specify how the probabilities $P(e)$ are to be assigned. But any such assignment rule would restrict the use of the probability concept to one particular case; it is precisely this incompleteness that makes his axioms so useful. We have here another case of a 'hole' in a high-level theory, and we have already discussed the need for such holes to achieve generality; we have also seen the case of the force in Newton's second law.

Kolmogorov's axiomatisation is not the only one that has been given; it is, however, the only one that fits neatly into both mathematics and physics. Mathematically, it might be summed up by saying that a probability is a nonnegative measure function, bounded by 1, on the Borel field of subsets of the phase space; its physical implications will be brought out in the next section. Kolmogorov's work has been usefully extended by Rênyi, who used the notion of *conditional probability* as basic (it is a derived concept for Kolmogorov).

The question has been raised whether the Kolmogorov axioms (or an equivalent axiom set) can justifiably be taken to determine what can be called a probability and what not. To some extent this is a terminological question merely; all of probability theory is a consequence of Kolmogorov's axioms, and anything to which this theory does not apply is clearly a different concept. The uses of probability in physics and almost all the ordinary-language uses do have the properties implied in probability theory and spelt out by Kolmogorov; we follow tradition and continue calling the concept involved by the name of probability. What is important is not to conflate it with other concepts having certain similarities or simply confused with it. Quite a different matter is the extension of the probabilty concept to fields other than the Borel field of standard set theory; such extensions have been proposed for various purposes and studied extensively, but their use in physics has not yet justified itself by producing significant new insights. We mention certain attempts of this kind in connection with quantum mechanics.

(vi)

The inadequacies of the two major views on probability are, in a sense, complementary. Propositionalism offers an internally consistent picture which, however, remains on the theoretical level and proves hard to link to any real phenomena; frequentism, on the other, by sticking myopically close to one particular phenomenon and framing its theory in rigid connection with that phenomenon, gives a partial account of it and is unable to rise above it. But we cannot solve the problem by adopting both interpretation, as Carnap tried to do, or formulating a compromise between them, as Braithwaite attempted. A solution is offered, however, by a concept from theoretical physics: that of an ensemble.

Popper's account perhaps comes closest to recognising that the fundamental phenomenon to be studied is that of *fluctuation*. A repetition of an experiment produces a somewhat different result every time; a particle, instead of moving in a straight line, shows irregular deviations from it; and so on. The fluctuation need not be actual: a boy is born – but it could have been a girl. The problem is to build a theoretical account of systems that fluctuate in this sense, and to maintain the needed degree of generality. It is in principle possible in each case to develop a model containing sufficient further detail so that the particular fluctuation is explained: the particle changes its direction because it collides with other, smaller ones, and we could include the details of their motion in the description. But such models are not general; when we deal with the next particles whose path deviates from a straight line, a different account of the background particles would have to be included. In fact, we would have to develop an enormous set of such models in order to be able to account for how any such particle moves.

But this plethora of model will still not give the answers we require. For in such situations we are almost never interested in the details of the motion but rather in properties that are much more general: do the particles keep in some degree to their original directions of motion, in spite of the buffeting from background particles? And if they start out from a certain point, do they stray indefinitely far away? None of the models individually can answer such questions; but if we take them jointly, and calculate an average over them, appropriate answers will be forthcoming.

This is even clearer if we consider the case of very many particles, as for instance in a vessel full of gas. We could, at least in principle, compute the trajectories of all the molecules as solutions of the Newtonian equations of motion and so obtain a complete description of how the gas behaves. Or apparently complete. For if we compute the solution for the molecules in an exactly similar vessel placed next to the first one, the solutions – starting from entirely distinct initial positions and velocities for the molecules – will be quite different. Yet the behaviour of the gas in the two vessels will be the same as regards pressure, temperature, and so on; such properties are independent of the initial conditions since they correspond to an equilibrium state, yet Newtonian mechanics knows no properties of solutions that do not depend on the initial conditions, and such concepts as equilibrium are wholly foreign to it. But if we average over all the relevant Newtonian models of molecular motion, the averaging process tends to suppress the differences between the models and brings out those features that they share; the initial conditions (and whatever varies with them) differ among the models and so disappear on averaging; what stands out are the common properties shared by all such systems – temperature, pressure, entropy, and so on. (There are non-trivial technical problems involved in identifying these properties with the appropriate averages; these problems are not relevant here.)

Note that it is not, as – to the confusion of the student – is so often stated in textbooks, the large number of molecules in the gas that makes a statistical treatment necessary; it is the need to determine quantities that are independent of the initial conditions that obliges us to abandon a purely Newtonian approach. With modern computers, surprisingly large numbers of molecules could be handled; but however many one has, their detailed trajectories are of no interest. Not until the computer has averaged over them do the interesting features emerge. At the other hand of the scale, systems with only a few particles also often require statistical treatment; in nuclear theory, the concept of a nuclear temperature, of just this type, works even for nuclei with only twenty or so nucleons. And when dealing with a stochastic process, Brownian motion being an example, even one particle is treated in this way.

(vii)

The plethora of models, following the pioneering work done by Einstein and independently by Gibbs at the turn of the century, is known as an *ensemble*. The ensemble concept is the basis of statistical physics, and although there are still some significant open problems concerning its justification, it has proved extraordinarily fruitful. It is, incidentally, important not to confuse an ensemble, a set of theoretical models, with a many-body system such as a gas. Among the members of an ensemble there is no interaction and indeed the concept of an interaction between alternative descriptions is meaningless; the molecules of a gas can and often do interact quite strongly. For a many-body system, the ensemble is one in which each member contains many particles. The confusion is possible when the interactions are very small, for then the many-body system can behave like a realisation of an ensemble – but of individual particles. Such cases are often of interest outside physics, for instance in epidemiology, or in the study of macroeconomics modelled as an ensemble of microeconomic enterprises.

Taking an ensemble average permits us to pass from one level of description to a different one. An average is a new quantity, with properties that differ from those of the variables averaged over; for instance, an average over a series of integer values is a rational fraction, or even a real number if the series is infinite. When we apply this process to an ensemble, it eliminates one set of variables and introduces another, so that we have created a different epistemic model, suitable for describing a different type of property. Among the new properties the ensemble concept provides us with, probability is significant in that it appears in every possible ensemble.

The original definition of an ensemble explicitly used in the probability concept; what is interesting is that the construction leading to the ensemble concept can be inverted, so that the ensemble is defined without using any probabilities, and then in its turn serves to define probability. This is what we proceed to do.

If the ensemble contains a finite, discrete number of models, then an average is calculated simply by summing the values the quantity we are interested in takes over all the models, and then dividing by the number of models; in the usual notation, if n is the number of models and $Q(i)$ the quantity of interest in model number i, then the average will be

$$\langle Q \rangle = \sum_{i=1}^{n} Q(i)/n \ . \tag{9.2}$$

Mathematical subtleties attend the limit $n \to \infty$ of a countable infinite set of discrete models; fortunately it is rarely needed. It is much more usual to find that a model is described by one or more continuous parameters; the initial positions and so on of the molecules in the case of a gas, for instance. Calling these ω and letting this symbolic variable range over the space Ω (for this is exactly the phase space introduced above), we have, instead of Eq. (9.2), for the average of Q

$$\langle Q \rangle = \int_{\Omega} Q(\omega) dn(\omega) / \int_{\Omega} dn(\omega) \ . \tag{9.3}$$

here the function $n(\omega)$ "counts" how many models there are "before" ω; we have placed quotes around the verb 'count' because here again there are some mathematical subtleties involved in counting over a continuous set of objects. We need not enter into these; it is sufficient for our purposes to think of the function $n(\omega)$ as a weighting function, and we immediately see that it must be specified in order to have a fully defined ensemble. The specification for $n(\omega)$ can sometimes be derived from the theoretical basis on which the ensemble is built. But if this cannot be done, or not fully, it is adjusted by comparison with experiment. One guesses an initial form, derives predictions from it, makes the corresponding experimental measurements, and their comparison with the predictions show how to modify our guess at $n(\omega)$. We have here an epistemic cycle, of a type usual in scientific research.

Probability is one particular kind of ensemble average. Given a property A that the members of the ensemble can either have or not have (for example, the property that $Q(\omega)$ exceeds a fixed value q), we can define what is called an indicator function

$$\chi_A(\omega) = \begin{cases} 1 & \text{if member labelled } \omega \text{ has property } A \\ 0 & \text{if not .} \end{cases} \tag{9.4}$$

The probability of A in the ensemble is then simply defined as the ensemble average of $\chi_A(\omega)$, namely

$$\Pr(A) \equiv \langle \chi_A(\omega) \rangle = \int_\Omega \chi_A(\omega) dn(\omega) / \int_\Omega dn(\omega) \,. \tag{9.5}$$

The meaning of Eq. (9.5) is simple: the probability is simply the relative weight in the ensemble of those members that have property A. Writing the equivalent of (9.5) for the case of a finite ensemble,

$$\Pr(A) = \sum_{i=1}^{n} \chi(i)/n \,, \tag{9.5'}$$

it is clear that the sum merely counts those members that have A, so that the probability is the relative frequency of the members having A in the ensemble. Eq. (9.5) is merely the generalisation of (9.5').

(viii)

The ensemble picture thus bears a family resemblance to the frequentist view; it is the theoretical analogue to an experimental relative frequency. Unlike von Mises' collective, however, the ensemble is, from beginning to end, at the theoretical level; nor does it need the remarkable mathematical properties of the collective. Its members need not be countably infinite, nor need they be ordered. The labelling given by ω is quite arbitrary and can be redefined without difficulty; it is usually chosen to have some physical significance, or at least to simplify finding a good counting function $n(\omega)$, but this is a matter of convenience only.

The probability concept defined by this use of an ensemble is, as I have just said, a theoretical construct; its family resemblance to experimental relative frequencies immediately suggests that these constitute the experimental aspect of a physical probability; or, since probability can often also be measured in other ways, its central element. Now a relative frequency, as measured, almost never agrees exactly with what one finds from a suitably chosen ensemble. But this is precisely what is to be expected, as was explained earlier when the general idea of a concept in physics was discussed. Moreover, probability theory – which can indeed be based on the concept as just defined – offers us an estimate of what is an acceptable difference between the theoretical value found from the ensemble and the experimental result found as a relative frequency; if the actual difference is too large to be acceptable, then clearly our theoretical model – the ensemble we have constructed and in particular the counting function $n(\omega)$ – will have to be revised (provided, of course, that we can have confidence in the experimental work). Here then we find just the sort of firm but flexible link between the two aspects of probability that is needed, without either having to suppose that an infinite set of experimental data can be obtained or that we must accept some further postulate such as the principle of indifference. We note also that the link is not established in general through a postulate or by definition, but must be found afresh in each case, as a result of the deliberate effort to fit the theoretical ensemble to the physical system it is intended to describe.

(ix)

We proceed to outline some properties of the probability concept as described in the preceding section.

Taking as the property A that ω is in a subset s of Ω, we find from (9.5) that

$$P(s) = \int_\Omega \chi_{\omega \in s}(\omega) dn(\omega) / \int_\Omega dn(\omega) = \int_s dn(\omega) / \int_\Omega dn(\omega) . \qquad (9.6)$$

Thus a well defined probability may be derived from the counting function $n(\omega)$ by renormalising it. It is this probability that has traditionally been used to define ensemble averages; Eq. (9.6) shows that it is entirely consistent to use the ensemble in order to define probability, and then find a probability function in terms of which ensemble averages become

$$\langle Q \rangle = \int_\omega Q(\omega) dP(\omega) . \qquad (9.7)$$

(We write $dP(\omega)$ for the probability of an infinitesimal increment $d\omega$ in s.) Eq. (9.6) also shows that the counting function $n(\omega)$ must have the mathematical properties of a non-negative measure; it need not, however, be bounded so that the integrals in (9.6) remain finite (though that is the simplest case), for one can also take the integrals in (9.6) over a subset $\mathcal{E}$ of Ω and let $\mathcal{E}$ tend to Ω after taking their ratio. In this way a probability is found which corresponds to Rênyi's extension of the Kolmogorov axiomatisation. This case is of some practical importance; it also makes it clear that the counting function need not, but can be, a probability function.

(x)

A much more significant property of our probability concept is this: in most applications in physics, the quantities Q should be written $Q(\omega, t)$, for the systems move and change with time. Therefore we get probabilities that depend on time, such as $P(A; t) = \langle \chi_A(t) \rangle$. The time is a parameter in the probability and not itself a random variable (the probability is that of finding A at a previously fixed time t, not that of finding A and t). Such probabilities, as we noted above, are hard to accommodate in the main philosophical accounts of probability; yet their importance can hardly be exaggerated. They underlie statistical mechanics, and form the basis of the theory of stochastic processes, applications of which go well beyond physics. An example will indicate how time-dependent processes are of relevance.

The time evolution of the ensemble may lead to a state of equilibrium, in which the ensemble averages no longer change with time (though the individual members of the ensemble keep on evolving, of course); for this to happen the ensemble must be ergodic, and the ensemble average at a fixed time will coincide with the time average taken over the evolution of an individual member in it:

$$\langle Q(t)\rangle = \overline{Q(\omega)} \equiv \lim_{T\to\infty} \frac{1}{T}\int_t^T Q(\omega,\tau)d\tau\ . \qquad (9.8)$$

(The bar over the Q indicates the time average; it depends only on ω, the member in question of the ensemble; Eq. (9.8), it should be noted, is valid for almost all members of the ensemble, but there may be a statistically insignificant number of exceptions.) But for a non-ergodic ensemble (9.8) will not necessarily hold, and a sequence of measurements on a system could well stray far from the equilibrium value, *and stay away*. In such cases the Urphänomen that von Mises postulated will not be observed – yet the probability concept developed here remains perfectly applicable. Non-ergodic ensembles appear, for instance, whenever the member systems are not only not closed but affect their environment in a way that influences their future interactions with it. Biological species, to take an interesting case, often change their environment quite profoundly; perhaps the most outstanding case is that of greeen plants, to whom we owe our present atmospheric composition. Biological evolution does not therefore tend to equilibrium states, yet probability concepts are extremely useful for understanding its theory. (I am not here defending any kind of reductionism; the theory of evolution is in no sense 'merely' physics; but that physical concepts can usefully be employed in order to clarify *biological* thinking is not to be doubted.)

The study of chaos in deterministic systems has proved astonishingly fruitful and rich in unexpected insights. While the standard philosophical conceptions of probability could only with some difficulty account for these new uses of probability, the ensemble-based concept appears entirely suitable. Instead of a general review of this question, too technical to be attempted here, I consider only one aspect: the random-number generators used in computers. Such a generator is a short program, that is to say a predetermined sequence of instructions, and its output is predictable with certainty from the input. Yet the sequence of numbers it produces when each output is fed back again as a new input appears random under all statistical tests; indeed, for a well written generator its statistical quality is much better than the data from a genuinely random physical system such as the pulse sequence from a radioactive decay process. When such algorithmic generators were introduced, von Neumann said that anyone who speaks of the algorithmic generation of random numbers is in a state of sin; this comment tells us more about its author than it does about what explains the remarkably random appearance of such sequences. The term 'pseudo-random sequence' has often been used, but so far with little discussion to justify it.

If the probability concept presented here is to apply, it must be possible to construct the relevant ensemble. For the random-number generators used in computers, this presents no difficulty. Every possible sequence of random numbers obtained from them is determined by the first input – the 'seed'– from which the entire output sequence derives by the repeated application of the generator to the preceding output. (Some generators depend not simply on the preceding value that was generated but on several earlier ones; for these the event space over which the ensemble is defined possesses several dimensions.)

Since every generator has a finite Poincaré cycle, at the end of which the seed is generated, the "time" evolution is given by a finite number of discrete steps; similarly, since the computer can only represent a finite number of different numbers, the event space itself also is finite and discrete. Thus Eqs. (9.5) to (9.8) must be written as sums, just as in (9.5') above, instead of integrals. But this does not affect their validity; at most it means that certain properties and values will only be approximate. This fact not only means that one can explain why the sequences of numbers generated by these algorithms should behave so much like random sequences; it is because they *are* random sequences, in the full sense of the word, and there is nothing 'pseudo' about them; it also means that we have all the tools of statistical and ergodic theory at our disposal in handling them; and it has proved invaluable in developing new random generators. We note in parenthesis that the ensemble point of view does not invalidate the approach developed by Kolmogorov and Marint-Lóf, who show that the randomness of a finite sequence of numbers can be expressed in terms of the length of the minimal program needed to generate the sequence.

10. The Ensemble Interpretation of Probability [*]

Abstract. Current philosophical interpretations are shown to be unsatisfactory when applied to problems of scientific research. An alternative, based on the work of Einstein and Gibbs, is proposed: probability as a scientific concept, with a theoretical and an experimental component, the former based on the ensemble and averages over it, the latter on relative frequencies in (finite) sets of experimental data. The two will agree only to the extent that the theoretical background of the ensemble is satisfactory. This interpretation extends in a natural way to the time-dependent probabilities of stochastic processes. The relevance of the concept in other areas of physics is exhibited.

10.1 Is Another Interpretation Needed?

It might be said that there already exist sufficient alternative interpretations of the probability concept to satisfy any possible need. An opinion common among physicists, perhaps induced by the present vogue of formalism, is that the mathematical theory of probability obviates any need for a further interpretation. However, Kolmogorov's axioms establish the mathematical properties of probabilities, but they do not tell us how to determine the elementary probabilities, i.e. how to measure them or derive them from a physical theory, nor do they tell us when talking of probabilities makes sense (see e.g. Suppes 1968). The problem of finding an interpretation is that of establishing the relationship between the real world in which we use probabilities and the mathematical structure which describes them. What this chapter will maintain, then, is that the presently available forms of interpretation do not satisfactorily solve this problem, and that the ensemble interpretation, described below, does so.

There are two main schools of interpretation. One identifies probability with the confidence that we place in a proposition; variants range from those that treat it as a logical relation like, but less certain than, an inference (Keynes 1921; or, in another vein, Jeffreys 1948) to those who, quite openly subjective, use the individual's estimations as their starting point (Ramsey 1926; de Finetti 1970). What they have in common is that for them probability describes a feature of our knowledge of some physical system, resumed

---

[*] E.I. Bitsakis and C.A. Nicolaides (eds.) (1989): *The Concept of Probability* (Kluwer Academic Publishers), 353–369.

in a proposition, and not an objective property of that system; if our knowledge were certain, this sort of probability would be superfluous. These views are thus based on the "ignorance interpretation": uncertain knowledge is expressed as probability. Popper (1959a) is therefore quite right in classing all these interpretations as subjectivist.

The other kind of interpretation prides itself on being objective because it refers probability to an event – something that quite objectively either happens or does not. But its chief and almost exclusive representative, the frequency interpretation, is better called positivist. Its principal exponents, v. Mises (1919, 1931) and Reichenbach (1935), were logical positivists, and their notion of probability shows significant – and problematic – traces of this philosophy. Quite characteristically, they confound the theoretical and the experimental when they define the probability of an event as the infinite limit (a theoretical notion) of the relative frequency (derived from necessarily finite and experimental data). Objectivity, in probability as elsewhere, requires the interaction of theoretical and experimental elements; conflating them in an unanalysable way does not contribute to objectivity.

These two schools of interpretation are subject to several well known philosophical criticisms; but what is relevant here is that neither of them is at all appropriate for a probability concept to be used in scientific research. Now scientific concepts have a double-sided nature, with a theoretical and an experimental aspect. A theory in which such a concept occurs will describe how it relates to other concepts; there may also be a more abstract background theory for the concept itself, particularly if the concept is common to several theories. On the experimental side there exist procedures and techniques for measuring the value taken by the concept in a concrete case; what characterises such measurements is that each is associated with certain limits on its precision. The interpretation of the concept – its physical meaning – provides for linking it to a feature of the real world, for relating its experimental and theoretical aspects; but this relation cannot be either automatic or exact. It is not automatic, for a theoretically derived value and the corresponding experimental one will agree only insofar as we have both an adequate theoretical model and a satisfactory experimental technique. It is also not exact, because besides those contemplated in the theoretical model, endless other factors will influence the experimental finding; and since these factors vary in complex and often unknown ways, measurements are repeatable only to within a certain range, technically known as the error limit. Agreement between theory and observation is then to be sought only to within these limits. Indeed, if it is notably better, it becomes suspect.

In the present case, we have a background theory of mathematical form, based on Kolmogorov's (1933) axioms; we also have specific theories for the objects of research, from which we derive values of the "elementary" probabilities to which probabilistic reasoning is then applied, to give the sought-for theoretical predictions. In the (apparently) simple case of coin tossing, we predict a probability of 1/2 for heads. If we actually toss the coin, we will mostly find something between 40 and 60 heads for every 100 tosses, while exactly 50 heads is rather improbable. With a bent coin the proportion might be quite far from this range, not because probability theory has failed but because we ought to have used a different physical model that applies to bent coins. Here then a theoretical probability will predict a relative frequency, but only with limited precision and only if the physical model is adequate. By keeping the two sides of the probability concept distinct, we can achieve a relation between them that satisfies the needs of scientific research.

For the subjectivist interpretations probability concerns a proposition, and more specifically our view of that proposition, and no experimental component is offered. The link to any observed frequency then poses an insoluble problem. To circumvent it, peculiar abstract principles, such as the principle of insufficient reason (Keynes 1921) or the stability of long-term frequencies (Hacking 1966), must be invoked. But they are apt to fail at the critical moment; insufficient reason gives rise to paradoxes, and probability can be meaningfully used even if there are no stable long-term runs. The profound divorce between theory and experiment in subjectivist conceptions appears again in their inability to incorporate the Kolmogorov axioms in a natural way: indeed, the actual use of the betting quotients that some of them take as starting point may not even conform to these axioms.

In v. Mises' conception, we have the opposite problem: the observed frequency serves actually to define the theoretical probability, and so the link between them is automatic, rigid and exact. Even though now the Kolmogorov axioms hold, essentially because they hold for relative frequencies, once again we have here a viewpoint that differs fundamentally from what is needed in the sciences. The discrepancy is shown up by the frequentist's inability to countenance probabilities for singular events, although these are constantly needed in statistical theory. Again, there is no specifically experimental component of probability.

These failures become serious in the less trivial applications of probability, e.g. when it depends on some parameter such as the time, or on whether some preceding probilistic event occurred. One's knowledge, and hence a subjectivist probability estimate, will not usually change merely because time passes; and v. Mises' basic tool, the collective, requires each event in it to be statistically independent of all the others. Hence on either view stochastic processes and ergodic theory (the theory of the connections between probabilistic and time averages) cannot be consistently developed. The recent exciting discoveries concerning chaotic behaviour in classical, deterministically described systems, are even more inexplicable: their deterministic theory means that full knowledge is available and subjective probability either 0 or 1, and no frequentist collective can be constructed since all events are wholly determined by their predecessors.

Other philosophical views of probability have been propounded, which are no more satisfactory; because of space limitations they will be ignored here (but see Rédei and Szegedi 1988).

We conclude that a new interpretation of probability, scientifically sounder, is needed.

10.2 The Physical System and Its Surroundings

When doing scientific research (and in everyday life) we never deal with the universe as a whole, we select from it a segment designed to contain everything relevant to our purpose and nothing irrelevant; following the physicists' custom, we shall talk about the (physical) system. Such a system is not simply given geometrically, for even of the objects involved only those aspects that interest us are taken into account: in calculating the trajectory of the inkpot that Luther threw at the devil, we do not ask what the colour of the ink was, nor do we worry whether Newton's apple tasted sweet; for calculating the tide we include the positions of both sun and moon, although they are very far yet we do not consider the boats floating on that tide.

Nor is our choice of system final; half the battle in doing research is finding a satisfactory system, for only then can we build an adequate theoretical model. It is only on the basis of an appropriate system that we can even design and carry out relevant experiments. In daily life past experience is usually a good enough guide to lead us quite rapidly to a satisfactory system; in research it may take many repeated cycles of setting up a system, developing a theoretical model, making experiments, and from the discrepancies seeing how to revise the system.

We also use several systems in parallel, either to devote special attention to certain aspects, to provide several different levels of approximation, or to study a problem with different aims in mind. Thus there must be many hundreds of different models of our earth, some treating it simply as a mass point, others as a perfect sphere, still others as a flattened one, others yet taking into account its surface irregularities, and some even treating it as an infinite flat body. And in probability theory we contemplate an infinite number of systems.

A system is specified both theoretically and experimentally. Normally we begin with the theoretical side, with the mental image, which we shall call its model. From it we derive the experimental specification of the system (through the use of suitable equipment like thermostats and so on). This is intended to isolate the system as well as possible from the remainder of the universe; nevertheless, an almost endless number of interactions of diverse nature and strength still link them. It may be possible to ignore these "outside" factors; if the model explains the system's behaviour in terms of a finite set of factors with negligible outside dependence, then we call the system closed. Some outside factors act so as to stabilise the system and so allow us to use a simpler one; thus the force of gravity reduces the billiard table to a two-dimensional system, by not allowing significant motion in the third dimension. Many small interactions may be neglected at first and later be taken into account as corrections. Finally there are almost always very many outside factors whose individual effect is negligibly small but whose joint influence may be quite large and may even change the system's character. In such cases we can neither ignore these factors, nor can we include them explicitly, for individually they are quite irrelevant and we are interested only in their joint action. Thus we need a new approach: the probabilistic one.

10.3 Ensembles, Averages and Probabilities

To each concrete and tangible physical system there corresponds essentially only one model, for if we change the model, we include new features and drop old ones, so that a different system is described; but one model may describe many systems, differing from each other in features not contained in the model. We could have built lower-level models that make these differences explicit; but instead of actually building them, we think of them as having been averaged over, and now a higher-level model built using these averages takes what above we called the irrelevant factors into account – jointly but not individually.

This procedure was first developed for the kinetic theory of gases. Traditional thermodynamics gives us a description of a gas in terms of pressures, volumes, temperatures, and so on. Yet if one could treat a gas simply as a mechanical system made up of a great many loosely interacting molecules, the same results should appear. However, they do not; instead we have endless facts about the positions and velocities of all the molecules at different moments, all of them strongly dependent on their initial positions and velocities. The problem of how to remove the unwanted information that depends on the initial conditions while keeping what does not so depend was solved by Einstein (1902, 1903, 1904) and by Gibbs (1902), using the method indicated above; they considered not one model, with the appropriate initial values for positions and velocities, but a whole collection of models with all conceivable combinations of initial values. Averaging over this set of models technically known as an ensemble – is a trivial operation mathematically, but one has the power of creating new concepts of quite different characteristics: for instance, ensemble averages no longer depend on initial conditions, and in statistical mechanics possess the properties needed for temperatures and so on.

Now ensembles are usually discussed in terms of an already established probability concept (e.g. Tolman 1938 or Balescu 1975); but the ensemble notion and the averaging operation can be quite straightforwardly defined without this concept, simply by clothing the preceding discussion in somewhat more formal terms. Probability is then nothing but a particular kind of average over the ensemble. Indeed, statistical mechanics can be presented in terms of no more than these two notions, ensemble and average (Fowler and Guggenheim 1939). We indicate briefly how this more formal argument might run:

Consider a set of hypothetical physical systems, all described by a common higher-level model and each by a lower-level one which is not made explicit but merely labelled by variables collectively denoted by ω. This could simply be a numbering, but in statistical mechanics the initial conditions are used. Any property f of the systems will then depend on ω, and we write it $f(\omega)$. Its ensemble average is then

$$F \equiv \langle f \rangle = \int_\Omega f(\omega) d\mu(\omega) / \int_\Omega d\mu(\omega) \,. \tag{10.1}$$

Here Ω is the range of possible values of ω, the angle brackets $\langle\ \rangle$ are standard notation for an ensemble average, and the function μ characterises the ensemble. If we now consider various related functions f, then Eq. (10.1) will connect the corresponding averages F in a new theory, from which the underlying variables ω and any information explicitly dependent on them have disappeared. In the case of statistical mechanics, the ensemble concept has thus allowed us to pass from the level of initial-condition dependence, in the mechanical model, to that of a new theory, thermodynamics, no longer so dependent.

Probability appears as a particular average. Given a property A that some systems in the ensemble possess but others do not, we consider the indicator function $\chi_{A(\omega)}$, which is 1 if the system labelled ω has the property A and 0 if not; the probability of A is then simply the ensemble average

$$\text{Prob}(A) = \langle\chi_A\rangle = \int_\Omega \chi_A(\omega)d\mu(\omega)/\int_\Omega d\mu(\omega)\,. \tag{10.2}$$

If Ω is a finite set of points (or systems) and μ the same for all of them, this is equivalent to a (theoretical) relative frequency; but Eq. (10.2) is more general, and works for an infinite set or a continuous Ω. The definition (10.2) is easily seen to satisfy the Kolmogorov axioms, since

$$\begin{aligned} \chi_{A\cap B}(\omega) &= \chi_A(\omega)\chi_B(\omega) \\ \chi_{A\cup B}(\omega) &= \chi_A(\omega) + \chi_B(\omega) - \chi_{A\cap B}(\omega)\,. \end{aligned} \tag{10.3}$$

Therefore the probability defined by Eq. (10.2) is indeed that dealt with by standard probability theory. (This definition may be further generalised.)

The two fundamental quantities Ω and μ are not determined by probability theory; they must be derived from the particular theory (physical or other) that covers the application envisaged. In some cases they can be fully specified in this way; more commonly Ω is so determined, but μ must be found by research into the problem, usually in fact by trial and error. In other cases μ need not even be fully detailed: provided a central-limit theorem exists, all we need to establish is that μ belongs to the class to which it applies; the commonest such case gives rise to Gaussian distributions, and explains why this distribution occurs so frequently as to have earned the epithet "normal". In certain cases it is even sufficient to know that μ exists in principle.

It may however happen that Ω has a structure for which no μ can be found, or Ω itself may not exist. The notion of probability does not then apply. The truth of a scientific theory is a case in point. The ensemble here would consist of situations where the theory is true and others where it is not, and it is easily established that there is no consistent way to formulate such an ensemble. One cannot therefore speak meaningfully of the probability that a theory is true, and induction is better forgotten about.

The construction of a suitable ensemble – when possible – provides us with the theoretical aspect of the probability concept. At least one experimental method derives from such intuitive views as saying that if the probability of a coin falling heads is 0.5, then about one half of the coin tosses will come up heads. In other words, a relative frequency (obtained on a finite data set) is an experimental correlate of a theoretical estimate found from an ensemble. But these two will coincide only exceptionally; and the relative frequency will have error limits. For short data series these limits will be dominated by the so-called statistical errors, i.e. the errors due to the fluctuations in the numerous factors outside the model but still of some influence on the system; statistical theory describes how to estimate and treat these errors. But as the data series grows, other error sources become increasingly important, until at length the coin we are tossing becomes too worn to be treated as symmetrical or the two faces as distinguishable.

Determining a relative frequency is the most general method of estimating probabilities. If the successive events in the data series accumulate very rapidly, we may only be able to observe their rate of occurrence: thus the intensity of a light source is really the rate at which individual molecules emit and are deexcited. In other cases neither the relative frequency nor a corresponding intensity are accessible, but we can measure the value of a function like the F defined in Eq. (10.1), which by using the discrete equivalent of Eq. (10.2) may be rewritten as

$$F = \sum_i f(\omega_i)\mathrm{Prob}(\omega_i) . \tag{10.4}$$

We cannot of course invert Eq. (10.4) to find the probability; but if several different such functions have been measured we can often make a reasonable guess at the probabilty distribution.

Clearly all methods of measuring a probability except that of counting instances to give a relative frequency depend rather strongly on the particular application. And all of them have finite error limits, a fact which if forgotten causes numerous confusions, some of which still haunt the philosophical discussion of probability. Many of the paradoxes surrounding the "principle of indifference" spring from this source.

Having briefly indicated what the ensemble interpretation of the probability concept consists in, I should add that I am in no sense its inventor (Brody 1975). It is, if anything, part of the common culture of physicists, and I suspect of other scientists too. It does not correspond, of course, to what scientists are taught in university courses, but their daily practice forces it on them. Analogous ideas appear to have been in the minds of some very outstanding mathematicians; thus Kolmogorov (1969) comes very close to an explicit statement of the ensemble view presented here, and Kac (1959), after presenting some very illuminating examples, sums up as follows:

At this point it should have become clear to the reader that probabilistic reasoning consists in imbedding a particular situation in an ensemble of like situations and replacing statements about individuals by statements about the ensemble.

Could the viewpoint outlined here be presented more succinctly?

10.4 Probability in Stochastic and Chaotic Systems

We turn now to the problem, mentioned in Sect. 10.1, of the time dependence of probabilities. A probability may vary in time through three mechanisms: under an external time-varying influence, because it belongs to a stochastic process and therefore depends on preceding events, or because a deterministic system has entered a region of chaotic behaviour.

A good example of the first case is found in the incidence of an epidemic that is being combatted by an effective public-health programme, so that the probability of falling a victim to the disease diminishes. Such a time dependence creates a difficulty for the frequentist, since at any given time there is only a finite sequence of events available, and lengthening the sequences by admitting different times gives series that do not converge to a well-defined limit. For the subjectivist the difficulty is rather that now probabilities alter without any further evidence being adduced. No problem arises in the ensemble view; any ensemble that represents reality adequately may be expected to evolve in time, and so also its probabilities.

In a stochastic process all probabilities are conditional on what has happened earlier to the system; their time evolution will therefore be different for different members of the ensemble, being given by a subensemble defined by the common prior history. As a result, the average along a trajectory (for a given system over its history) and over the ensemble (all systems, whatever their history, at a given time) will in general differ. But there is a special type of ensemble for which they coincide: the so-called ergodic systems. This concept (and the corresponding theory) plays an important role in much of modern physics; it is also relevant to the understanding of probability, in that non-ergodic systems (now known to be much the commonest case) will not show any long-term stability of relative frequency, such as many authors (e.g. Hacking 1966) try to use as underpinning for their probability views. The ensemble view provides a probability concept to be used even in such cases; more in general, it is the only one in which stochastic processes and ergodic theory can be given a consistent treatment.

In the last few decades it has been found that non-linear systems, even though they are described by perfectly deterministic equations of evolution, can exhibit rather startling behaviour of random character (Hao 1984, Cvitanović 1984). These phenomena are so complex (and as yet so little understood) that I only discuss one of the simplest and oldest cases, the generation of random numbers in a computer. The chief method used is the congruence method (Lehmer 1951): given suitably chosen integers a, c and m, one chooses a starting integer μ_0 and calculates successively for all i

$$u_{i+1} = (au_i + c)\bmod m \,, \tag{10.5}$$

where "$x \bmod y$" denotes the residue after division of x by y. Dividing u_i by m now gives a random number between 0 and 1. Other algorithms (e.g. Brody 1984) are often advantageous.

Now a computer is a deterministic device and no algorithm should, according to traditional views, produce a sequence of numbers with probabilistic behaviour. However, if the algorithm is well designed, the sequence has statistical properties notably closer to random than natural processes; one should not then sidestep the philosophical problem by talk of "pseudo-random" numbers, as is often done. But an ensemble can be constructed here: the event space is, for the congruential generator (10.5), simply the set of integers from 0 to m; the "systems" are the different sequences generated and may be labelled by their initial value u_0, which determines them completely. A good generator can be proved to be ergodic – averages along a sequence coincide with averages over the ensemble for fixed i, which here plays the role of a time. And even if we know the values of the three parameters in (10.5), we cannot from a number in the sequence deduce its predecessor, since there are many different possible ones. Not that our ignorance is necessary: the computer could easily be programmed to keep a record; but we do not normally do so, because the algorithm can make no use of it. Just as in the case of statistical mechanics, in other words, we deliberately ignore part of what is knowable; and by ignoring the individual trees, here irrelevant, we are able to see the wood. To put this point more generally: the probability concept is not needed because of our ignorance; on the contrary, it permits us to eliminate unwanted infromation. To link probability with ignorance, as so many authors do, is thus a mistake; the paradox that we need probability to deal with an excess of information comes closer to reality.

Neither the subjectivist views nor the frequency interpretation are useful in understanding random-number generation or indeed the chaotic behaviour of deterministic systems; the best they can do is summed up in the quip that "anyone who considers arithmetical methods of producing random numbers is, of course, in a state of sin" (v. Neumann 1951).

10.5 Probability and Quantum Mechanics

We have discussed the interpretation of probability in statistical mechanics where it is the physical application that has helped to clarify the philosophical concept; in quantum mechanics, inversely, the philosophical confusion has become reflected in the conceptual confusion surrounding this theory. Soon after the development of matrix mechanics, its founders saw the need for an interpretation, which they based on the idea that a quantum description (in terms of a state vector, or equivalently a Schrödinger wave function) refers to a single physical system only. The square of the wave function then gives the probability of finding the system in the relevant state; but for a single system, what does such a probability signify? Each expectation value derived from it has a certain spread, and Heisenberg has shown that for a pair of conjugate variables the product of the spreads has to be greater than a certain limit; neither of them can thus be zero. Heisenberg interpreted the spreads as the theoretical minima of the experimental errors, which thus become irreducible.

Does this mean that the variables are, to this extent, indeterminate? And how is it possible for the extent of the indeterminacy to depend on what else is being measured? It seemed necessary to conclude that it is the measurement procedure that determines the value of what is measured, which therefore is not really a property of the physical system. Indeed, one might say that it is the measurement which creates the quantity measured. The spiral into steadily more subjective views was impossible to stop, and even outstanding physicists fell into fully subjectivist and in the end solipsist world views (Bohr 1936, 1948, Jordan 1938, Heisenberg 1951, Pauli 1954, Houston 1966) or the acceptance of quantum mechanics as incomprehensible (Dyson 1958).

That a probability may be stipulated for a single physical system is only possible, as our previous discussion has shown, within one or another of the subjectivist views, where probabilities belong to propositions; the attempt to reach consistency then leads to these undesirable consequences. The alternative approach, that the state vector represents not one but an ensemble of systems, is conceivable for the frequency interpretation and the only possible one for the ensemble interpretation. In quantum mechanics this approach was adopted first by Slater (1929) and developed further by Einstein (1936, 1953), Blokhintsev (1953), Lamb (1969, 1978), and others; but much work remains to be done on it, a fact which has given rise both to criticism of various kinds, often conceptually confused, and to the refusal of most textbooks even to mention that there are alternative interpretations. Yet the ensemble view offers many advantages among which its conceptual straightforwardness stands out; it eliminates the endless confusions, misunderstandings and paradoxes which afflict the "orthodox" view (see e.g. Ballentine 1970, Ross-Bonney 1975).

One criticism levelled at the ensemble viewpoint is that the EPR argument (Einstein, Podolsky and Rosen 1935) that quantum mechanics is incomplete would have to be accepted. Without entering into technicalities, what the argument implies is that quantum mechanics, being a theory whose predictions take the form of ensemble averages, is the "thermodynamics" to a statistical mechanics which remains still to be formulated. But is this not precisely the case? In the last few years, an increasing number of physicists have come round to the idea that work in this direction is worthwhile (Brody 1983 and references given there).

Work along such lines might have been initiated many years ago had an adequate conception of what probability is been generally accepted. As it is, except for some less significant addition and a certain amount of largely formal rewriting, non-relativistic quantum mechanics has remained unchanged for half a century, while experimental knowledge has progressed by leaps and bounds. The single-system view has even misled many people into concluding that quantum mechanics is our last fundamental theory, so that physics, apart from improvements in technical detail and new applications, has essentially come to an end. Such views arise if one argues that quantum probabilities are basic and not further to be analysed (as the ensemble view of course permits and fosters).

10.6 Philosophical Conclusions

The survey of the preceding sections leads us to conclude that traditional views on probability have not generally been helpful in physics and have sometimes done damage; the ensemble interpretation, on the other hand, aids in untangling conceptual confusions. There still remain some points of philosophical import to be mentioned.

(i) A probability concept based on ensembles might, on a superficial view, be considered subjective, since ensembles are thought constructions. But we cannot use just any ensemble: like any other scientific concept, an ensemble must, by repeated cycles of reformulation and comparison of experimental reality, be made to describe that reality fittingly, in accord with our purposes; and like any scientific concept, it acquires objectivity in the course of reformulation and adaptation. Note that objectivity in a concept can only mean that it correctly represents a selected aspect of the world. Objectivity is, therefore, a property we confer on the concept through our work in formulating and reformulating it. Correspondingly, no concept is purely objective; it is precisely its subjective aspect that allows us to capture, understand, adapt and use it. What is wrong, then, with the subjective views of probability is not that they possess a subjective aspect; it is that they possess nothing else and are therefore subjective where they should not be. Concretely, therefore, the objectivity of each particular application of the probability notion must be achieved through research work and must be critically examined; there cannot be a global solution. There may be numerous and difficult problems in validating the particular probabilistic approach taken; but these problems are not in essence distinct from those facing any other kind of theoretical model.

There is also a much more general sense in which probability is objective: probability arises as a way of recognising the non-isolated character of the system we are dealing with, without describing the "outside" influences explicitly. Probability connects the level of full description (outside influences made explicit) with that of statistical description, of description via the elements of behaviour common to all systems; these two, as we have seen, may be of entirely different character. The two-level structure is a consequence of the fact that physical systems are only partially separable from the rest of the universe, and so is probability, the tool we use to describe it; the probability concept is objective inasmuch as it yields an account of this partial separability.

(ii) Probability is also relative, and that again in a double sense. Each probability is meaningful when referred to its ensemble; probabilities belonging to different ensembles are not comparable, and such comparisons give rise to well known absurdities. And for a given event we can find a large number of different probabilities, according to the various ensembles into which this event will fit; the choice between them is guided by our purpose, but given that purpose is essentially determined (to within indefinitions not discussed here). Probability is relative in the sense in which distances are relative: given a reference point, they are perfectly objective, without it they are meaningless.

Points (i) and (ii) are adequately catered for by the ensemble conception, but at most with great difficulty by other views.

(iii) The ensemble view of probability allows the construction of an ensemble each of the members of which are ensembles in their turn; indeed, constructions of three or more levels are conceivable (though they do yet seem to have been used). This is of practical utility: in statistics, where it underlies the notion of the probability of a probability; in quantum mechanics, where it is needed for quantum statistical mechanics; and in the biological theory of evolution. In the last, the top level concerns the evolutionary process itself, and the individuals making up the ensemble are evidently the biological species, which are born, change without becoming a new species, change into a new species, divide into two or more species, or become extinct. It is important to note that this process involves a time-dependent ensemble whose (stochastic) evolution is highly non-ergodic; there is thus no equilibrium state to which most trajectories tend, and final states, in themselves of low probability, are "frozen in". Each member of this ensemble must itself be an ensemble, however, in order to be able to account for the evolution of the species in terms of individual histories. If this double structure is not recognised, it becomes impossible to see what is actually the bearer of evolution, and sterile and confusing debates have hence arisen. But the ensemble-of-ensembles construction makes it clear that it is the species that evolves, not the individual, and least of all the gene (indeed, since an individual living being has one and only one set of paired genes, it seems unlikely that a third level of ensembles could be useful).

(iv) Finally, as the probability concept developed here suggests and the study of chaotic systems demonstrates, the philosophical dichotomy between probabilistic and deterministic world views is very inadequate. Neither is applicable to the universe as a whole but only to specific systems; and in every case, given a deterministic model we can build a probabilistic one for the same system, and inversely. The two model types complement each other. It is possible that instead we face a much more fundamental distinction (so far ignored by philosophers): in a non-chaotic system, only the precision and completeness of our knowledge of the initial conditions limits the time range over which we can predict future behaviour; a chaotic system, on the other hand, possesses an inbuilt limit to predictability. In such a system, trajectories starting at almost coincident conditions will quite rapidly diverge so widely that improving our knowledge of the initial conditions will hardly help at all. A good example is that of weather forecasting; although the atmosphere is a rather well understood "deterministic" system, it is capable of very varied chaotic behaviour (Lorenz 1963, Mason 1968), so that even with ideally precise and abundant data we could not foretell the weather for more than a few days. Since this distinction between the predictable and the unpredictable characterises the physical system rather than its theoretical model, we have here an ontological difference; that between a deterministic and a probabilistic description being that between two aspects, possessed by almost every system, but brought out by different models.

References

Balescu, R. (1975): *Equilibrium and Non-Equilibrium Statistical Mechanics* (J. Wiley, New York)

Ballentine, L.E. (1970): Rev. Mod. Phys. **42**, 358

Blokhintsev, D.E. (1953): *Grundlagen der Quantenmechanik* (Deutscher Verlag der Wissenschaften, Berlin)

Bohr, N.: Erkenntnis **6**, 263

Bohr, N. (1948): Dialectica **2**, 312

Brody, T.A. (1975): Rev. Mex. Fis. **24**, 25

Brody, T.A. (1983): Rev. Mex. Fis. **29**, 461

Brody, T.A. (1984), Comp. Phys. Comm. **34**, 39

Cvitanovic, P. (1984): *Universality in Chaos* (Adam Hilger, Bristol)

Dyson, F.J. (1958): Scientific American **199** (9), 74

Einstein, A. (1902): Ann. d. Phys., IV. Folge, **9**, 417

Einstein, A. (1903): Ann. d. Phys., IV. Folge, **11**, 170

Einstein, A. (1904): Ann. d. Phys., IV. Folge, **14**, 354

Einstein, A. (1936): J. Franklin Inst. **221**, 349

Einstein, A. (1953): in *Scientific Papers Presented to Max Born* (Oliver & Boyd, Edinburgh) p. 33; Einstein, A., Podolsky, B., Rosen, N. (1935), Phys. Rev. **47**, 777

Finetti, de B. (1970): *Teoria delle probabilitá* (Einaudi, Torino)

Fowler, R.H., Guggenheim, E.A. (1939): *Statistical Thermodynamics* (Cambridge University Press, Cambridge)

Gibbs, J.W. (1902): *Elementary Principles in Statistical Mechanics* (Yale University Press, Yale, Connecticut)

Hacking, I.M. (1966): *The Logic of Statistical Inference* (Cambridge University Press, London)

Hao Bai-Lin (1984): *Chaos* (World Scientific, Singapore)

Heisenberg, W. (1951): Naturwissenschaften **38**, 49

Houston, W.A.: Amer. J. Phys. **34**, 351

Jeffreys, H. (1948): *The Theory of Probability* (Oxford University Press, Oxford)

Jordan, P. (1938): *Die Physik des zwanzigsten Jahrhunderts* (Vieweg, Braunschweig)

Kac, M. (1959): *Probability and Related Topics in Physical Sciences* (Interscience, London New York) p. 23

Keynes, J.M. (1921): *A Treatise of Probability* (Macmillan, London)

Kolmogorov, A.N. (1933): *Grundbegriffe der Wahrscheinlichkeitsrechnung* (Springer, Berlin Heidelberg New York)

Kolmogorow, A.N. (1969): Probl. Inf. Trans. **5**, No. 3, 1

Lamb, W.E., Jr. (1969): Phys. Today **22** (4), 23

Lamb, W.E., Jr. (1978): in S. Fujita (ed.) *The Ta-You Wu Festschrift* (Gordon & Breach, London) p. 1

Lorenz, E.N. (1963): J. Atmos. Sci. **20**, 130

Mason, B.J. (1968): Contemp. Phys. **27**, 463

Mises, R.v. (1919): Math. Zeits. **5**, 52

Mises, R.v. (1931): *Wahrscheinlichkeitsrechnung und ihre Anwendung* (F. Deuticke, Wien)

Pauli, W. (1954): Dialectica **8**, 112

Popper, K.R. (1959): *The Logic of Scientific Discovery* (Hutchinson, London)

Ramsey, F.P. (1926): in R.B. Braithwaite (ed.) *Truth and Probability in the Foundations of Mathematics* (Routledge and Kegan Paul, London)

Rédei, E., Szegedi, P. (1988): in E.I. Bitsakis, C.A. Nicolaides (eds.) *The Concept of Probability* (1989) (Kluwer Academic Publishers)

Reichenbach, H. (1935): *Wahrscheinlichkeitslehre* A.W. Sijthoff (Leiden Wissenschaften, Berlin)

Ross-Bonney, A.A.: Nuovo Cim. **30B**, 55

Slater, J.C. (1929): J. Franklin Inst. **207**, 449

Suppes, P. (1968): J. Phil. Sci. **65**, 651

Tolman, R.C. (1938): *The Principles of Statistical Mechanics* (Oxford University Press, Oxford)

11. The Philosophy of Ensemble Probability

Chapter 10 outlined an interpretation of the probability concept that is based on the notion of an ensemble of theoretical replicas of the system being studied. The replicas represent all possible variations of those factors influencing the system which we do not want to have included in the description given by the ensemble averages. As we noted, the ensemble description does not automatically yield satisfactory values for these ensemble averages, including the particular average which is called probability; to achieve this, the specification of the ensemble and in particular the counting function $n(\omega)$ of Eq. (9.5) must be made to fit the segment of nature under study. This fit is obtained through the operation of an epistemic cycle. A first epistemic model (usually found from a general theory within the framework of which the ensemble is being formulated) provides predictions for the behaviour of the system which are used to set up an experiment and compared to the outcomes. This comparison may not be satisfactory to within the tolerance required for the purpose for which the ensemble is being built; if so, the comparison will furnish indications towards improving the specification of the ensemble, of $n(\omega)$ in most cases. In this process, several features of philosophical interest become apparent.

Firstly, in many cases incomplete specifications of the ensemble are sufficient. There are even cases where it is enough to know that an ensemble description is possible, for then one can use probability arguments. The opposite case also occurs, that the probability concept is inapplicable because no ensemble can be formulated: we will examine such a case below. In other cases certain aspects of the ensemble description are adequate; of these the commonest is the one where only the joint result of very many small factors, acting independently or approximately so, is of interest, for in such a situation one or another form of the central limit theorem of statistics applies, and provides the distribution function which can be used as $n(\omega)$.

Secondly, where one ensemble can be formulated, others that are relevant to the problem always exist. They can come about in three different ways: (i) Ensembles may be specialised by restricting the factors included in the models that are their members; for the example of a volume of gas, we can ignore the positions of the individual molecules and obtain the ensemble from which the Maxwellian velocity distribution is derived. (ii) Related ensembles may be built by altering the overall conditions; in the gas example, we have the microcanonical ensemble if the total energy of molecules is taken as fixed, but we obtain a different one, the canonical ensemble, if an exchange of heat with the surroundings is permitted, and still another one, the grand canonical ensemble, if matter can also be exchanged and the number of molecules is not fixed. (iii) The counting function $n(\omega)$ of many ensembles depends on parameters such as the total energy for the microcanonical ensemble; we then have a family of ensembles, rather than a single one, and since in general these parameters can have only one value at a time, the ensembles in such a family are not compatible with each other. A particularly interesting case combines possibilities (i) and (iii): under suitable experimental conditions it is sometimes possible to convert a variable $x(\omega)$ for the members of an ensemble into a parameter that has the same value for all members and so is a parameter of the ensemble. Thus an ensemble has been converted into a whole family of ensembles. This usually corresponds to selecting a subensemble; for instance, from the ensemble of all married couples we may pick out those who have exactly k children; k is then a parameter that creates a family of mutually incompatible ensembles. This point, rather obvious in itself, nevertheless will become of importance below, in connection with recent discussions about the foundations of quantum mechanics.

(ii)

A more directly philosophical question concerns the extent to which the probability concept is objective. The frequentist view is often termed objective, while at least some of the versions of the propositional one must be called subjective. Where does the ensemble view propounded here stand?

For a concept, it is referential objectivity that is significant, that is to say the extent to which it represents correctly the reality of an aspect of the segment of the world being examined; the number of chairs in a room is objectively seven if there really are seven chairs in it, for instance. In this sense, it is a matter of some doubt whether the frequentist conception really is objective; as has been pointed out, the definition given by von Mises not only begins with a confusion between the experimental and the theoretical aspects but is also based on a notion, the collective, which is only very loosely, if at all, linked to reality. But what makes both these conceptions of probability unsuitable for use in science is that the degree to which they are objective is a matter to be decided once and for all. The ensemble conception, on the other hand, is like other scientific concepts in that its objectivity – its fit to reality – must be achieved individually for each application of it through the iteration of the appropriate epistemic cycle until the fit is good enough, in a sense we have already insisted upon.

Inevitably, since this fit is never perfect but only good enough, the objectivity of probability is like that of any other scientific concept: it contains an ineradicable admixture of subjective elements. Some of these are necessary; we have seen that in making a model of a segment of reality one must distort certain features in order to be able to isolate and bring out those other features which are relevant to the purpose in hand. Other subjective elements could be less innocent. To reduce either, a further epistemic level would be needed; but even once that has been achieved, the probability concept will not lose its usefulness.

As several times before, we find that this is not a yes-no matter; objectivity is not automatic but must be achieved by hard work.

12. On Errors and Approximations

In physics, as in other sciences, approximation is the rule of the day. The theoretical work involved in preparing an experiment and the interpretation of the experiment's results are approximate in nature. Much purely theoretical work also involves approximation of one sort or another, while laboratory work yields results that are reproducible at best to within certain error limits (in a sense to be discussed below). The choice between different theoretical models for a given application depends critically on the degree of approximation required, and publication of an experimental datum without appropriate error limits is strongly discouraged as being essentially meaningless. It is therefore rather surprising that philosophy of science has so far paid scant attention to what is evidently a central matter in scientific practice.

Why have these evidently fundamental matters been so neglected by philosophy?

Part of the problem is no doubt that many philosophers of science keep too remote from the day-to-day work of scientific research. It is no help that even good scientific popularisation tends to present only exact results, often accompanied by graphs that show beautifully adjusted curves and no error bars. But a possibly more significant factor is an attitude inherited from 19th-century physics: the fluctuations and incertitudes of experimental results used to be considered a matter of laboratory technique merely, to be eliminated by simple statistical techniques (and even tricks), so that the average could now be considered an almost exact result and plugged into the theory. This is how physics could be dubbed an exact science. But we are well aware now that physics and the other natural sciences are exact only to the extent that their inexactitude can find quantitative expression; it is high time that this awareness should be elaborated into a viable philosophical account. The present paper is an attempt in this direction, necessarily incomplete because of the complexity of the issues and because the centrality of the problem involves many epistemological and even ontological matters. Section (i) briefly outlines how the problem appears in "real-life physics", so as to ground the discussion in fact, and Sect. (ii) discusses one facet of the problem area, that of approximation, which must be distinguished from that of experimental errors. Section (iii) presents the only contemporary philosophical account of this problem area. Section (iv) attempts an outline account of what underlies the phenomena of experimental error, and Sect. (v) proposes a mathematical description for certain types of experimental errors, whereas others will be seen not to be susceptible to such treatment.

(i)

One difficulty in achieving a philosophical understanding of the problem of errors and approximations is the immense complexity inherent in a practical situation. The usual textbook illustrations are not unreal, in the sense that they do represent cases which occur; but the situation is rare where the raw observational datum is also the hoped-for result, and one merely has to account for repeated measurements which yield slightly differing results. Normally considerable and elaborate calculations must be carried out, aberrant observations must be detected and eliminated, and discrepancies between various measurements or between different but equally justifiable treatments of the data must be reconciled.

Indeed, one is often faced by the kind of situation which an outstanding physicist characterises as "very tricky" and as requiring "intuition of no ordinary order, combined with a careful and conscientious juxtaposition of very different calculations (perhaps even some semi-empirical ones)" before even tentative conclusions can be arrived at (Sutcliffe 1975, p. 89). This paper is in fact a storehouse of information concerning actual procedures for developing approximations and evaluating errors in its field; and from it and similar work it is possible to derive at least a preliminary list of the "facts of life" which philosophy of science will have to account for.

The first such "fact of life" is that errors (which are by no means mistakes, the term keeps its original Latin meaning of wanderings) are the fluctuations of successive measurements under the same circumstances. That such measurements yield varying results in an apparently inexplicable disorder is what makes necessary a statistical treatment. In the simplest case we find that the data have a Gaussian distribution, so that determining their mean and standard deviation gives us all the information needed; but less simple distributions occur not uncommonly. Moreover, we generally measure more than one quantity, and therefore need more complex statistical methods in order to find correlation coefficients and similar results. As the problem acquires greater complexity, it shades over into the realm of statistical physics, where the fluctuations themselves are the phenomena under study.

Secondly, experimental errors are of several kinds; three are worth distinguishing in the present context:

1. Instrument errors. These determine the precision with which individual data can be read; they are usually obtained from ancillary experiments or from a (usually highly phenomenological) theory of the instrument. But even with the best instrument theory the actual size of the instrument errors will depend heavily on the skill of the observer, on how consistently he observes the stipulated laboratory procedure, and on how well this procedure is designed; the "observer" here may be human, or it, may be a suitably programmed computer. The human experimenter also plays a significant role in weeding out, repeating or correcting faulty observations;

this requires a full understanding of the whole situation, and cannot really be made automatic.

2. Measurement errors. These are the fluctuations among repeated measurements. They are observable only if on the average they exceed the instrument errors, a situation that has become almost standard due to the tremendous strides in instrument technology of recent years; but unlike the instrument errors they are intrinsic to the physical system being studied, and therefore important to understanding it.

3. Derived and estimated errors, often called uncertainties. Almost always different data are combined (in ways determined by the theoretical model used) with each other, with data from other experiments, or with approximations obtained from other theories. Because the original data have errors associated with them, the resulting quantities also do; these errors must be estimated through various sophisticated statistical techniques. Such errors are expressed as "error limits" – a misleading term, since they are statistical estimates of the corresponding standard deviations, so that for a Gaussian distribution, for instance, there is a probability of about 34% of their being exceeded. No actual upper bounds for a possible error can be set if the distribution is Gaussian, since even very large values have a finite though small probability of occurring.

The term "experimental error" generally refers to any or all of these three types, and of course there are other ways of classifying experimental errors, relevant in other circumstances.

Thirdly, together with the statistical errors one commonly observes deviations which are systematic, i.e. non-statistical. They spring from two essentially different sources, experimental inadequacies and theoretical deficiencies; from inadequately controlled factors in the experimental set-up and from discrepancies between the behaviour of the system and the model predictions. But to conclude that a systematic deviation is of this second origin, we must first be confident that we have eliminated all relevant systematic errors of experimental origin, by improving the experimental control, by measuring the perturbing factor and correcting the experimental datum, or by a combination of both; and even then the deviation will be visible only if it is large compared with the standard deviation of the statistical errors.

One fundamental conclusion is that at no point is it really possible to follow cookbook recipes in evaluating experimental errors. It is rather a matter of using one's critical judgment in balancing conflicting considerations in order to arrive at an acceptable compromise; differences of opinion are common, and no reasonable optimum may be available. Nor, once we have found error limits for our final results, can we state, on that basis alone, that they are satisfactory or not. A comparison must always be made with theoretical models; if only one is available, we can at most say that it is compatible with the data, because there could well be and often is a whole set of models that fit more or less equally well (or even equally badly!). Such a comparison is always made by treating the predictions yielded by the models as approximations to the experimental data, and then proceeding as discussed below. The comparison is rarely clean and decisive, because mostly each of the models considered more or less fits the data but only in a certain restricted region; and these regions are generelly distinct and many not even overlap.

Secondly, the interplay between experimental results and underlying theory is constant and complex. Experimental data serve to complete theoretical models, and these in their turn are needed to provide interpretations for the data from which the quantities used in theory-experiment confrontations are derived.

Thirdly, this confrontation is not a matter of one given theory against one given set of laboratory results. Instead, we have a carefully chosen set of models, built from several theories, with suitable approximations and empirical values thrown in; and we compare them with a number of sets of data, derived from different results obtained in sometimes very different circumstances. It is exceptional if the comparison favours one model exclusively; in most cases each model reproduces certain data rather well but not others, and the type of conclusion drawn is often to assign to each model (or at least to a sizable number among the models) its role, carefully stating the ranges of situations or parameter values where they work. Thus even if a systematic discrepancy attributable to theoretical insufficiency appears, large enough to stand out above the statistical fluctuations, it is often not evident just which model or models have to be adapted.

Lastly, the use of statistical techniques (often highly sophisticated ones) is indispensable; it calls for philosophical examination, because in the process we pass from a set of experimentally measured values to a prediction, based on statistical theory together with the model of the physical system, of the error in the final results, i.e. of the fluctuations we should expect in them if the entire series of measurements and calculations were to be repeated a sufficient number of times.

I do no propose here to offer a philosophical account of all these points; but however incomplete any account may be, it must at least not conflict with them; it is this which will guide the discussion.

(ii)

We turn now to approximations; this is a term applied to several somewhat different situations: finding an approximate analytical representation for a set of experimental data (or data from a computer simulation), finding a "simple" function to represent another one, and finding a whole model that is in some sense more appropriate than a given one.

Fitting a function to data of experimental origin is clearly a combination of error analysis, in the sense discussed, and approximation proper; it is exceptional in involving statistical fluctuations. Much confusion exists in the philosophical literature about this matter (as an example, see Quine 1960) mostly concerning what sort of approximating function should be used, and how close the approximation should be; but statistical theory offers quite adequate solutions to these problems (see, e.g. Green & Margerison 1978). Thus the nature of the function is, ideally, suggested by the theoretical model which is intended to fit the data; polynomials (unless suggested by the model) are used only as a solution of last resort, when we know too little. Furthermore, the experimental error of the data sets the limit to which the approximation must be good, and any parameters in the function are adjusted to achieve this limit, and no more: too high a degree of approximation is as bad as one too low, since its precision is spurious and the curve may have a shape that changes wildly from one set of data to another one taken under the same circumstances.

The second kind of approximation, the functional one, is the problem of finding a "simple" function (one, that is to say, that we know how to handle) to approximate a more complicated one; a special case is that of calculating one or more numerical values, because very few of the mathematical functions used in physics can be evaluated exactly. This type of approximation exhibits an apparent symmetry in that we are attempting to bring two theoretically derived functions close to each other. But the symmetry is only apparent: one of the two functions is always "given" (in that particular context), while the other is being adjusted. And in the process of deriving the approximating function from the "given" one information is lost, so that it is no longer possible to recover the "given" function if we only know the approximating one. The asymmetry we observe here in fact characterises all types of approximation.

It should be stressed that the approximating function, though perhaps suggested by the approximating theory, can in no wise be deduced from it. The function is always a free invention, to be tested against its possible applications, and commonly one has to try out several different ones before hitting on a satisfactory version. Since functional approximations of all kinds are much used in building a model to fit a given experimental situation, it is evident that describing this latter process as "logico-deductive" is inappropriate.

Mathematical theory provides a simple and straightforward abstract description for approximations in the form of the O notation (see e.g. de Bruijn 1958). This is in fact more than a notation; thus $f(x) = O\left(g(x)\right)$ for $x \in S$ indicates that $f(x)$ belongs to a whole class of functions which have the property that there exists some positive, finite constant c such that $|f(x)| \leq c|g(x)|$ for all x in the set S. And $O\left(g(x)\right)$ which defines our approximation condition, an approximating function $a(x)$ such that

$$b(x) = a(x) + O\left(g(x)\right),$$

where $b(x)$ is the function to be approximated. Note that in spite of the really rather inappropriate (but customary) = sign this relation is not symmetric; $b(x)$ is rarely an approximation for $a(x)$ in the sense stated. Both this asymmetry and the existence of an infinite class of approximating functions are very suitable to the needs of the physicist. The $O(g)$ in the definition of the class allows him to state even very complex approximation conditions in a clear and simple way, and the asymmetry is always present in applications. In the theory-model situation it is obvious enough, but even when comparing theories there is not much room for doubt: thus it is Newtonian mechanics that approximates quantum mechanics, and not the other way round. Note here that the O scheme describes the general character of the approximation but does not give numerical values for the discrepancy between the two function at stated values of x: for this we must determine the value of c, which is not given by the mathematical theory but has to be found from the details of the specific application. Note, furthermore, that what we have called "x" here may in fact be a set of variables, and not all of them need appear in both a and b. A particularly important case is that of some parameter which is present only in b: it provides an acceptable version for the assertion that "classical mechanics is the limit of quantum mechanics when h tends to zero", a statement that on the face of it is meaningless since h is a universal constant and cannot tend to zero.

Procedural approximations are a subclass of this kind of approximation. They arise from the use of mathematical procedures beyond their range of validity. On occasion this gives rise to a new mathematical structure, so that the approximation turns out to be exact; thus the theory of distributions in the sense of L. Schwartz came to justify the use of the mathematically shocking Dirac delta functions (Lighthill 1966) But much more commonly such an a posteriori justification is known to be impossible, as when we use a finite basis set to approximate the solutions of a Schrödinger equation whose eigenvalues lie in the continuum. However, justifying an approximation means, for a physicist, determining how good it is in the needed range of parameter values; and the difficulty with most procedural approximations is that generally we do not know how to do this. For the philosopher, on the other hand, it must be noted that this type of approximation, like the others, is not deductively describable.

The fourth and most complex kind of approximation is the structural one. Here we approximate the whole of a theoretical model, whose mathematical or conceptual structure is inappropriate (mathematically intractable, or containing quantities whose values are inaccessible, or because it is to be combined with another model of incompatible conceptual structure), by means of another model. The approximation may be made by eliminating or simplifying some of the concepts involved, i.e. on the theoretical level; or one eliminates or simplifies the representation of certain objects, working on the level of the concrete information with which the model is built from the theory. Thus the earth, a sizable object of roughly spherical shape, is sometimes treated as if it were an infinite plane, and sometimes as if it were an extensionless massive point.

In some cases it is possible to reduce the problem to that of approximating one or a set of basic equations in theoretical model A by those of model B; if one can then take a further step and say that the B equation approximates the A equation in the sense that the B solutions are functional approximations to the A solutions, then this kind of approximation has been reduced to the earlier one. But only to a certain point: only if every B solution approximates an A solution. This need not be the case, and never less so than when A and B are whole theories rather than different models for a specific situation. Newtonian mechanics, for instance, is in many ways an approximation to special relativity; yet its differential equations have solutions that in no wise approximate any relativistic orbits. Moreover, the concepts in one theory are never simply translatable into those of the other, however closely related they may be. Witness here the question of celerity vs. velocity, or the discussion whether it is the rest mass or the dynamic mass that is analogous to the constant Newtonian mass. Both are possible and meaningful, but the choice between them presumably depends on the application and may contain a significant conventional element. An even more extravagant structural approximation is the renormalisation technique used in quantum electrodynamics, which is a mathematically very sophisticated way of putting divergent (i.e. infinite-valued) integrals equal to 0; its phenomenal success justifies it, but we do not at all understand yet why it works. What is needed here is a notion of conceptual approximation; but this clearly depends on a much more fundamental matter, namely how far – conceptually speaking – a given theory may be said to approximate to the reality it pictures. This constitutes a fourth type of approximation; but it is too different from the others to discuss here.

Several conclusions may be drawn. Firstly, though it is clear that approximations are in some sense the theoretical counterpart of experimental error fluctuation, the two are very different. Experimental error is a phenomenon of statistical fluctuation; the errors are random both in sign and in magnitude (this is true also for data from numerical simulations, trivially so for the so-called Monte Carlo calculations, but also for other kinds, because of the accumulated rounding errors). On the other hand, the difference between the approximated function and the approximating one is (except for the rare case when one or both of these functions is stochastic) regular and non-random, and can often itself be shown to be analytic. Only in the first situation described above is there a fluctuating component due to the errors of the data being used; it can occasionally be reduced to the second one by first fitting some fairly arbitrary function, derived from an ad hoc phenomenological model, to the data and then approximating this function by means of another one that possesses a more satisfactory theoretical background.

Secondly, approximations are always asymmetric; the approximating function of model or whatever (the one that is adjusted) is always of theoretical nature.

Thirdly, approximations exhibit a degree of nearness which can generally be expressed in numerical terms: the maximum difference between two functions in the physically relevant region, or a suitable average over these differences, or a maximum relative difference, and so on. The choice of criterion depends on the particular situation; but whether the approximation is satisfactory depends on comparing the value of the criterion with the limit dictated by the application to be made of the approximation. Clearly, a crude approximation is sometimes entirely adequate, while in other cases a much better one is needed. We return to this matter below.

Fourthly, different approximations can produce models that are wholly incompatible with one another, even though the same theory was used as a starting point; thus the two models of the earth, mentioned above. This is one of the most fertile sources of the contradictions that appear in physics, for two or more such models are frequently combined to yield yet a third one. For instance, the two models of the earth could be combined in order to calculate the combined effect of the earth's and the moon's gravitational fields on a pendulum: the point-mass model is used to determine the moon's motion, and the infinite plane one gives the basic behaviour of the pendulum which is then perturbed by the moon's action.

(iii)

The structuralist philosophers of science have the great merit of having recognised the central importance of experimental errors in physics and other natural sciences, and of having devoted considerable effort to the philosophical elucidation of this problem area (see Przelecki 1969, 1974, 1976; Moulines 1976a,b, 1980; Ludwig 1978; Hartkämpfer & Schmidt 1981), quoted below as HS). This account is not yet fully unified; certain divergencies remain among the mentioned authors. We shall not here enter into their differences, since they do start from a common point of view and reach very closely connected conclusions, and it is rather with these that we must deal here. Nor need we discuss in much detail the elaborate formalism these authors have developed; we shall be more concerned with what this formalism expresses.

The structuralist analysis starts from the presupposition that a physical theory is a highly formalised structure or at least can be converted into one without significant loss or distortion. Purely theoretical terms in the formal structure are sharply set off, and any relevant experimental information is not thought of as linked to the theory except by way of confirmation or disconfirmation; it is, moreover, presented as a set of observation sentences, which some structuralist authors think of as atomic. The observations described by these sentences are said to be imprecise (unscharf: Ludwig 1978) or vague (Grafe, Balzer, in HS), or the corresponding theoretical concepts are said to require blurring (Moulines, op. cit. and in HS), but in fact it is never made clear in what this imprecision consists, or what characteristic of the observation sentences manifests it. Ludwig – who is a physicist – explicitly distinguishes "imprecision between theory and reality" (which in fact covers the discrepancies usually found rather than any imprecisions) from "imprecision of measurement"; the latter he describes (HS, p. 7) as something every student of experimental physics has to learn how to estimate, and later on comments (ibid., p. 14f):

There is much confusion about the concept of imprecision of measurement. The notion that there are precise quantities in nature but that we cannot avoid making errors in measuring them is in my opinion a fiction. This fiction must be replaced by a theory of the measuring apparatus.

Only such a theory of the measuring apparatus can explain what we mean by an imprecision of measurement. Such theories show that there are two sources producing imprecisions of measurements.

The first source is the imprecision between the mathematical picture and reality discussed above. There are such imprecisions also in the theory of measuring methods. The second source is the finding that measuring apparatuses (in most cases) do not measure what one "wanted" to measure. It is in general difficult enough to construct apparatuses which measure something which is not far away from the quantity one has in mind. These last "imprecisions" are consequently differences between the real measurement and the desired measurement. This last fact plays an important role in quantum mechanics if one "tries" to measure together quantities which can't be measured precisely together.

It is our opinion that all these imprecisions can be described by uniform structures in the theory without going back in all details to the complicated theories of measurement. It is only necessary to give a value for the so-called "error" of measurement, a value gained from the theories of measurement.

At least with the first sentence we need not quarrel; but the remainder of this long quote calls for some comments.

We note, in the first place, that Ludwig talks of "mathematical pictures" where we would expect "physical theory"; yet these are rather different objects. Secondly, the notion of model (more fully described in the following section) does not appear. Thirdly, the two sources of instrument errors mentioned here not only do not produce errors of measurement so much as systematic discrepancies between the actual properties measured and our theoretical representations of them, discrepancies which have little to do with experimental errors. Moreover, these two sources of instrument error are not fully distinct, because the first corresponds to the discrepancy between what is being measured and the fully worked-out theory of the measurement, the second to the discrepancy between what is measured and the intuitive notions on which the theory is (or, perhaps better, should be) based. Fourthly, as we saw above, the instrument error is not really derived from the theories of measurement, or only in part, but depends heavily on the appropriate laboratory procedures and on the skill and patience of the observer. Fifthly, the instrument errors do not provide more than a rough lower bound for the measurement errors, and these are not considered. Ludwig is the only author in HS who seems aware of the need for statistical treatment of the data; but he does no more than mention it, so that the most significant form of experimental error, predicted error limits, brought out in such a treatment thus remains beyond the purview of his approach.

This is also, presumably, the origin of another difficulty we have to comment upon: his use of uniform structures to describe experimental errors. The term belongs to set theory and Ludwig's use may be briefly described as follows: For any two values α and β of a physically measurable variable ν we establish a criterion $\gamma(\alpha, \beta)$ which decides if they are indistinguishable; for instance $|\alpha - \beta| < \varepsilon$, where ε is a suitably chosen parameter, though many others are conceivable, and Ludwig cites examples. The imprecision set is then the set set of all indistinguishable pairs.

Since several imprecision sets may be used in different circumstances, a "uniform structure" is taken to be a set which includes all possible imprecision sets. This uniform structure must satisfy certain postulates and can be interpreted as a set of nested intervals, including both the limit of infinite precision (given by $\emptyset$) and the opposite limit of null precision – the infinitely

long interval. From the "theories of measurement", in some not very clearly
specified way, an instrument error is derived, and this now determines which
of these intervals is taken to give the bounds within which experimental data
are acceptable, outside of which they indicate disagreement between theory
and experiment. Other kinds of error are not mentioned.

There are several fundamental problems with this approach:

(a) The distinction between different types of fluctuations or statistical errors is neglected; in fact, as the passage above from Ludwig shows, as well
as others in HS, only the instrument error is considered, but is taken as
given by external sources. In fact the instrument error is the (statistical)
experimental error obtained from other theory-experiment situations and
should therefore be explainable within the structuralist framework. Moreover, the instrument error is commonly of little importance; we saw above
that if the actual error used by the physicist, obtained from the fluctuations in repeated measurement, exceeds it, then the instrument error can
be ignored. Thus the types of experimental error of significance in physics
are neglected in the structuralist account.

(b) The use of the uniform structure as descriptive tool precludes examining
the central role of statistics in the whole problem of handling experimental
error. In particular, the fact that errors have different sorts of distribution (in the statistical sense), with different kinds of parameters, is swept
under the carpet; while the use of a fixed interval within which all values
must fall is quite inappropriate: since very large errors have a small but
non-zero probability of occurring, the only satisfactory interval would be
the infinite one. (Note that in practical work a fixed-interval technique
of this sort is often used as a crude approximation when better methods
are not needed or not achievable; but this cannot be justified except as
a rough approximation.) Moreover, the width of the error distribution is
not in general arbitrary, to be chosen according to (not very well specified) external criteria. The process of building the physical model from
one or more basic theories, together with other, ancillary ones and with
input from independent experiments, generally yields information about
the errors to be expected; and if the observed errors are very different,
even though the mean of the observations comes close to the theoretical
expectation, some explanation is called for. Thus the real nature of experimental error is both far more complex than the structuralist account
allows, and in many ways rather different from it.

(c) Because they ignore the random, statistical character of experimental error, the structuralists also ignore the distinction between errors and approximation. Quite analogously, they also ignore the distinction between
different types of experimental error, and between different types of approximation. The essential asymmetry of the approximation is also not
well represented.

(d) The structuralists seem also to be unaware of the fact that theoretical
models are used for concrete applications in daily life, far more than

in purely academic research, and it is here that predicted uncertainties are judged acceptable or not, according to the specific limits of precision needed in the application. If the model works to within these limits, fine and well; if not, another one is sought. Thus what is important in actual use (and therefore calls for philosophical study) is not the uniform structure but a specific uncertainty range or distribution width; and both the infinite and the null case must be excluded as unphysical.

(e) Associated with this problem is the fact that precision and accuracy are not distinguished; yet it is accuracy which decides whether a physical theory is "good", and not precision.

(f) The formalism necessary for the uniform-structure description is needlessly complex. Part of this complexity arises because the uniform structures are defined on the space of models of the theory (for details see Moulines (1976): Erkenntnis 10, 201; Critica 24 (1976) 25); note that "model" here is used in Sneed's sense, i.e. a logico-mathematical structure that satisfies the theory's axioms, and not in the physicist's sense. Now questions of approximation arise only in the context of a specific (physicist's) model, where a function in theory A is approximated by another one from theory B, and here this superstructure is excess baggage. There is furthermore considerable doubt whether the model space is measurable: if not, the distance function used as the set-defining criterion will not be definable. At the same time, the flexibility of the O scheme, which allows the approximation criterion to vary from point to point, is not available for the uniform-structure method, so that the much needed notion of the region where the approximation is satisfactory cannot be developed in it. And where experimental errors are concerned, entirely analogous comments may be made.

(g) Finally and most fundamentally, the structuralists give us just a description, with no explanatory power. At no point is an attempt made to discuss the origin of experimental error or to examine its philosophical implications; ontology does not exist, one might say. Hence we cannot expect from them an explanation of why certain types of measurements (e.g. counting – but only when the number of objects to be counted is sufficiently small!) have no experimental error limits, or why some errors are large and others small; nor do they tell us what the existence of errors means for the concept of a physically measurable quantity, or what conclusions we may draw for epistemology from their almost universal presence and their essentially random nature. Yet such conclusions are of obvious importance.

The last two points may be illustrated from Moulines' example (in HS) of the approximation of Newtonian gravitational theory by Kepler's laws. He only treats the two-particle case, where both theories give rise to elliptic orbits with the sun at one focus; in this case nearness of a Kepler ellipse to a Newtonian one is equivalent to nearness of the respective initial conditions, that is to say the starting points and starting velocities. But in the case of more than two particles this is no longer true: the Newtonian orbit will deviate steadily further from the Kepler ellipse (it is in general not even a closed curve), yet the Kepler approximation is good to within a limit, ε say, for a time t which is a function of ε. It is this fact which lies at the base of the computations in positional astronomy; but there is no room for it in the Moulines scheme. Indeed, his formalisation of the Kepler and Newton approaches is incomplete in that he has not included any representation of the initial conditions.

The analysis of the Kepler-Newton relation by Mayr (HS, p. 55) is still less satisfactory; for instance, though he recognises that the Kepler approximation to Newton is valid only during finite time intervals, his abstract formalism distorts this by applying a uniform structure to the time axis, – thus confusing the finite time limit with an uncertainty in the time measurements.

These authors' complex formalism hides the reason why the Kepler approximation works, namely that as the planets' masses get smaller, their acceleration due to the sun's attraction remains constant (because of the equivalence of gravitational and inertial masses), while their mutual interactions decrease. Thus the physics is obscured as the analysis becomes more complicated.

It is clear that the structuralist approach remains too close to its neopositivist ancestry and too far removed from the actual situation to be satisfactory; an alternative approach must therefore be developed.

(iv)

We shall now examine what appears to underlie the observed facts concerning experimental errrors; and though we shall need a number of ontological and epistemological notions that go beyond what is commonly accepted in present-day philosophy of science, we cannot here justify them more than rather briefly.

We begin with the concept of "system"; the term will be applied exclusively to any segment of our world whose behaviour it is proposed to study. It may have been subjected to suitable control by experimental preparation in the laboratory, or else it may be defined by an appropriate selection of observations, as in the astronomical case.

Repeated measurements of the same quantity on similarly prepared systems differ. "Similarly prepared" means, of course, that the experimenter has reproduced a certain small set of conditions, judged to be the decisive factors; but even if this judgment is correct (which it will be only insofar as the theory behind it is correct), there are very many, perhaps infinitely many, other factors that individually have an almost vanishing influence but which jointly have a detectable effect. Thus the concept of a closed system, exhaustively described by a limited set of parameters, is an extrapolation. Of course, the degree to which we can approximate to a closed system has improved enormously over the years, as experience has accumulated and as new experimental techniques have become available; but an experimental system cannot ever be strictly closed, if only because we could not even detect the existence of such a closed system (by definition, it has no interaction whatsoever with its surroundings), let alone measure any of its properties. In fact, a system may even be acted upon by very strong forces which are nevertheless not taken into account; thus in many laboratory set-ups we completely ignore the earth's gravity, not because of its weakness but because the gravitational potential acts equally on all parts of the system and is equally counterbalanced by supporting structures.

What constitutes the system is determined by the experimental procedure that establishes its (relative) isolation and determines the initial conditions; this procedure in its turn is derived from our ideas about the system, – to begin with very often vague and nebulous, so that our procedures for establishing the system are clumsy and indadequate. But as our knowledge of the system improves so do our experimental measures, and the system's properties correspond more closely to our description of it. This description will be based on just a few characteristics of the system, deemed to be essential on the basis of past experience with the system; the process of improving our description is basically one of picking out the right characteristics, often also of reducing their number, i.e. of eliminating the irrelevant ones, and of determining and describing (where possible, in mathematical terms) their interrelations. For a great many kinds of systems this process has now gone far enough that we can make very effective use of "our ideas about the system", i.e. of our theoretical knowledge.

What we have described so far is the content of two related basic assumptions. The first, that of the partial separability of the world, is ontological in character: it says that the world may be split up into systems, i.e. segments that are largely but never wholly isolated from the rest of the world. Note that the systems are not mutually exclusive: they may overlap, or be contained one in another; nor is the division into systems in any way unique. Note furthermore that this splitting-up is our work: and it is not theoretical, though both the abstract possibility and the specific features of isolating a certain system are first conceived theoretically, but is the result of our active intervention by means of appropriate experimental procedures.

The second basic assumption is epistemological and concerns the cyclic process of building theoretical pictures, applying conclusions from them to make alterations in the system, and using any discrepancies between the conclusions and the observed behaviour to modify the theoretical pictures, – what we shall call the epistemic cycle; the assumption is that by repeating the epistemic cycle it is possible to reduce the discrepancy to any required level. Note that nothing like convergence to an absolutely faithful theoretical picture is postulated; nor do we assume that the discrepancies will diminish in every repetition of the epistemic cycle; nor, finally, do we take the physical system to remain constant during the process. Indeed, it is a matter of experience that not only do we modify the system during each cycle but that we redefine and correspondingly alter the system between cycles.

It is important to observe here that our theoretical knowledge is not formed at the level of individual pictures of individual systems. For in carrying out each epistemic cycle we need criteria to determine the significant discrepancies, to indicate how to modify our theoretical picture, to make predictions from the picture, and to turn these predictions into suitable modifications of the experimental set-up. Thus behind the epistemic cycle there are epistemic cycles of a higher order that allow us gradually to build up these criteria; indeed there is a whole hierarchy of them. In fact, our theoretical knowledge mimics the world in that it has an underlying unity while at the same time we may isolate parts of it, – but, again, this isolation is only incomplete. Each part depends on one or more higher-level parts, and in its turn gives rise, together with others, to a lower-level part. The image of a hierarchy is not exact: not only are there interactions in both directions, but what may be high-level at one point becomes low-level at another in the structure. And, of course, the entire structure evolves and grows and is restructured in the course of time. In order to describe these interrelations we shall adopt the terms "theory" and "model" but give them a relative meaning only: each element of the whole edifice of theoretical knowledge is a theory relative to the models built with it, and is a model relative to the theories from which it derives. A model, moreover, is not simply deduced from a particular theory, rather it is derived by combining elements of a central theory with elements from others, together with data obtained directly from the experimental level or from other models; information concerning our intentions is also relevant, and the "glue" holding this structure together is a series of appropriately chosen approximations. This implies that a formally consistent logical description of the structure will be to some extent inadequate; at the same time, the fact that it is joined by approximations give the edifice of theoretical knowledge a flexibility which makes it possible to alter it at one point without having immediately to jettison the whole of it.

The picture of a theory as a wholly independent part of our knowledge and therefore as formalisable in terms of a finite set of concepts is only what a physicist would call a zero-order approximation; it is sometimes useful, but often too restrictive. Theories are linked both by common concepts (mass, length and time, for instance, occur in almost all physical theories) and, as mentioned, by various approximations, of which the structural kind discussed above is a very important particular case. Qualitative approximation between concepts is another that so far is not well understood; Moulines, in the work cited above, has the great merit of at least having recognised its significance, but unfortunately his formalism places the approximation at the level of the whole theoretical structure and not at that of the concept and so is not quite what is needed.

There need not then be strict logical coherence between different parts, and so we find – to take an example – that quantum and classical arguments are combined and relativistic corrections thrown in to make a theory for atomic spectroscopy, in spite of the obvious inconsistencies between the three theories.

Successive executions of the epistemic cycle do not, strictly speaking, affect the same system; they work either on different though similar systems, or on a given system at different times so that the system has had a chance to evolve into something slightly different. The model involved in the cycle cannot therefore be applicable exclusively to one single system, it must apply to the whole class of systems which share the same values of the controlled characteristics but may differ in any of the others; in other words, the model must have some degree of generality. If it did not, we could not even confirm that it is, at least in part, correct – let alone make use of it for understanding how other instances of this type of system function. Higher-order models have correspondingly greater generality, for they must model not just one but a set of lower-order models in order to be satisfactory. Thus the hierarchy of knowledge starts at the bottom with models of specific experimental (or, more generally, experiential) systems and ends in the clouds of abstract speculation, in ontology and logic and mathematical methods. The process of generalisation involved may, of course, be mistaken in some respect; for instance, if it is not recognised as part of the epistemic cycle, then the opposition of specificity in the system to generality in our picture of it gives rise to the ancient quarrel of realism and nominalism. And if the fact of generalisation is relegated to laboratory practice instead of being seen as central in our knowledge, then we arrive at 19th-century physics, where experimental fluctuations are "eliminated" by taking the average of a sufficiently large number of data, and the theory is fitted to this average as if it corresponded to a single system which, moreover, is conceived as closed in order to be able to neglect all factors not explicitly included in the theory. Note that in very many cases this is an excellent approximation, but grossly misleading in other situations.

With these background concepts, the nature of experimental error should be clear: it arises from the fact that in repeated measurements either on the same set-up or on similarly prepared set-ups the system (in the sense indicated above) is not quite the same and therefore the value of what is being measured will also differ a little. These differences arise from the factors not taken into account in the model which we expect describes the system adequately; but it does not describe the experimental error, in the sense that it gives no prediction for its value. It is, of course possible in principle to build a model that includes some of these factors, and sometimes it is even necessary to do so; but if we keep on increasing the size of the model it becomes more and more unwieldy, while its generality decreases to the point where it is no longer possible to test it adequately. And it proves impossible to include explicitly all conceivable factors that might cause fluctuations in the measured values. There is one significant exception to this: we have tacitly supposed our physical variable to take on values that can be much closer to each other than any instrumental error at present achievable, so that a continuous description of it in the model is appropriate; but if it only takes discrete values, integers for example, then it is often possible to discriminate with certainty between different integer values. For instance, we can count the number of people present in a room at a given moment, provided there are not too many of them. Thus experimental error originates in actual experiments; but as we saw, it propagates upwards into theories of much higher levels.

Note that instrument errors are no more than the experimental error of the instrument-observer combination, rather than of the system being studied. It is therefore sufficient if they can be made small enough to reveal the fluctuations of the system itself. They constitute then errors in the determination of the system's errors, second-order errors, so to say, and are commonly negligible.

(v)

A model, we have said, cannot predict each fluctuation explicitly. It is nevertheless possible to construct theories and models that describe the statistical properties of these fluctuations in great detail without making any prediction for individual fluctuations.

This is achieved by making explicit the generality necessarily inherent in any model. A closed-system model refers to a single (abstract) system but faces an indefinitely large class of actual systems, differing in the many excluded factors. We cannot explicitly include them all in the model; but we can conceive a set of models, each of which has been extended to include at least a very large number of these factors. We need not construct these models in detail, we only need to consider the whole set; and we build a theory that describes this whole set. A great many of the factors would, if explicitly treated, be represented by continuous variables, so that the set of all possible detailed models is, at least in principle, infinite. Such a set of models, of theoretical replica of our basic model completed with all possible variations in the factors not contained in the basic model, is known as an ensemble. The concept, central to statistical mechanics, was introduced, independently and almost simultaneously, by Einstein and J.W. Gibbs.

The different models belonging to an ensemble, its members, are distinguished from each other by specifying the values of certain variables; these variables commonly cannot be measured in the laboratory, but we can certainly talk about them and their values. In statistical mechanics one uses the initial values of the position and velocity of the atoms of neon making up a sample of that gas, for instance; from them mechanical theory (either classical or quantum according to how the ensemble is specified) permits calculating the state of the model system at any other time. These initial values, or whatever other parameters differentiate between the models, may be used as the coordinates of a space known as phase space, and the model's time evolution then gives rise to a trajectory in phase space which is different for each model.

The phase space for the ensemble is measurable, and it is moreover possible to formulate a bounded measure function for it which remains unaffected by the time evolution of the ensemble; if this does not hold, the ensemble – which is after all a theoretical construct – is altered until it does. At this point a basic assumption is added: that the individual members of the ensemble are statistically independent of each other; in statistical mechanics, this is the so-called hypothesis of molecular chaos. It is possible to confirm experimentally that this assumption is justified. Several statistical tests can be applied; for instance, the nearest-neighbour correlations in the data series should not alter appreciably if the series is randomly reordered. If the hypothesis is not verified, then successive measurements influence each other through some factor which should have been taken into account in the model or else controlled experimentally. A common case of this leads to the use of thermostats to control the temperature of the laboratory set-up.

With this hypotheses, we have, firstly, that the bounded measure function becomes (upon normalisation) a probability, the probability of finding a member of the ensemble in the state given by the point in phase space which is the argument of the measure; secondly, we may now study the statistical behaviour of the ensemble as a whole under time evolution. This is the subject matter of what is known as ergodic theory. We cannot enter into its details here (see e.g. Onicescu and Guiasu 1971 or Penrose 1970). But if the ensemble is ergodic, then the time average of a property of a member is almost always equal to the ensemble average, i.e. the average over all members of the ensemble at the same time; in that case, and it is a common one, the measurement result will be well represented by the ensemble average provided the time it takes to carry out the measurement is sufficiently long, and often a fraction of a second is "sufficiently long". In such circumstances it suffices to formulate a degenerate ensemble, with all its members identical, as our theoretical description, and we recover the approximation of classical physics, i.e. using a single model which corresponds to the ensemble average.

But many ensembles are not ergodic or even close to it; then a member's properties will not be well represented by the ensemble average, there will be observable fluctuations, and we can no longer simply substitute the ensemble average and use it as if it came from an individual model. Here then a proper statistical theory must be developed, in which the average properties will generally differ both quantitatively and qualitatively from those of an individual member.

These two situations are the extremes. The case of experimental errors falls in between, for we explicitly need the ensemble concept to explain and describe the properties of the errors; but one we have dealt with them, we may – if the distributions have relatively small standard deviations – ignore the ensemble and deal only with its averages.

Thus the ensemble is the concept that best represents the actual physical situation; since it simultaneously provides an adequate basis for the interpretation of the probability concept it also explains why statistical techniques must appear in dealing with experimental error: statistics is the inverse theory to probability in that it shows how to derive probability distributions, that is to say ensemble specifications, from experimentally determined frequencies.

From the measured data we obtain the statistical properties of the ensemble used to describe them. From the ensembles corresponding to the errors in various measurements we can determine the ensemble that should describe a derived error. The ensemble needed for determining the uncertainties applicable to a prediction from the theory is deduced from one or both of these kinds of ensemble. The actual mathematical deductions are often not possible; but one can generally obtain at least the first few moments of the distribution function characterising the ensemble, and this information is generally sufficient. In many situations it is enough to know that the central-limit theorem holds (so that the distribution is Gaussian and the mean and standard deviation contain all needed information); this is the commonest case in the treatment of experimental errors. And there are even cases where the mere fact that an ensemble exists is sufficient.

In this way the use of ensembles provides for all three kinds of statistical errors. Instrument errors can, of course, be dealt with in the same way, since they are experimental errors in another physical system. But systematic errors are quite evidently not susceptible to this treatment; the basic assumption that the individual observations, embodying the realisation of a sample from the ensemble, are statistically independent is invalid for them. Their occurrence indicates a discrepancy between model and system, and may require reformulating the theoretical description of the system, or modifying the experimental set-up, or both. The decision cannot be made on general grounds but must depend in each particular case on what is possible and on what the aim pursued is. For instance, in many cases maximum experimental precision is not needed, and a systematic errror below the threshold of utility, though detected, may simply be ignored.

References

Bruijn, N.G. de (1958): *Asymptotic Methods in Analysis* (North-Holland, Amsterdam and Noordhoff, Groningen)

Green, J.R., Margerison, D. (1978): *Statistical Treatment of Experimental Data* (Elsevier, Amsterdam)

Hartkämpfer, A., Schmidt, H.J. (eds.) (1985): *Structure of Approximation in Physical Theories*. Colloquium on Structure and Approximation in Physical Theories (Osnabrück, 1980) (Plenum Press, New York)

Lighthill, M.J. (1966): *Introduction to Fourier Analysis and Generalized Functions* (Cambridge University Press)

Ludwig, G. (1978): *Grundstrukturen einer physikalischen Theorie* (Springer, Berlin Heidelberg New York)

Moulines, C.U. (1976a): Erkenntnis 1:0, 201

Moulines, C.U. (1976b): Critica 2:4, 25

Moulines, C.U. (1980): Synthese 4:5, 387

Onicescu, O, Guiaşu (1971): *Mécanique Statistique* (Springer, Wien and New York)

Penrose, O. (1970): *Foundations of Statistical Mechanics* (Pergamon Press, Oxford)

Przelecki, M. (1969): *The Logic of Empirical Theories* (London)

Przelecki, M. (1976): Erkenntnis 1:0, 228

Quine, W.V.O. (1960): *Word and Object* (MIT Press, Cambridge), p. 19

Sutcliffe, B.T. (1975): in G.H.F. Diercksen et al. (eds.) *Computational Techniques in Quantum Chemistry and Molecular Physics* (D. Reidel, Dordrecht), p. 89

Part III

The Philosophy
of Quantum Mechanics

13. Problems and Promises of the Ensemble Interpretation of Quantum Mechanics[*]

Abstract. After a brief outline of the ensemble interpretation, its advantages and promises are described, including both the elimination of the puzzles that beset the Copenhagen interpretation and newer lines of research such as the application of ergodic theory to quantum mechanics. Three problems facing the ensemble interpretation are discussed, of which that of joint probability distributions is seen to require further research; some possible lines are indicated. Certain philosophical problems are discussed, including that of the relation between formalism and interpretation, where it is suggested that the formalism neither implies nor is indifferent to the interpretation. The hiddden-variable question is then considered; the von Neumann theorem is seen to be a special case of a very general theorem, and is interpreted to mean that only stochastic hidden-variable theories are acceptable. The outline of a possible such theory is given.

(i)

The puzzles and paradoxes of quantum mechanics are, as is well known, rather closely associated with the Copenhagen interpretation; this designation will here be taken generically to cover the wide variety of views which has in common the conception that the wave function describes a single system. The list of these difficulties is long and has not yet ceased to grow – witness the recent discovery of a Zeno-type paradox in quantum mechanics (Misra and Sudarshan 1977). They may usefully be classified according to the particular aspect of the Copenhagen view (or views!) to which they relate. Most of them will be found to fall into one of three groups:

a) Those associated with the basic question concerning the nature of the wave function, e.g. the problems of the wave-particle duality. This variety has given rise to a great deal of heated discussions but is usually not susceptible to precise statement in mathematical language; this fact does not diminish their importance but does complicate their analysis.

b) Those derivable in one way or another from the work of Einstein, Podolsky and Rosen (1935); these include the well-known puzzle of Schrödinger's cat and its ramification at the hands of Wigner's friend. Here the mathematical formulation was made clear from the very beginning; unhappily it cannot be said that the large number of papers that over the years have

* Presented at the *Symposium on the philosophical aspects of quantum theory*, Dubrovnik, 2–6 April 1980. First published in: Revista Mexicana de Física **35** Suplemento (1989) p. 19–45

attempted to analyse these problems have shed any significant light on them.

c) Those linked to certain specific points in the Copenhagen interpretation which may be doubted without affecting the main part of that structure. Of this nature are the questions raised by the projection postulate.

The third group is not very relevant to the present purpose, for obvious reasons, and I shall not further discuss them. My aim here is rather to show, in the first place, that the ensemble interpretation of quantum mechanics successfully deals with the difficulties in groups a and b; then to ventilate certain philosophical questions concerning the relationship between the mathematical framework and the interpretation of any physical theory; thirdly, in the light of these considerations, to examine certain problems in the ensemble interpretation and the attempts made to solve them; and finally, to consider the outlook for the future work. But before these points are raised I will briefly outline the nature of the ensemble interpretation.

What I call here the ensemble interpretation is essentially what is commonly known as the statistical interpretation; but I mark the difference for two reasons. One is that the term "statistical" may be misleading, since it has sometimes been used to refer to Born's conception of the wave function as a probability amplitude (Heisenberg 1930, Messiah 1959). The other reason is that the statistical interpretation is usually left rather unfinished, while I propose to study the problems involved in rounding it out and so making it the basic interpretation of all quantum mechanics.

The ensemble interpretation was adumbrated by Slater (1929) and further developed by quite a number of authors, including Einstein (1949, 1953), Margenau (1958, 1963 a,b) and Blokhintsev (1953). A recent review (Ballantine 1970) provides a good account of the arguments leading to this interpretation, so that here I need only touch on the central points (see also Ross-Bonney 1975).

In the ensemble interpretation the wave function is taken to refer not to one simple system but rather to an ensemble of such systems, in the sense of statistical mechanics: a (possibly infinite) set of theoretical replicas of the system under study; this set is described by a measure function $\mu(d\mathbf{x})$ which yields the relative abundance within the set of systems that lie in the region $d\mathbf{x}$ of the underlying sample space, usually taken to be the phase space of classical mechanics. In such an ensemble the theoretical prediction for a function $f(\mathbf{x})$ is the expectation value,

$$\langle f \rangle = \int_\Omega f(\mathbf{x})\mu(d\mathbf{x}) , \qquad (13.1)$$

where Ω is the volume of the space,[1] and f may, of course, also depend on the time t.

---

[1] Eq. (13.1) assumes that the measure μ is normalized. A very clear discussion of ensembles will be found in Balescu (1975)

One point must be stressed: the ensemble is a *theoretical* construct, and as always its use does not necessarily imply that only averages over many measurements of the same kind can be compared to its predictions (and even less that it can only deal with systems composed of many particles); I return below to this question.

What the ensemble interpretation maintains is, in other words, that the ensemble average (13.1) coincides with the quantum-mechanical expectation value, provided a suitable ensemble is chosen. But as will be seen below, it is precisely this choice of an ensemble that the quantum formulation does not determine uniquely: it is here that we rediscover the incompleteness of quantum mechanics first pointed out by Einstein, Podolsky and Rosen (1935). As will be seen in Sect. vi, this incompleteness points beyond quantum mechanics and opens up a fruitful field of research. Within the quantum framework the nonexistence of well-defined criteria for choosing an appropriate ensemble creates no problem: the quantum formalism has precisely the aim of allowing us to calculate expectation values without specifying explicitly the ensemble used for this.

But this does not mean that all is plain sailing: there are still certain problems raised by the ensemble interpretation, among which the issue of joint probability distributions occupies pride of pace; accordingly, it will be reviewed in Sect. iv. This, however, requires the clearing up of some philosophical matters, and these, together with some related points, will also be discussed below, together with some possibilities for resolving the joint-probability problem.

(ii)

Now what does the ensemble interpretation achieve?

In the first place, conceptual clarity. As has been stressed, for instance by Penrose (1970), the notion of an ensemble is fundamental not only for statistical mechanics but is a basic concept for the cognitive process we call scientific research. A particular aspect of this situation is the role it can play in adequately explicating the much confused idea of probability (Brody 1980); this explains its relevance to quantum mechanics, where it has been clear for some time that the two chief interpretations are closely linked to the two most prevalent views concerning the nature of probability (Popper 1967; see also Ballentine and Brody et al. 1979).

The conceptual clarity I refer to is particularly evident in the elucidation the ensemble interpretation offers of Heisenberg's so-called uncertainty principle: if $\hat{a}$ is the (Hermitian) operator corresponding to an observable a (i.e. a quantity for which a measuring process is known) such that the values observed for a state ψ yield a mean that, within experimental error limits, corresponds to $\langle\psi|\hat{a}|\psi\rangle$, then

$$(\Delta a)^2 = \langle\psi|(\hat{a} - \langle\psi|\hat{a}|\psi\rangle)^2|\psi\rangle \tag{13.2}$$

is the theoretically predicted statistical dispersion of the measured values of a. Similarly for the operator $\hat{b}$. And if a and b do not commute,

$$[\hat{a}, \hat{b}] = i\hbar \hat{1} , \tag{13.3}$$

for instance, then

$$\Delta a \Delta b \geq \tfrac{1}{2}\hbar . \tag{13.4}$$

Here Δa, Δb can only be interpreted as standard deviations in the usual statistical sense, and the fact that for Eq. (13.4) to hold they must both be defined with the same ψ only means that we must have a state-preparation procedure which can generate an indefinitely long sequence of systems belonging to the ensemble described by ψ; on some of these systems we measure a, on others we measure b. It is not necessary that a and b should be measured on the same systems. Thus Δa and Δb bear no relation to the experimental errors δa and δb; fortunately so, for otherwise it might prove difficult to provide experimental proof for the validity of Eq. (13.4), as is very clearly explained by Ballentine (1970).

Similar conceptual simplifications arise in the consideration of the measurement problem. This has become a problem only because, in a view that associates the wave function with an individual system, the prediction that different measured values occur with nonzero probabilities is incompatible with the fact that only one of these values is obtained in each measurement while the others do not occur at all. What privileges these values? We do not know; hence the need for von Neumann's projection postulate and all its undesirable consequences. The difficulties are often further compounded by the quite unwarranted assumption that the measurement process does not alter the value of the measured quantity (of course, if ψ is an eigenstate of $\hat{A}$ with eigenvalue α, then $\hat{A}\psi$ is an eigenstate with the same eigenvalue; but no principle underlying quantum mechanics allows us to identify the mathematical effect of $\hat{A}$ with the physical interaction between the systems described by ψ and external systems) and by the neglect of the distinction between state preparation and state measurement. Any physical experiment begins by suitably preparing the system under study; the system is then submitted to whatever interaction is the object of the experiment, and finally suitable measurements are carried out. Therefore a state preparation is a process that leaves the system in a known state; while state measurement determines, at least partially, what state the system is in as it enters the measurement. The system's state before state preparation and after measurement are of no interest, and in the second case may be meaningless when the measurement is destructive. There is thus a symmetry under time reflection between preparation and measurement; but the symmetry is not complete, for not every measurement procedure can be turned into a preparation method. The distinction between state preparation and measurement, established clearly by Margenau (1958, 1963a,b), is vital. Among other points, it permits defining the proper significance contained in the projection postulate, namely that after state preparation by means of a procedure describable through an operator $\hat{A}$, the system will be in an eigenstate $|\alpha_i\rangle$ belonging to the eigenvalue α_i of $\hat{A}$, if the subensemble i is selected. The last phrase is essential, for without a selection a mixed state containing all eigenstates of $\hat{A}$ is obtained; in a Stern-Gerlach apparatus, for instance, we obtain a beam of particles with all spin projections directed upwards only if we select that part of the split beam which goes through the upper slit. On the other hand, if we replace the slit by a counter, the particles are absorbed: we have turned a preparation procedure into one of measurement, but the projection postulate is now meaningless. It is the confusion between preparation and measurement which generates the many absurdities apparently derived from the projection postulate.

Once these unnecessary confusions are eliminated the discussion of measurements in the ensemble interpretation is very simple (once more follow Ballentine 1970): the expectation value is found from the basic rule

$$\langle \hat{A} \rangle = \langle \psi | \hat{A} | \psi \rangle \, , \tag{13.5}$$

and corresponds, to within experimental error, to the mean of a sufficiently large set of experimental determinations of A on similarly prepared systems (i.e. all belonging to the ensemble described by ψ). Suppose $|S; \alpha_i\rangle$ to be the eigenstate of $\hat{A}$ for the system S to be measured, and $|M; \mu_j\rangle$ to be those states of the measuring apparatus M which are macroscopically distinguishable, $|M; \mu_0\rangle$ being the initial state of M. Then the initial state of the system $S + M$ will be $|S; \alpha_i\rangle |M; \mu_0\rangle$; if we now write V for the evolution operator of the interaction between S and M, the final state will be

$$V|S; \alpha_i\rangle |M; \mu_0\rangle = |S; \phi_i\rangle |M; \mu_i\rangle \tag{13.6}$$

say. The state ϕ_i of S may but need not coincide with α_i and in general is a functional of it (Arika and Yanase 1960). If instead of an eigenstate of the quantity to be measured we have a state

$$|S; \psi\rangle = \sum_k \langle \alpha_k | \psi \rangle |S; \alpha_k\rangle \, ,$$

then

$$V|S; \psi\rangle |M; \mu_0\rangle = \sum_k \langle \alpha_k | \psi \rangle |S; \alpha_k\rangle |M; \mu_k\rangle \, , \tag{13.7}$$

and the probability of the macroscopic observation μ_i is $p_i = |\langle \alpha_i | \psi \rangle|^2$, meaning that if we repeat the state preparation yielding ψ and the measurement of A a sufficient number of times, the relative frequency of our finding μ_i is just p_i. If, then, the states $|M; \mu_i\rangle$ are distinguishable at the macroscopic level, the system S must have had the value α_i for A with just that probability p_i. It is this which makes M into an appropriate measuring system. No peculiar phenomena such as the "collapse of the wave function" is involved. Schrödinger's cat presents no problem: there is now a multiplicity of them, half of them being dead and the other half alive; which one of them we look at is not, however, implicit in the wave function.

The only remaining question concerns the simultaneous measurement of more than one quantity. The problem appears, as expected, when the operators for these quantities do not commute. It is surprising that the textbooks make at best confusing reference to this problem, for simultaneous measurement is one of the most widely employed experimental techniques; indeed, in bubble chambers the momentum of a particle is derived from a set of position measurements which yield the curvature in the magnetic field.[2] Surprisingly little work has been done on this question since the pioneering work of Park and Margenau (1968). They showed that the non-commutativity of operators does not imply the incompatibility of the corresponding measurements; they offer a not implausible conjecture that there exists a type of joint measurement (which they term historical because its results depend on earlier states of the system), that is the only one always feasible, and they furnish examples of this type (the curvature measurement on bubble-chamber tracks belongs to it, as do time-of-flight experiments). This conjecture has not to my knowledge been fully validated yet.

It is worth adding, since there exists much confusion about the matter, that there is no connection between non-commutativity and correlation. Thus position and momentum of individual particles are *in some experimental situations* so strongly correlated that we can deduce one from the other; while two perpendicular spin components are quite uncorrelated.

A quite different direction in which the ensemble interpretation has shown notable promise is in the study of the ergodic properties of quantum states. The ensemble represented by the wave function ψ has, in general, a time dependence, though for a stationary state of energy E this takes the relatively trivial form $\exp(-iEt/\hbar)$; it makes sense, therefore, to ask under what circumstances ensemble averages may replace the temporal average for a single system – which is the most fundamental of the properties that ergodic theory (see e.g. Arnold and Avez 1968) has shown to be relevant. It is surprising that such questions should never have been asked until quite recently; the first ones to do so were Claverie & Diner (1973, 1975), who defined the correlation function of an operator $\hat{A}(t)$ (in the Heinsenberg picture) as

$$B_A(t,t') = \langle\psi|\tfrac{1}{2}[\hat{A}(t)\hat{A}(t') + \hat{A}(t')\hat{A}(t)]|\psi\rangle , \qquad (13.8)$$

and then showed that, firstly, for a stationary quantum state it depends only on the difference $\tau = t - t'$ and so is stationary in the stochastic-process sense also; and secondly that if the quantum state is non-degenerate, then the variance of the quantity $a(t)$ corresponding to $\hat{A}(t)$ in such a way that its ensemble mean is the quantum-mechanical expectation value $\langle\psi|\hat{A}|\psi\rangle$ tends to zero as t^{-2}: thus we have a strong ergodic property. Numerically, it is found that ergodicity is reached in a surprisingly short time; for the average for any

---

[2] On more than one occasion, a bright student has interrupted my discussion of bubble-chamber techniques and stated that this was impossible, because the laws of quantum mechanics forbid the simultaneous determination of non-commuting quantities. This is explicitly stated to be the case even by reputable authors, e.g. Roman (1965), Sect. 1–2: "the necessary and sufficient condition of the simultaneous measurability of two or more observables on any system is that the corresponding operators commute".

operator is within one millionth of its ensemble average when t exceeds 10^{-10} sec. On the other hand, an unconfined system is never ergodic.

These results are highly relevant to the interpretation of quantum mechanics. Among other points, they go a long way to clearing up the source of many confusions; for instance, it is evident that measurements on individual hydrogen atoms (and similar confined non-degenerate systems) are well represented by ensemble averages for they usually take a much longer time than that needed to achieve ergodicity. This is the reason why the single-system interpretation of the Copenhagen type could develop and achieve a persuasive series of successes; it is also the reason why they create so many paradoxes for situations like the famous double-slit experiment, where ergodicity fails and therefore the predictions of the quantum-mechanical ensemble correspond only to averages over many measurements: the passage of a single electron, whichever slit it goes through, never creates a diffraction pattern.

Much further work remains to be done on these and related questions; but it is already evident that such investigations will shed light on some very obscure corners of quantum mechanics, as for instance the unsatisfactory description we have of processes like the separation of H_2^+ into $H^+ + H$; this problem is well known to quantum chemists (see e.g. Claverie and Diner 1976), but ignored by physicists.

(iii)

The ensemble interpretation thus offers the advantages of conceptual simplicity and clarity, of freedom from paradox (so far as we know), of a quite natural fit to the experimental situation, and of great possibilities for further research. Why then has it not simply displaced the Copenhagen interpretation? There exists a philosophical bias which I return to below; but there are also difficulties more directly linked to the physics of quantum phenomena, and these may be examined under three headings.

The first one concerns the existence of discrete states. If we accept the ensemble interpretation, quantum mechanics has a basic structure rather like that of statistical mechanics, and one might therefore expect that distribution functions which are Dirac δ's (or sums of them) would appear only as limiting cases; in quantum mechanics, however, such distribution functions appear to be fundamental. The difficulty vanishes once it is noticed that even in quantum situations discrete states (in the mathematical sense) can only be an extrapolation: for such states appear only in systems that are essentially confined to a finite volume (described by a square-integrable wave function) and have no interaction whatsoever with anything outside it, and this is in fact an idealization which real systems at best approximate. From a slightly different point of view, an energy eigenstate, for instance, must have a certain finite width, otherwise its lifetime is infinite, it can never decay and therefore cannot be observed; we should not know about its existence and if the theory predicted it should judge the theory wrong. Here, then, the ensemble interpretation suggests that the basic elements of the quantum formalism be extended so as to consider primarily open systems in interaction with their surroundings, and closed systems only as limiting cases. This point will be touched on again below.

The second difficulty to be looked at here arises from the conclusion many physicists have arrived at that joint distribution functions for non-commutating observables cannot exist in the quantum formalism. The argument may be set out in the following points, which we number for their discussion below; they are largely adapted from Cohen (1966a).

(1) The ensemble of quantum mechanics are characterized by probability distributions over a classical phase space, augmented where necessary by variables that account for spin etc. (In what follows we have no need of this generality and therefore restrict ourselves to a phase space (p, q) with one degree of freedom, where q is the position and p the momentum of a particle.)

(2) The probability distribution[3] $f(p, q)$ must satisfy the following conditions:

$$f(p, q) \geq 0 \quad \text{almost everywhere} \tag{13.9}$$

$$\iint f(p, q)dpdq = 1 , \tag{13.10}$$

$$\int f(p, q)dp = |\psi(q)|^2 \tag{13.11}$$

and

$$\int f(p, q)dq = |\phi(p)|^2 , \tag{13.12}$$

where $\psi(q) = \psi(q, t)$ is the coordinate wave function for the state under consideration and $\phi(p) = (2\pi\hbar)^{-1/2} \int \psi(q) \exp(ipq/\hbar)dq$ is the corresponding momentum wave function. Conditions (13.9) and (13.10) are needed so that $f(p, q)$ is a proper density function for a probability distribution, while condition (13.11) stipulates that it should have the correct marginal distributions for the probabilities of finding values for the position or the momentum. Condition (13.10) is redundant, being implied by the normalization of ψ or ϕ, but is noted for completeness' sake. All integrals from $-\infty$ to ∞.

(3) For each quantum-mechanical operator $\hat{A}$ there should exist a function $a(p, q)$ such that the expectation value

$$\langle \hat{A} \rangle = \langle \psi | \hat{A} | \psi \rangle = \iint a(p, q)f(p, q)dpdq . \tag{13.13}$$

(4) Furthermore, if another operator $\hat{B}$ is a function $F(\hat{A})$, then the corresponding function $b(p, q)$ which enters into (13.13) should satisfy

$$b = f(a) . \tag{13.14}$$

---

[3] For the sake of simplicity, I have written probability densities $f(p, q)$ to discuss distributions here, though the differentiability of the distribution $F(p', q') = Pr(p \leq p', q \leq q')$ is not a requirement.

It can now be shown that

(5) There exists no *unique* rule for deriving a distribution $f(p,q)$ for a given wave function ψ that will satisfy all the conditions of points (2) and (3); and moreover,

(6) There exists no function $f(p,q)$ that satisfies point (4) for any F as well as (2) and (3).

Conclusion (5) was first brought out in a paper by Shewell (1973); both are proved (though differently stated) in Cohen (1966b, 1966c).

These conclusions are all the more surprising in that joint distribution functions have found wide use in various applications, and are particularly useful in quantum optics (Agarwal and Wolf 1970). Indeed, their history goes back to 1932, when Wigner (1932) introduced a distribution

$$f_{\mathrm{W}}(p,q) = \frac{1}{2\pi\hbar} \int \psi^*(q - \tfrac{1}{2}\tau\hbar)e^{i\tau p}\psi(q + \tfrac{1}{2}\tau\hbar)dq \ , \qquad (13.15)$$

as a phase-space representation of ψ. This is still the best known of the distribution functions, and it has formed the basis of a serious attempt to restate quantum mechanics as a phase-space theory (Moyal 1949). Yet the Wigner function of Eq. (13.15) is not a probability function for all of quantum mechanics; it fails in three aspects: it satisfies condition (13.9) above at best for the ground state, but is non-positive already for the first excited state of the harmonic oscillator; for functions $a(p,q)$ which cannot be written as $a'(p) + a''(q)$ it can give the wrong expectation value; and – a special but important case – it predicts finite widths for the excited states of systems. Concerning the first point, it has been shown (Urbanik 1967, Hudson 1974, Piquet 1974) that the Wigner function is non-negative for the coherent states first introduced by Glauber (1963); but however useful these have in practice proved to be, it is not possible to reduce quantum mechanics to them, and therefore several authors, among others Mehta (1964) and Margenau and Hill (1961), have proposed alternatives to Eq. (13.15). That none of these can be satisfactory is the result of Cohen's work, who was able to introduce a general form of which earlier proposals are particular cases:

Suppose, in classical statistics, we are given a probability density $f(x)$ for a set of variables $\mathbf{x} = (x_1, \ldots, x_n)$ and wish to determine the density for a set of variables $\mathbf{y} = (y_1, \ldots, y_n)$ which are functions $\mathbf{y}(\mathbf{x})$ of the $\mathbf{x}$. A convenient way to carry out the transformation is to find the characteristic function for the y, i.e. the Fourier transform of f in the y space

$$\chi(\theta) = \int f(\mathbf{x})\exp(i\theta \cdot \mathbf{y}(\mathbf{x}))d\mathbf{x} \ , \qquad (13.16)$$

and then to recover the looked-for density $f(\mathbf{y})$ as the inverse Fourier transform

$$f'(\mathbf{y}) = \frac{1}{(2\pi)^n} \int \chi(\theta)\exp(-i\theta \cdot \mathbf{y})d\theta \ . \qquad (13.17)$$

In quantum mechanics $\mathbf{x} = (p, q)$ and the $\mathbf{y}$ we are interested in are the operators $\hat{p}$ and $\hat{q}$. This prescription therefore implies that (writing θ, τ for the two components of θ).

$$\chi(\theta, \tau) = \int \psi^*(q) \exp i(\theta \hat{q} + \tau \hat{p}) \psi(q) dq$$

$$= \int \psi^*(q) \exp(\tfrac{1}{2} i\theta\tau\hbar) \exp(i\theta\hat{q}) \exp(i\tau\hat{p}) \psi(q) dq$$

$$= \int \psi^*(q - \tfrac{1}{2}\tau\hbar) e^{i\theta q} \psi(q + \tfrac{1}{2}\tau\hbar) dq \ . \tag{13.18}$$

The Fourier inverse of this is the Wigner function (13.15); but the notion that we must replace p, q by $\hat{p}, \hat{q}$ in the exponential is, as Cohen (1966b,c) shows, insufficient, and a more general correspondence

$$\exp i(\theta q + \tau p) \longleftrightarrow g(\theta, \tau) \exp i(\theta \hat{q} + \tau \hat{p}) \tag{13.19}$$

satisfies the requirements (2) and (3) provided

$$g(\theta, 0) = g(0, \tau) = 1 \ , \tag{13.20}$$

and

$$g^*(\theta, \tau) = g(-\theta, -\tau) \ . \tag{13.21}$$

Equation (13.20) is needed to ensure the quantum-mechanically prescribed marginal distributions (13.11) and (13.12), while Eq. (13.21) ensures that all operators $\hat{A}(p, q)$ are Hermitean and so have real eigenvalues. Wigner's function now corresponds to the case $g = 1$.

It is not difficult now to show that there exists no function $g(\theta, \tau)$ such that the conditions (2) and (3) are satisfied for all ψ – which proves conclusion (5) – and that there cannot exist any $g(\theta, \tau)$ for which condition (4) is satisfied – which proves conclusion (6). These two conclusions not only create foundational difficulties for the extensive applications of distribution functions (and the Wigner function in particular); they are also awkward for the ensemble interpretation. Since their mathematical background is irreproachable, we must critically examine the points (1) to (4) on which they are based. But this requires the elucidation of certain philosophical points, a matter to which we now turn.

This will at the same time create a basis for discussing the third difficulty of the ensemble interpretation, namely that in it Einstein's theorem (Einstein, Podolsky and Rosen 1935; Einstein 1949) must be taken seriously and the consequence must be faced that the quantum-mechanical description of nature is incomplete.

(iv)

The origins of quantum mechanics are closely linked with strongly antimaterialist philosophical positions. This is not the place for exploring them, a task which has already been well carried out by others (e.g. Jammer 1966, 1974);

what is relevant here is that this created an evident bias against the ensemble interpretation because of its open association with the view that the physical world has a reality which is independent of and both logically and chronologically anterior to our ideas about it. Such conceptions have been repeatedly expressed; the first to state them explicitly was, perhaps, Jordan (1936). As a consequence the precise arguments of, for instance, Einstein (1949, 1953) have been either ignored or misunderstood, and the ensemble interpretation has remained underdeveloped, its problems stressed rather that studied, while the peculiarities and paradoxes of the Copenhagen interpretation have been taken, with a naive pride, as signs of its revolutionary character. It is time, I feel, for us to abandon such attitudes and to behave as physicists rather than as blind defenders of our respective Weltanschauungen. If, therefore, I proceed from a materialist point of view (in the sense indicated above), this is to be taken as a postulate whose justification is to be found in its successes, no more – and no less.

This position has significant consequences for the concept of probability used in the ensemble interpretation. It can evidently not correspond to any of the so-called subjective views, whether Keynes or de Finetti, for these can be consistent only with a rejection of objective reality as the starting point for the philosophy (of science as of anything else). Unfortunately the frequentest viewpoint also creates difficulties, due largely to its positivist origins which lead it to ignore the subtle but fundamental distinction between theory and experiment; nevertheless this view (often called objective) should have led the first generations of quantum physicists to something like the ensemble interpretation.

Yet this did not happen. Those founders of quantum mechanics that thought along Copenhagen lines tended towards one or another subjective view of probability; Heisenberg's interpretation of it as some sort of Aristotelian potentia is well known (Heisenberg 1955). Those that followed the ensemble interpretation on the whole accepted von Mises' formulations (von Mises 1931), and then ran into trouble. For some quantum mechanics became a kind of theory for many particles when they interpreted the frequentest conception of probability too strictly as an experimental prescription; it is to avoid such misunderstandings that I have preferred here to speak of the ensemble interpretation instead of the statistical one, as it is traditionally known. For most the smoothing-over of the theory-experiment distinction in the positivist tradition underlying von Mises' work made the agreement of theoretical prediction and experimental result almost automatic, and therefore "automatically" eliminated the whole region of problems that would have led to an ergodic theory of quantum mechanics;[4] we saw above that the first fruits of such a theory already provided useful insights.

---

[4] This is particularly striking in the work of J.v. Neumann, who made significant contributions to the development both of quantum mechanics and of ergodic theory, who carefully formulated the postulates of quantum mechanics in terms of statistical ensembles, and who yet did not realize how neatly one part of his work would apply to the other (v. Neumann 1932).

It was necessary, therefore, to develop a conception of probability that could fit better into an interpretation of quantum mechanics based on the idea of an ensemble. In fact, the ensemble concept turned out to provide precisely the required structure. This idea is discussed elsewhere (Brody 1975, 1979). Here we need only say that a full justification can be found for the use of ensembles on the basis that a physical theory must cover a certain range of similar situations and must therefore be an approximation for each particular situation; that averaging over the ensemble provides a way for selecting the common features among all the situations covered by the ensemble; and the probability is then introduced as the ensemble average of one particular kind of property. It will be clear that since the ensemble is a theoretical construct, the agreement of its theoretical predictions with experimental results is not automatic; it must be striven for, by adjusting and improving the ensemble until the fit is adequate. Lastly, if the ensemble is to describe physical situations, it will have a time evolution; thus ergodic concepts can appear naturally in this picture and they turn out to be very relevant to understanding the role of probability in our descriptions of nature.

On such a basis the ensemble interpretation of quantum mechanics is quite natural, and the conceptual difficulties and misunderstandings I have mentioned above do not develop. But because things did not in fact happen in this way, certain further points require discussion before going on to consider possible solutions to the difficulties of the ensemble interpretation.

The first one concerns the relationship between a theory's structure and its interpretation, where two opposing viewpoints can be found. One is that these two elements of a physical theory have a unique connection, so that the formalism implies the interpretation; this widely held view is clearly stated by Rosenfeld (1957). Such a view raises several problems; even if it were true, it would not be helpful until we actually knew how to deduce the interpretation from the formalism; and since, above all for quantum mechanics, no one has been able to carry out such a programme of deducing the interpretation, we cannot do better than to continue comparing the relative merits of different interpretations. But the view is not even true; this is obvious when we consider that the interpretation of a formalism, i.e. the connection we propose to establish between the various concepts of the formalism and elements of physical reality,[5] involves notions that do not at all appear in the formalism. Moreover, historically the first vague ideas of what later becomes the interpretation precedes the construction of the formalism, and it is precisely because the connection between the two is not unique that the business of scientific research requires that element of creativeness all the great scientists have insisted upon. Also, it must not be forgotten that scientific theories are not static; they change and evolve; sometimes it is the interpretation that

---

[5] The language here is deliberately borrowed from Einstein (Einstein, Podolsky and Rosen 1935); for the problem is a real one only in his materialistic philosophy. In the phenomenalism of Mach or the conventionalism of Poincaré it reduces to a methodological question of little fundamental interest, and for a subjective idealist it vanishes completely.

changes, as when the ideal gas laws are reinterpreted in the light of the microscopic models of statistical mechanics; sometimes the formalism is renewed,
as when classical mechanics is rewritten by Hamilton and Jacobi. In quantum
mechanics, in particular, the formalism has not remained at all static, and we
have a Hilbert-space version a C^*-algebra one, and lattice-theory one – not
forgetting the two original formulations of Schrödinger and Heisenberg-Born-
Jordan. Do we then have to find a slightly different interpretation for each of
these versions?

But if the Rosenfeld view of a unique connection between formalism and
interpretation is not tenable, neither is the view apparently held by many
critics of the Copenhagen interpretation (I say 'apparently', for its absurdity
would be patent if it were fully spelt out, and so it can at best be glimpsed as
implicit in their writings) that these two components are largely independent,
so that one can remove the interpretation from a theory and simply plug in
another one. This view might be termed a 'Meccano' one; household appliances
and motorcars can be built on such a principle, but scientific theories lie
beyond its pale. We need not labour the point.

Clearly the actual situation lies in between these extremes. Formalism and
interpretation do not imply each other, and are not mutually deducible; but
neither are they independent, to be changed at the whim of the scientist. They
have a strong influence on each other. This fact has an obvious implication
here: if the ensemble interpretation of quantum mechanics is to be made fully
workable, we may expect that some change in the formalism will be required.
Just what changes are in fact needed is still an open question, requiring much
further work; in the next sections some relevant ideas are discussed. It might
be said, of course, that with such changes in the formalism we no longer
have quantum mechanics, we have a new theory. This is to some extent a
terminological question to be settled by convention, though only a radical
change in basic notions, methods and results could really justify speaking of
an entirely new theory; the changes that can at present be foreseen, however,
are hardly more important than for instance the introduction of superselection
rules. But a more significant answer is that the general acceptance of what we
have called the ensemble interpretation would surely mean the equally general
relinquishment of the Copenhagen interpretation: this is not what would be
expected if it were to constitute a new theory, since new theories (pace Kuhn
& Co.) do not replace earlier ones except in special, limiting situations.

It is perhaps more important to observe that if we abandon Rosenfeld's
view then we must look for a criterion of choice between the alternatives for an
interpretation. One should expect, on general grounds, to derive experimental
tests; after all a difference in interpretation signifies a difference in the links
between formalism and observable fact. Unhappily, in the present case nothing of the kind seems possible. Two factors combine to bring this about: on
the one hand, in all interpretations of the probability concept – and we have
seen how central this is to the formulation of the two main interpretations of
quantum mechanics – the experimental estimate of a probability is derived
from a relative frequency; and on the other, it is quite usual to find experi-

mental physicists who in writing adhere to the Copenhagen interpretation but whose experimental practice corresponds to the ensemble one. There is thus no way experimentally to decide between the two interpretations.[6] Since conceptual simplicity is at best a confused and subjective criterion (Bunge 1963), there remains only one possibility: fruitfulness in suggesting further lines of research. In the last section we shall show that on this criterion the ensemble interpretation wins hands down.

Another point concerns the "return to outmoded classical models" which has been a common reproach directed at critics of the Copenhagen interpretation. Now it is true that the ensemble interpretation makes no such demands of ontological renewal as the Copenhagen viewpoint does – no doubts about the underlying reality, no holistic implication uniting object and measuring device, no renunciations of causality (which, moreover, is generally confused with determinism). But it is not at all true that it reduces quantum theory to a special if elaborate case of classical mechanics, for two good reasons: as the comparison with statistical mechanics makes clear, in a statistical (or better, ensemble) theory, new concepts and qualities appear (e.g. temperature or entropy), others disappear (the positions and momenta of the microscopic components) and even new basic principles (irreversibility and ergodicity) can arise which not only have no counterpart in the underlying mechanics but can even contradict it; in the present case, moreover, the ensemble interpretation is not a complete statistical theory, for it lacks the required mechanical theory that it would be based upon.

Precisely this is the last point to need making here: in the ensemble interpretation Einstein's theorem on the incompleteness of quantum mechanics acquires the specific meaning just mentioned; quantum mechanics, then, requires an underlying physical model of which it will be the statistical theory. In a sense we shall have here a hidden-variable theory and we must therefore explore the question to what extent such theories are conceivable, and how far it has been possible to construct them. Here we anticipate the discussions below to underline that the ensemble interpretation provides both the motivation for research in this direction which has proved to be very promising, and the link between the resulting theoretical constructions and quantum mechanics. The Copenhagen interpretation, on the other hand, leads to the conclusion that this line of work is impossible; it has even been used to turn quantum mechanics into the definitive fundamental theory of physics, no longer susceptible of further modification (Born and Heisenberg 1928); we need not refer to the many historical precedents of similar predictions that further research has falsified.

---

[6] It was hoped at one time that Bell's inequality (Bell 1965; see also Clauser and Shimony 1978) might indirectly provide relevant evidence, in that its experimental confirmation would eliminate all but macroscopically non-local hidden-variable theories. As we shall see below, the concept of hidden variables is, though not implied, at least suggested by the ensemble interpretation. But recent work (Brody and de la Peña, 1980; Brody 1980) has dashed this hope.

(v)

On the basis of these general considerations it is possible now to examine the problem of the joint probability distributions.

The first observation to be made here concerns point (1) in Sect. (iii). An ensemble theory needs appropriate probability distributions over a suitable sample space; but that this sample space should be the phase space of classical mechanics – augmented or not – is a matter that all writers on the subject have simply taken for granted. It is at first sight plausible, particularly in view of an early paper by von Neumann (1931); but it springs from the notion that quantum mechanics should in some sense be derivable from classical mechanics. If this is not so, if classical mechanics appears as a limiting case of quantum mechanics but not also as its basis, then there is no special reason to accept the classical phase space as the sample space relevant to the ensemble interpretation. No work appears to have been done on this question, so that no more can be said here beyond pointing out an open problem.

With this reservation we may accept classical phase space as the basis for the ensemble interpretation, and go on to consider point (4). The meaning of the stipulation that Eq. (13.14) should hold becomes clear if we take the simplest case, $b = a^2$ corresponding to $\hat{B} = \hat{A}^2$. The square of an operator will correspond experimentally either to the repetition of a measurement on the same system, or alternatively to a different device that measures the observable A^2. In the second case there seems no conceivable reason to suppose that Eq. (13.14) should always hold; only the first implies this, under the condition that we accept the projection postulate (without which the two measurements might yield different eigenvalues of $\hat{A}$). But we have already noted that the projection postulate is untenable in the majority of experimental situations. Thus (4) is not in general valid in quantum mechanics, and conclusion need no longer be taken as standing in the way of a consistent ensemble interpretation.

Conclusion (5) presents us with a different situation. It is not in itself a very plausible requirement that there should be a unique rule for deriving the quantum mechanical distribution function, in the sense that Eq. (13.19) admits only one function $g(\theta, \tau)$; on the one hand, such requirement has no analogue in classical statistical theories and would therefore need a specific justification which it has never received; and on the other hand, as is well known (see e.g. Cohen 1966a–c), each $g(\theta, \tau)$ generates a particular correspondence rule between classical quantities and quantum operators, while no single such rule can have general validity (Shewell 1959). This last point is obvious enough, for the existence of such a privileged correspondence rule would tie quantum mechanics to the apron strings of classical mechanics in a very unacceptable way.

Unfortunately, this does not dispose of the matter. As is already clear from the example of the harmonic oscillator, a joint distribution function would have to be state dependent, as well as problem dependent.

That it should depend on the particular problem what $g(\theta, \tau)$ is appropriate seems eminently reasonable; that g should also be a functional of the quantum state is clearly less so. We must conclude that the conditions given in Sect. (iii) must be reformulated, but it is not yet completely clear in what terms; though condition (4) can be eliminated, something must be added that will allow an appropriate freedom of choice for $g(\theta, \tau)$ while at the same time a physically plausible picture is created.

Research on these questions has been going on now for some time, with useful and interesting results, but no definitive solution as yet. Curiously enough, though various groups have proceeded on the basis of quite dissimilar notions, their conclusions converge. Apart from those authors (e.g. Shankara 1967, Leaf 1978) who attempt to remain within the framework of traditional quantum theory, two lines of thought appear.

One is due to Prugovečki (1976, 1979; Ali and Prugovečki 1977; Boisseau and Barrabes 1978) and is based on the concept of a "fuzzy" sample space, made up of "fuzzy" sample points defined as a point in phase space together with a certain "confidence function" giving the certitude of how near an experimental observation actually is to the point. The unsharpness of such points raises a number of problems, among others certain conceptual difficulties with the confidence function. For the present purpose, however, the main awkwardness of the "fuzzy" approach is that fuzziness is taken as a primitive concept, not further explained; thus the approach sidesteps the problems rather than elucidating them, and though on this basis it is possible to define probability functions that are positive semidefinite and fit into the quantum-theoretical picture, they are not ordinary probability distributions in the sense of Kolmogorov, say, and their use would require more fundamental study that they have yet received. We shall not further discuss this approach here.

A second approach is represented by the very different formulations of Bopp (1956), Ruggeri (1971) and Kuryshkin (1972a,b, 1973). Bopp's paper broke new ground and in fact anticipated the later work of Shewell (1959) and Cohen (1966) in many ways. His proposal was – in the terminology used here – to introduce a particular function $g(\theta, \tau; l)$ that depends on a parameter which possesses many of the characteristics of a fundamental length. Now $g(0, \tau; l) = 1$ only in the limit $l \to \infty$; hence we do not here satisfy condition (iic) except approximately; condition (4) is likewise not satisfied. Bopp's work has been rather unjustly neglected and incorrectly described as wrong (Kuryshkin 1973); it is so only in the sense of breaking with the Copenhagen interpretation. The choice that Bopp made for g was a little too specialized, and the papers by Ruggeri and Kuryshkin attempt to remedy this; their methods are at first sight quite unrelated but work in course by my collaborator J.L. Jiménez establishes the fundamental identity of these three approaches and their connection with related work.

Further possibilities exist that have not yet been worked out. One promising idea is to take explicitly into account the comment made above that eigenstates of zero width are idealizations. This can be done by eliminating pure states from "real" quantum mechanics and only admitting density matrices such that $\text{tr}\{p^2\} \leq \text{tr}\{p\} = 1$. Whether this gives rise to a feasible theory is at present under study.

There is, however, one problem with such approaches. We noted that Prugovečki's fuzziness concept is not really satisfactory, essentially because it makes it impossible to reach the underlying physics; the Bopp-Ruggeri-Kuryshkin method does not do this, but as yet it lacks any background model that would make its various assumptions sufficiently plausible and remove their rather indefinite generality. After all, the ensemble interpretation is not looking for a new formalism; it looks for better physics and it necessarily adapts the formalism.

Both these approaches involve functions $g(\theta, \tau)$ that do not satisfy Eq. (13.20); thus the marginal distributions of q and p will not quite equal $|\psi(q)|^2$ and $|\phi(p)|^2$ respectively; this discrepancy with the prediction of usual quantum theory requires discussion. Quantitatively, the difference can be explained as the effect of experimental uncertainty, as has been shown by Cartwright (1976) and Yoshihuku (1979); if one considers that the experimental errors in the measurement of q and p obey Gaussian distributions $\chi_q(\xi)$ and $\chi_p(\eta)$ with dispersion σ_q^2 and σ_p^2 respectively, then the experimental joint distribution will be given by the joint distribution of $q + \xi$ and $p + \eta$, that is to say the convolution of $f(q, p)$ with χ_q and χ_p; and the resulting distribution function is non-negative provided

$$\sigma_q^2 \sigma_p^2 \geq \tfrac{1}{4}\hbar^2 . \tag{13.22}$$

It is evident, then, that a non-negative distribution function $f(q, p)$ can come as near as one wishes to reproducing one or the other of the quantum-mechanical marginal distributions, which are thus seen as the limiting cases of infinite experimental precision; and even when both dispersions have to be taken into account, the uncertainty-like inequality (13.22) imposes no restriction that we can as yet achieve experimentally. For the fact is that the experimental validity of $|\psi(q)|^2$ and $|\phi(p)|^2$ for the experimental distributions has not received anything like as solid an experimental confirmation as one could wish. The best data available have been obtained from experiments with particle beams, where $\phi(p)$ is well defined but, because the system is not confined, $\psi(q)$ is not normalizable and the relation between the two is not simply that of Fourier transformation as required for the theory that interprets them as marginal-distribution amplitudes;[7] on the other hand, for confined systems measurements of sufficient precision seem to be very difficult. Since so far the exact form the marginal distributions given by a positive joint distribution function have not yet been worked out, the matter must be considered one more open problem; but it might be added that the theory of stochastic electrodynamics, to be discussed, should differ slightly from the quantum prediction.

In summary we have more problems calling for future work than answers in this matter of the joint distribution of q and p; yet it may fairly be concluded

---

[7] The ordinary Fourier transform must be generalized to cover this case, and the function $|\psi(q)|^2$ must be interpreted as a probability distribution in the sense of Rényi. Certain conceptual problems arise that have not yet been elucidated.

that our failure so far to find fully satisfactory joint distributions cannot be ascribed to a fundamental weakness of the ensemble interpretation.

(vi)

We have so far discussed two of the three difficulties mentioned above that arise in the attempt to make the ensemble interpretation complete and consistent; but the third is in a way the most interesting. The EPR theorem (Einstein, Podolsky and Rosen 1935; Einstein 1949) leads to the conclusion that quantum mechanics is incomplete. For the ensemble interpretation this conclusion is inescapable and therefore raises the problem of how to complete the theory. The incompleteness of quantum mechanics takes a specific form in the ensemble interpretation, as we have mentioned: the formalism provides no guiding lines for the choice of an ensemble on which to base the theory. Any attempt to formulate a suitable physical model which would give rise to an appropriate ensemble must therefore lead us beyond quantum mechanics. We shall outline in the last section a possible theory that does just this; here we have first to face the third difficulty alluded to above: the existence of proofs that a completion of quantum mechanics, of the sort contemplated here, is impossible. In other words, we have to deal with the hidden-variable problem. Much of the extensive literature concerning it seems to obfuscate the matter, and I enter on it here only because I consider it to be much simpler that it is generally thought to be.

The first proof of the impossibility of hidden variables was given by von Neumann (1932) as a straightforward corollary of his derivation of the density-matrix formulation for quantum mechanics. For many years this proof was taken to be conclusive, though many people felt misgivings about its implications (e.g. de Broglie 1956); then certain weaknesses in the derivation of the density-matrix results were discovered, and for some time these appeared to justify a search for hidden-variable theories. The question had acquired importance because of the appearance of a fully worked out hidden-variable theory, due to Bohm (1952); this theory has important weaknesses from the physical point of view (in particular with an implausible space dependence), but it provided a counter example to von Neumann's theorem – or so it seemed.) But further work, in particular by Kochen and Specker (1967) and by Gleason (1957), then reestablished the validity of von Neumann's density-matrix theorem, and therewith also the hidden-variable corollary. A good summary of these developments may be found in Bell (1966).

Yet, interesting though Gleason's theorem certainly is, it is irrelevant to the issue of hidden variables; for the validity of von Neumann's result was never in doubt, only his methods of proof. This, though obvious once it is pointed out, seems to have been generally overlooked. From the ordinary Hilbert-space formalism, based on pure states, the density-matrix formalism may be derived though an additional postulate which seems unchallengeable; inversely, a pure state takes the form of a special density matrix. Both connections can be found fully worked out in von Neumann's book. As a consequence, if the density-matrix theorem had to be given up in order to allow the introduction of hidden variables, the rest of quantum theory would also have to be given up, which is just what hidden-variable theories are intended to prevent.

Fortunately the way out of this dilemma is not difficult; in fact, it could have been found in von Neumann's book itself, for he carefully specified the hidden variables he proposed to exclude as deterministic, i.e. dispersionless. His corollary does not apply to stochastic hidden variables. This becomes clear when it is observed that the corollary is in fact the quantum-theoretical case of a much general result, valid for any statistical theory. The general case is a simple consequence of probability theory: consider the dynamical variable x of statistical theory, with a distribution function[8] $P(x)$. This theory is to be embedded in a broader theory which contains the hidden variable h as well as x; this means that we must find a joint distribution function $Q(x, h)$ such that $P(x)$ is the marginal distribution for x in it. Let $R(h)$ be the distribution of h and $P_c(x|h)$ the conditional distribution of x given h. We have, from standard probability theory

$$Q(x, h) = \int_{-\infty}^{h} P_c(x|h')dR(h') . \tag{13.23}$$

If h is to be a hidden variable of deterministic type, with the fixed value $h = h_0$, then

$$R(h) = \begin{cases} 1 & h \geq h_0 \\ 0 & h < h_0 \end{cases} \tag{13.24}$$

and thus

$$dR(h) = \delta(h - h_0)dh . \tag{13.25}$$

Substituting (13.25) in (13.23) we have

$$Q(x, h)0 = \begin{cases} P_c(x|h_0) & h \geq h_0 , \\ 0 & h < h_0 . \end{cases} \tag{13.26}$$

Therefore

$$P(x) = Q(x, \infty) = P_c(x|h_0) . \tag{13.27}$$

Using (13.24) and (13.27) in (13.26), we have

$$Q(x, h) = P(x)R(h) . \tag{13.28}$$

We conclude from (13.28) that x and h are statistically independent. Since this argument can obviously be carried through for the set x of the dynamical variables and the set h of "hidden" variables to be added to them, the variables of the original theory are statistically and thus also functionally independent of the h's which therefore are irrelevant to any explanation of the behaviour of the x's. Nothing is gained by adding the h's. If we allow h to vary over a small interval, from h_0 to h_1, say, then

$$R(h) = \begin{cases} 1 & h > h_1 \\ 0 & h < h_0 \end{cases} \tag{13.29}$$

---

[8] Here we use distribution functions (which are integrals over probability densities when these exist), to make the argument simple and general.

and the argument is still valid outside this interval. Only if this interval is large enough to be significant within the original theory can there by any genuine connection, statistical dependence or otherwise, between x and h.

Note that the conclusion depends entirely on the requirement of maintaining the character of the original theory, i.e. that the introduction of the hidden variable h does not alter $P(x)$; if this requirement is dropped the conclusion that h must also be a stochastic variable no longer follows. Nor does the argument apply to the introduction of h as a new parameter in $P(x)$; it must be a new dynamical variable, in the sense of being at least statistically linked to the other dynamical variable.

Returning to the particular case of quantum mechanics, the situation with regard to von Neumann's corollary is, then, that its validity need no longer be disputed but its meaning has to be reinterpreted. Stating it positively, we draw from it the conclusion, not that hidden-variable theories are impossible, but that they must be stochastic theories. Seen in this way, the corollary provides a useful hint for further research, rather than figuring as an obstacle to it, as seen by the Copenhagen school. The hint is borne out by the Bohm theory and indeed all other hidden-variable theories that have achieved some sort of consistency: without exception their hidden variables are stochastic in nature and are not dispersionless.

It is sometimes said that if hidden variables cannot be dispersionless then they are useless. But it should be clear that this view is inspired more by a desire to return to a fully deterministic, Newtonian kind of physics, and while quantum mechanics (or rather the immense range of experimental results that it accounts for satisfactorily) should not lead us to abandon the ontologically fundamental status of reality, it should convince us of the limitations of a purely mechanistic physics. Moreover, the idea that a stochastic theory cannot provide the basis for completing quantum mechanics and so yield a deeper understanding of it ignores the lesson of statistical mechanics: here we have an entirely stochastic theory that has enormously enriched and broadened our understanding of thermal physics and has gone well beyond the limits of applicability of the thermodynamics it was intended to underpin.

But statistical mechanics, as already noted in Sect. (iv), holds the further lesson for us that the underlying classical mechanics differs markedly from it in character, concepts and main quantities. This will be relevant below.

(vii)

To sum up, we have seen that, from the point of view of the ensemble interpretation, quantum mechanics is incomplete, that any completion should be stochastic in nature, and that the resulting theory will likely be of very different character.

Several theories along such lines have been suggested in the past; the first would appear to have been due to Fényes (1952). They have not, on the whole, been very successful, and for this there are two reasons. One is a technical problem: the best-known stochastic process is, of course, Brownian motion, and this misled many workers into identifying the process underlying quantum mechanics with a Winger process (the mathematical model for Brownian motion); that the two, though related, are different was first shown by de la Peña and Cetto (1977a). There are now reasons, as we shall note, for believing that the process is not even Markovian. The other reason is that without a physical model to serve as starting point, background and touch-stone, a mathematical description of a stochastic process, however ingenious, will be somewhat arbitrary and even ad hoc. Of course one has the great advantage of free choice of a nice, simple, tractable mathematical problem if one ignores physical models; but one runs the risk that the nice easy problem can be quite misleading.

I propose therefore to outline here a theory that is far from having a finished form as yet, but does possess a plausible physical conception to base its mathematical structure on; and however provisional the present form of the theory may be, it has already had some significant successes. This theory, stochastic electrodynamics, owes its inception to Braffort and coworkers (Braffort, Spighel and Tzara 1954; Braffort and Tzara 1954; Braffort, Surdin and Taroni 1965; Surdin, Braffort and Taroni 1965) and independently to Marshall (1963; 1965a,b) who gave it its name; it has been further developed by Santos (1975), Boyer (1975), de la Peña and Cetto (1977b), and Claverie and Diner (1976); the last three include a review of the earlier work.

The underlying physical conception is simple: consider a charged particle, such as an electron; in its movements it emits electromagnetic waves described by radiation-reaction terms; if it is considered in isolation, it would therefore lose energy and, in the case of an orbital electron, fall into the nucleus. This is the classical picture, often considered as an argument for quantum mechanics. But – and this is the central point of stochastic electrodynamics – the electron is not isolated; all the other charges in the universe also emit radiations through the same mechanism, and since these radiations are evidently incoherent, the electron being considered is bathed in a stochastic radiation field. The problem to be solved is thus the motion of a classical charged particle under the joint effect of a stochastic electromagnetic field, the radiation reaction and any external force (e.g. the Coulomb attraction of the nucleus that may be present.

The radiation reaction is of course well known; it is given by the Liénard-Wiechert potential, and for non-relativistic speeds is usually well approximated by a term proportional to the third derivative of the particle's position vector. In the same approximation only the electric component of the stochastic electromagnetic field need be taken into account. But because this is a stochastic force, we can at best write a Langevin-type equation, and can derive conclusions from it only if the probability distribution of this force is known. Here we see the significant advantage of stochastic electrodynamics over earlier theories as a physically grounded conception; for not only is the physical model plausible, but the stochastic properties of the background radiation field can be derived quite independently, from considerations of relativistic invariance (Marshall 1963; Santos 1974) and others (Jiménez, de la Peña and Brody 1980). The spectrum so predicted coincides with the quantumelectrodynamical one (ω being the frequency),

$$\rho(\omega) = \frac{\hbar}{2\pi^2 c^3} \omega^3 \,, \tag{13.30}$$

but $\hbar$ is merely a proportionally constant that yields the amplitude of the fluctuations in the stochastic background field; the use of the symbol $\hbar$ anticipates that eventually ordinary quantum mechanics has to be recovered. This is the one point in stochastic electrodynamics where this constant is introduced.

The equation of Langevin type for a particle of mass m and charge e now becomes

$$m\frac{d^2 x}{dt^2} = f(x) + \tau m \frac{d^3 x}{dt^3} + eE(t) \,, \tag{13.31}$$

the so-called Marshall-Braffort equation, in which

$$\tau = \frac{2e^2}{3mc^3} \tag{13.32}$$

and $E(t)$ is the component along x of the random electric field (for simplicity we only consider the one-dimensional case, for which the expectation value is

$$\langle E(t) \rangle = 0 \,, \tag{13.33}$$

and the spectrum is given by Eq. (13.30). We shall not enter here into the details of how phase-space and configuration-space probability distributions for particles obeying Eq. (13.31) are derived, since they are complex and have already appeared in the papers cited. Certain points, however, are relevant here:

a) The stochastic process considered here is markedly non-Markovian; but as equilibrium is approached, the importance of the memory terms diminishes. Very close to equilibrium, the radiative terms in (13.31) tend to cancel out, and the configuration-space amplitude then satisfies the Schrödinger equation: quantum mechanics is thus the equilibrium limit for this theory.

b) If we assume that an equilibrium state exists (this has not yet been proved for all relevant cases) then it will be reached rather rapidly; for instance, the relaxation time for an atomic orbital electron is of the order of 10^{-23} s. Before equilibrium has been reached, the configuration-space distribution and the moment-space one are not necessarily Fourier transforms of each other and so the Heisenberg uncertainty relations may be violated (in the ensemble-interpretation sense, evidently, that the product of the statistical dispersions may be less than $\frac{1}{4}\hbar$), In a particle beam, no equilibrium state in the strict sense can exist, if we take the beam to be infinitely long; but even in a finite beam only quasi-stationary states should be expected, and so the actually observed distributions of q and p could slightly diverge from $|\psi(q)|^2$ and $|\phi(p)|^2$, respectively; this purely qualitative argument was made use of above.

c) As the detailed analysis of the harmonic oscillator (de la Peña and Cetto 1979) in this theory shows, the quantum-mechanical discrete states are recovered. That there exists a ground state of finite energy, at which radiation reaction and absorption from the stochastic field balance each other, and that this state is stable, could already be seen by simple "hand-waving" arguments (Claverie and Diner 1976); that the energies of the excited states should also be correctly reproduced was not so clear; but that these states should be shown to have widths (in the sense that the instantaneous energy fluctuates and is equal to the quantum value only in the mean) was surprising. This last result, it should be noted, fits in well with the ensemble interpretation but not with the orthodox Copenhagen views.

d) A considerable number of detailed predictions have already been derived from stochastic electrodynamics; these range from the Planck black-body spectrum to the non-relativistic Lamb shift, and generally agree very well with the predictions of standard quantum theory. But the theoretical structure is far from complete, largely because no general mathematical formalism has yet been developed, and problems have to be tackled piecemeal. Nevertheless, for a great many questions rough qualitative arguments are available that show that at least in principle it should be possible to answer them satisfactorily within the framework of stochastic electrodynamics.

e) Stochastic electrodynamics does not agree everywhere with quantum mechanics; unlike the ensemble interpretation of the latter, it is a new theory. However, so far it has not proved possible to make predictions from it that differ in an experimentally accessible way from the standard quantum results.

There is one feature of stochastic electrodynamics to which attention should be drawn: it is from its very inception a theory of open systems. In more than one sense this constitutes perhaps a more decisive break with classical physics than the Copenhagen interpretation ever achieved. We are very far from understanding as yet all the implications of this fact; but a highly significant consequence is that in this theory there are no exact conservation laws, – only statistical ones. Hence all symmetries will hold only on the average, and we may guess that conceivably this will provide a mechanism to comprehend "spontaneous" symmetry breaking; but this is speculation. What seems clear is that the ultimate consequences may prove even more fundamental than the appearance of irreversible phenomena and hence of "time's arrow" when thermodynamics first – and as we now see, rather timidly – broke through the restriction of fundamental physical theories to closed systems.

Thus we have here the beginnings of a hidden-variable theory that is intended to underpin but also to go beyond quantum mechanics; in fact, the direction in which such a theory should be sought for was pointed to by that interpretation. The Copenhagen interpretation, on the other hand, never encouraged such a development and in the view of some even forbade it.

(viii)

By way of a general conclusion we may say that the ensemble interpretation, so far from being nonexistent as has sometimes been stated (Hanson 1959), forms a consistent body of ideas that removes or at least permits clearing up the peculiar paradoxes arising in connection with the Copenhagen interpretation, that sheds light on certain questions otherwise not even touched, and that rather strongly directs further research along promising new lines. Nevertheless, not all its problems have yet been satisfactorily settled; much work remains to be done, though there is little reason to expect significant conceptual breakthroughs from such investigations: it is rather a matter of adequately completing what has already been outlined.

The future of stochastic electrodynamics is another matter; here even fundamental conceptual problems may have to be solved before it is possible to say that this theory has firm foundations.

But what is to be stressed is that the philosophical problems felt to be peculiar to quantum mechanics in the past simply dissolve in the ensemble interpretation; this leaves room for tackling the genuine problems.

I would like to express my gratitude to J.L. Jiménez to whom many of the results of Sect. (v) as well as other points are due, and to Luis de la Peña, for numberless useful discussions and thorough reading of the manuscript.

References

Agarwal, G.S., Wolf, E. (1970): Phys. Rev. **D2**, 2161, 2187, 2206
Ali, S.T., Prugovečki (1977): J. Math. Phys. **18** , 219
Allcock, G.R. (1969): Ann. Phys. (N.Y.) **53**, 253, 286, 311
Arika, H., Yanase, M.M. (1960): Phys. Rev. **120**, 622
Arnold, V.I., Avez, A. (1968): *Ergodic Problems of Classical Mechanics* (Benjamin, New York, N.Y.)
Balescu, R. (1975): *Equilibrium and Nonequilibrium Statistical Mechanics* (Wiley, New York, N.Y.
Ballentine, L.E. (1970): Revs. Mod. Phys. **42**, 358
Bell, J.S. (1965): Physics **1**, 195
Bell, J.S. (1966), Revs. Mod. Phys. **38**, 447
Blokhintsev, D.I. (1953): *Grundlagen der Quantenmechanik* (Deutscher Verlag der Wissenschaften, Berlin)
Bohm, D. (1952): Phys. Rev. **85**, 166, 180
Boisseau, B., Barrabes, C. (1978): J. Math. Phys. **19**, 1032
Bopp, F. (1956): Ann. Inst. H. Poincaré **15**, 81
Born, M, Heisenberg, W. (1928): *5ᵉ Conseil de physique Solvay* quoted in Jammer (1966), p. 358
Boyer, T. (1975): Phys. Rev. **D11**, 790
Braffort, P. Tzara, C. (1954): C.R. Acad. Sci. Paris **239**, 1779
Braffort, P., Spighel, M., Tzara, C. (1954): C. R. Acad. Sci. Paris **239**, 157
Braffort, P. Surdin, M., Taroni, A. (1954): C. R. Acad. Sci, Paris, **261B**, 4339
Brody, T.A. (1975): Rev. Mex. Fís. **24**, 25
Brody, T.A. (1979): *6ᵗʰ Int. Cong. Logic, Methodology and Philosophy of Science* (Hannover) Sect. 7, p. 222
Broglie, L. de (1956): *Une tentative d'interprétation causale et non-linéare de la mécanique ondulatoire* (Gauthier-Villars, Paris) p. 68f
Bunge, M. (1963): *The Myth of Simplicity* (Prentice-Hall, Englewook Cliffs, N.J.
Cartwright, N.D. (1976): Physica **83A**, 210
Clauser, J.F., Shimony, A. (1978): Reps. Prog. Phys. **41**, 1881
Claverie, P., Diner, S. (1973): C. R. Acad. Sci. Paris **277B**, 579
Claverie, P., Diner, S. (1975): C. R. Acad. Scie Paris **280B**, 1

Claverie, P. Diner, S. (1976): in O. Chalvet et al. (eds.) *Localization and Delocalization in Quantum Chemistry* (Reidel, Dordrecht) p. 395
Cohen, L. (1966a): Philos. Sci. **33**, 317
Cohen, L. (1966b): J. Math. Phys. **7**, 781
Cohen, L. (1966c): Ph. D. thesis (Yale University)
Cohen, L. (1973): in C. Hooker (ed.) *Contemporary Research in the Foundations and Philosophy of Quantum Mechanics* (Reidel, Dordrecht) p. 66
Einstein, A. (1949): in P.A. Schilpp (ed.) *Albert Einstein Philosopher-Scientist* (Library of Living Philosophers, Evanston, Ill.) p. 665
Einstein, A. (1953): in *Scientific Papers Presented to Max Born* (Oliver and Boyd, Edingburgh) p. 33
Einstein, A., Podolosky, B., Rosen, N. (1935): Phys. Rev. **47**, 777
Fényes (1952): Zeits. f. Physik **132**, 81
Glauber, R.J. (1963): Phys. Rev. **131**, 2766
Gleason, A.M. (1957): J. Math. Mech. **6**, 885
Hanson, N.R. (1959): Am. J. Phys. **27**, 1
Heisenberg, W. (1930): *The Physical Principles of the Quantum Theory* (Dover Publications, New York, N.Y.)
Heisenberg, W. (1955): in W. Pauli (ed.) *Niels Bohr and the Development of Physics* (Pergamon, Oxford) p. 11
Hudson, R.L. (1974): Reps. Math. Phys. **6**, 249
Jammer, M. (1966): *The Conceptual Development of Quantum Mechanics* (McGraw-Hill, New York, NY.)
Jammer, M. (1974): *The Philosophy of Quantum Mechanics* (Wiley, New York, N.Y.)
Jiménez, J.L., Peña , L. de la, Brody, T.A. (1980): Am. J. Phys. **48**, 840
Jordan, P. (1936): *Anschauliche Quantenmechanik* (Springer, Berlin)
Kochen, S., Specker E.P. (1967): J. Math. Mech. **17**, 59
Kuryshkin, V.V. (1972a): Ann. Inst. H. Poincaré **17**, 81
Kuryshkin, V.V. (1972b): C. R. Acad. Sci. Paris **247B**, 1107, 1167
Kuryshkin, V.V. (1973): Int. J. Theor. Phys. **7**, 451
Leaf, B. (1968): J. Math. Phys. **9**, 65
Margenau, H. (1958): Philos. Sci. **25**, 23
Margenau, H. (1963a): Ann. Phys. (N.Y.). **23**, 469
Margenau, H. (1963b): Philos. Sci. **30**, 1
Margenau, H., Hill, R.N. (1961): Prog. Theor. Phys. **26**, 722
Marshall, T.W. (1963): Proc. Roy. Soc. **A276**, 475
Marshall, T.W. (1965a): Proc. Camb. Phil. Soc. **61**, 537
Marshall, T.W. (1965b): Nuovo Cim. **38**, 206
Mehta, C.L. (1964): J. Math. Phys. **5**, 677
Messiah, A. (1959): *Mécanique Quantique* (Dunod, Paris) Vol. I, Chap. iv
Mises, R. von (1931): *Wahrscheinlichkeitsrechnung* (Franz Deuticke, Vienna)
Misra, B. Sudarshan, E.C.G. (1977): J. Math. Phys. **18**, 756
Moyal, J.E. (1949): Proc. Camb. Phil. Soc. **45**, 99
Neumann, J. von (1931): Math. Annalen **104**, 570
Neumann, J. von (1932): *Mathematische Grundlagen der Quantenmechanik* (Springer, Berlin)
Park, J.L., Margenau, H. (1968): Int. J. Theor. Phys. **1**, 211
Peña, L. de la, Cetto, A.M: (1975): Found. Phys. **5**, 355
Peña, L. de la, Cetto, A.M: (1977a): J. Math. Phys. **18** , 1612
Peña, L. de la, Cetto, A.M: (1977b): Int. J. Quantum Chem. **12**, Suppl. 1, 23
Peña, L. de la, Cetto, A.M: (1979): J. Math. Phys. **20**, 469
Penrose, O. (1970): *Foundations of Statistical Mechanics* (Pergamon, Oxford)
Piquet, C. (1974): C. R. Acad. Sci. Paris **279A**, 107
Prugovečki, E. (1976): J. Math. Phys. **17**, 517
Prugovečki, E. (1979): Found. Phys. **9**, 575

Roman, P. (1965): *Advanced Qauntum Theory* (Addison-Wesley, Reading, Mass.)

Rosenfeld, L. (1957): in S. Körner (ed.) *Observation and Interpretation* (Butterworth, London) p. 41

Ross-Bonney, A.A. (1975): Nuovo Cim. **30B**, 55

Ruggeri, G.J. (1971): Prog. Theor. Phys. **46**, 1703

Santos, E. (1974a): Nuovo Cim. **B19**, 57

Santos, E. (1974b): Nuovo Cim. **B22**, 201

Santos, E. (1975): An. Fís. (Esp.) **71**, 329

Shankara, T.S. (1967): Prog. Theor. Phys. **37**, 1335

Shewell, J.R. (1959): Am. J. Phys. **27**, 16

Slater, J.C. (1929): J. Franklin Inst. **207**, 449

Surdin, M., Braffort, P., Taroni A. (1966): Nature **210**, 405

Urbanik, K. (1967): Studia Math. **21**, 117

Yoshihuku, Y. (1977): Chubu Inst. Technology preprint 77-2

14. Probability and the Way Out of the Great Quantum Muddle [*]

Abstract. An interpretation of the probability concept is presented which possesses a sound physical basis and avoids the usual pitfalls. Applied to quantum mechanics, this concept leads to the ensemble (Slater-Einstein) interpretation and the expectation that the incompleteness (revealed by the EPR Theorem) is resolved by an underlying stochastic process. The most successful theory of this kind is stochastic electrodynamics; the recently proposed modification in which quantum states are obtained as metastable states of the stochastic process appears as the most promising theory yet envisaged.

14.1 Probability

It is to Sir Karl Popper (Popper 1980) that we owe the insight that confusion over the probability concept lies at the origin of the "Great Quantum Muddle" – a most appropriate term for the interpretation debate in quantum mechanics which of course is also due to him. Here I begin with an overall version of this confusion, where the fairly obvious solution points us towards a view of probability that makes sense in physics and also goes a long way towards getting us out of the slough of the Great Quantum Muddle.

Consider a coin that we flip once, to find that it has come down heads. Before our flipping it, the probability of heads for this particular flip is generally accepted to be $\frac{1}{2}$. What is it afterwards? The more common school of thought judges that it is now 1, but others think that it remains at $\frac{1}{2}$. The thought behind these two views is, on the one hand, that the experimental evidence now shows it to be one, and on the other hand, that the theory of coin flips indicates it to be $\frac{1}{2}$; both are right, in a sense. But both are wrong in holding theirs to be the only view. What they overlook is that probability is not just an abstract philosophical notion, but a concept within the natural sciences, within physics in particular. But such concepts are characterized by having both a theoretical and an experimental component, and the relation between the two is non-trivial and often dangerously oversimplified. When the two sides can be stated numerically, the best we can hope for is that they will agree to within a certain degree of approximation; and this degree of approximation itself requires both a theoretical and an experimental study which in their turn

---

[*] First published in: A. van der Merwe, F. Selleri and G. Tarozzi (eds.) (1988), *Microphysical Reality and Quantum Formalism* (Urbino, Sept. 1985), (Kluwer Acad. Publ., Dordrecht, p. 443–456

must agree reasonably well. Here the theoretical estimate is obtained from a model created on the basis of the general theory by introducing various additions, simplifications and approximations taken from other theories and from the specifications of the set-up used in the experiment, and then inserting the appropriate data for initial conditions and the like. In this process the estimated degree of approximation is worked out step by step; but not only does it in its turn need experimental verification, several elements in its structure usually come from other experiments – or sometimes are just guessed. The result will be a theoretical estimate for our quantity together with an error limit on it; and this must be compared with the specific experimental observation, which commonly also has an error limit to it. The two (almost) never agree exactly, because the theoretical model we have used applies not only to this particular case, but also to an indefinitely large set of other cases. Indeed, the model does not contain any information that would allow us to discriminate between them. Thus, when we repeat the measurement a few times, the identical model will apply to the repetitions; but the values measured will differ, even if only slightly. The model, in other words, does not apply just to one specific case, but to all physical situations of the *same* type. Each of these will differ from the others to some extent (sometimes even negligibly), and so the measurement results will also differ among each other. These differences may not need any attention beyond being expressed as a statistical estimate of experimental error limits, or they may be relevant to the physical theory of the phenomena. Just what is meant by the words I have stressed above is defined by the theoretical model; the set of possible cases "of the same type" have in common all those properties the model is explicit about (again, up to experimental error limits) while all the factors neglected in the model can fluctuate in any way that is consistent with physical law. This set is known as an *ensemble*. The notion was introduced at the turn of the century by Gibbs and by Einstein to provide a foundation for statistical mechanics; it will be generalized here to apply to all the situations of comparison between theory and experiment.

The basic point of this generalization is that the ensemble may be defined and constructed without at any time using the concept of probability, either explicitly or even implicitly. This is so because the ensemble, being a set of theoretical constructs, can be given a structure such as to possess a measure. This is evident where the set is finite or denumerable (the Laplace definition of probability derives from a special case of a finite ensemble) and may also be shown to hold when the topology is not so simple. The measure need not even be normalizable, though I shall not discuss this extension here.

The measure function of the ensemble is then used to define averages over the ensemble; this is the second basic concept needed. It should be noted here that averaging, though a mathematically trivial operation, can profoundly alter the physical significance of what is averaged. The average kinetic energy of molecules becomes a temperature, with entirely different properties; and even the average number of children in the families of a country betrays its new character by no longer being an integer.

We can now use the ensemble to define probability: Given a property A which members of the ensemble may or may not possess, the probability of A is the expectation over the ensemble (or average) of the corresponding indicator function $\chi_A(\omega)$, which is 1 if the system labelled ω in the ensemble possesses A, and 0 otherwise. That is,

$$\Pr(A) = \int_\Omega \chi_A(\omega) d\mu(\omega) , \qquad (14.1)$$

where $\mu(\omega)$ is the measure function for the ensemble, and may be assumed to be normalized over Ω, the range of ω.

The normalization condition can be relaxed, in agreement with the Rényi (Brody 1975) formulation of conditional probabilities. A much more significant generalization may be found from the fact that the ensemble describes also physical systems that evolve in time. It must then be given by a time-dependent measure function $\mu(t, \omega)$ from which we obtain $\Pr(A, t)$ in terms of $\chi_{A(t)}(\omega)$. Such a time dependence of the probabilities characterizes stochastic processes, whose theory (Brody 1975) thus receives a new basis. This is relevant to the problem I began with, for stochastic processes may have various ergodic properties. Thus a non-ergodic process will not in general exhibit those stable frequencies which in some conceptions have to be specially postulated but here can be predicted to hold if the ensemble possesses sufficient ergodic properties. The tossing of coins and the rolling of dice belong to this type, as Hopf (1937) has shown. More in general, the measurement of a physical quantity takes time; if this time is larger than the relaxation time to ergodic equilibrium, the fluctuations will not be perceptible; the situation treated in classical physics is thus recovered. In an intermediate case, the fluctuations are small and may in general be neglected; here the traditional theory of errors applies. Further, if the ensemble is not only ergodic but has the mixing property, certain central-limit theorems apply and most distributions of interest will not depend on the exact nature of the measure $\mu(\omega)$. This explains why the Gaussian distribution dominates the theory of errors; on a different level, it explains why the various ensembles used in statistical mechanics make exactly the same predictions for several properties of thermodynamic nature.

The ensemble is the tool of probabilistic theorization; it remains to examine its experimental counterpart. This is furnished by the relative frequencies measured in actual and hence finite experimental series. Probability is a little unusual among physical concepts, because there is only this one direct method of measurement; there are, of course, numerous indirect ways of determining probabilities. In part this may be due to the fact that probabilities hardly ever occur in the theories of physics except as intermediate quantities; it is far more common to find an expectation value, an ensemble average of another quantity, than the simple yes-no property possessed by an attribute.

The probability given by Eq. (14.1) has also the properties associated with the mathematical concept; it is not difficult to show that it satisfies the Kolmogorov axioms. It follows that the extensive results of statistical theory may be applied to it; in particular, one can obtain all the well known theoretical estimates of the error limits on a measurement of a probability. This is the basis mentioned before for comparing theoretical probability estimates with an experimentally found relative frequency. If by this standard the discrepancy between the two is excessive, it may be necessary to reformulate the ensemble until a measure function is found that agrees well enough with experiment. But this cycle (which may have to be repeated) is in no way different in the case of probability than for any other physical property. It may sometimes be an exceedingly hard problem to find a theoretical probability distribution that agrees acceptably well with the known data; the general questions the cycle raises have by no means found a suitable solution; but since they are common to all of physics (and indeed to all of knowledge) they have no special relevance here.

Returning briefly to the coin-tossing situation, it should now be clear how each of the two views mentioned oversimplifies the issue. Those who cannot see anything but the theoretical estimate have left themselves without any way of connecting it to the experimental data as they accrue; while the other school, accepting the experimental side and simply declaring it to do duty as the theory as well, are left without a basis for their theory. This is evident if we ask what happens when tossing continues. Suppose the next toss is also heads, and then tails comes up. The experiment believers' probability then fluctuates, being successively 1, 1, and $\frac{2}{3}$. But whence this lack of stability? It is always the same penny that is being tossed, and presumably in the same way. If the relative frequency settles around 0.51, on the other hand, the theory believer has a problem: Is this evidence that the penny is slightly bent, so that he should revise the theory? And if so, when? After 100,000 tosses? And why not after 99,999?

To sum up, the view here proposed of probability is more reasonable for the physicist, among other reasons because exact agreement between theory and experiment is not expected. It is, I should add, the view that most physicists hold quite spontaneously, and I cannot claim any originality for it. It has also the advantage of escaping the usual paradoxical difficulties. Of these philosophical problems, I will here mention only that the probability concept is inapplicable when no ensemble can be defined; a particular case of this is the probability that a theory be true. On this view (Brody 1975) there can then be no such thing as induction.

Here I shall only explore the consequences of this view for the conceptual difficulties of quantum mechanics. I shall also ignore the many other schools of thought on probability that exist; since none of them can clearly draw the distinction, vital to physics, between theory and experiment, this seems justified for the present purposes.

14.2 Quantum Mechanics as an Ensemble Theory

It is universally accepted that probability figures centrally in quantum mechanics. But when probability is treated as a primitive concept in the quantum context, not requiring further elucidation, people inevitably attach their own meaning to it, consciously or not. Result, the "Great Quantum Muddle". Perhaps the single most devastating source of confusion has been to treat the wave function, understood since Born's work as a probability amplitude, as applying to the individual quantum system; this notion is the simplest physical interpretation of the theory believer's views.

The confusion could persist because quantum mechanics is a probabilistic theory for experimental results that in many cases show a statistical dispersion. Yet it is not a phenomenological theory such as thermodynamics, say, since it explicitly uses probabilities, nor a fully statistical one like statistical mechanics, since it admits no model for the individual fluctuating process. And this does raise a difficulty, namely that these processes often are accessible to experiment, and measurements must therefore be averaged before comparison with quantum predictions is possible. Postulating that probabilities are primitive notions then leaves the individual measurement results not yet averaged, unaccounted for.

But the discussion in the previous section opens the possibility of a more straightforward approach, since the notion of a quantum ensemble is natural to quantum theory. It was for instance used by von Neumann (1932) up to his well known discussion on hidden variables, and then dropped in favour of a single-system interpretation of the Hilbert-space vector. Had he fully developed the ensemble idea, he would have reached the conception first proposed by Slater (1929) and later developed by Einstein (1935, 1953) and others (Ballentine 1970; Ross-Bonney 1975); this is the ensemble (or, sometimes, statistical) interpretation, which explicitly denies that the wave function is the description of a single system and sees it, instead, as an ensemble property. From this single point there comes the resolution of most of the conceptual puzzles that have bedevilled quantum mechanics. Thus the position and momentum uncertainties Δq and Δp that enter into Heisenberg's relation, rather than limits on experimental precision, are the scatters of the values of position and momentum in the ensemble, as Popper (1980) observed half a century ago. And if we also maintain the distinction, not always clear in the Copenhagen view, between the preparation of a system and a measurement on it, we see that in many cases there is no such thing as a collapse of the wave function, while in others it reduces to the adoption of another ensemble in order to account for a change in experimental conditions. The projection postulate is thus superfluous.

The EPR problem (Einstein 1935) loses its paradoxical aspect. We cannot say of one particle in a dissociating pair that by itself it is described by one wave function and not another one; but if we have a large set of such particles, all at a certain position at a fixed time after separation, then they are a sample from the ensemble characterised by that position (or rather, by the corresponding eigenvalue of the position operator): If instead we choose those having a certain momentum, we have picked a different ensemble. Both are subensembles of the ensemble of all particles that the source emitted; they may have many common members; but they do not coincide. The subensembles formed by their partner particles going in the other direction will then also differ, and it is unsurprising if their wave functions are incompatible.

Note incidentally that one must transmit coincidence signals from one side to the other in order to realize the two subensembles experimentally; it is these signals that effect the particle selection according to the required criterion. Thus one need not suppose that a particle "knows" what wave function describes it; nor do particles mysteriously change their state when the experimenter does a different measurement on their partner particles (Paul 1985).

Note also that commonly more than one ensemble may be used to describe a given particle. Hence no measurements on a single particle suffice to define the wave function. This perfectly well known fact – well known in the laboratory, that is to say – is carefully ignored by Copenhagen theorists.

It is sometimes said that the ensemble interpretation cannot be made consistent with the ordinary formalism, in that there cannot exist a unique prescription for finding, given a quantum state, a non-negative distribution over phase space that correctly reproduces all expectation values. But in fact this requirement should not be stipulated, for two reasons:

(1) A unique prescription would mean also a unique correspondence rule for finding quantum operators that represent classical quantities (Shewell 1959); but no single rule can correctly reproduce the operator equivalents of all the classical quantities, and indeed, if it existed, quantum mechanics would be reducible by it to classical mechanics.

(2) The phase space to be considered is always taken to be the ordinary Liouville space of classical mechanics. That whatever phase space is used should go over to a Liouville space in the limit $\hbar \to 0$ is a reasonable condition; but that it should *coincide* with it has no such justification.

Taking into account these (and some other minor) modifications of the argument, Cohen's results (Cohen 1966, 1973; Cohen and Zaparovanny 1980) are seen to imply, as Suppes and Zanotti (1976) have argued on different grounds, that not only do the joint probabilities exist, but that quantum mechanics does not even provide sufficient information to specify them completely.

This is one aspect of a much more significant aspect of the ensemble interpretation: the EPR "paradox" has now become a perfectly good theorem, and it shows that quantum mechanics provides at best an incomplete description of physical reality. Indeed, any statistical theory is insofar incomplete as it does not furnish detailed trajectories of individual systems in their phase space but only averages over them. In classical statistical mechanics these trajectories are, at least in principle, described by Newtonian mechanics; no such underlying theory is as yet available for quantum mechanics. It would have to be a theory for time-dependent phenomena, and the probabilities in it would change with time; it would, in other words, have to be a stochastic theory.

A physically meaningful conception of probability thus leads to the search for a physical stochastic theory as a background to quantum mechanics; but it cannot well go beyond this point, and physical research has to take over. Before entering this field, we note that we have here an explanation of why hidden-variable patches plastered over the conceptual cracks offer little improvement compared to the Copenhagen view: not because von Neumann's theorem on hidden variable rules them out (see chapter 15) or because Bell's inequality forces them to be unacceptably non-local (Brody 1989b); simply, their use impedes the formulation of a proper physical picture. This is borne out by the fact that the simple acceptance of the stochastic character of quantum mechanics, as in Nelson's work (Nelson 1966, 1967) on a Markov-process approach, yields an explicit version of the ensemble view and so shows it to be entirely consistent, but offers little further physical insight. It merely reproduces all the already known results of quantum mechanics, where a theory with a physical basis would presumably disagree with some quantum predictions and in others go beyond it.

14.3 Stochastic Electrodynamics

Several theories that in one way or another approximate to this ideal have been proposed over the years. Of these, the only one with both a sound physical background and the capacity for further development has been stochastic electrodynamics; this we proceed to outline – in the briefest terms, since the literature concerning it is already too vast and the problem areas it has tackled too extensive for any detailed description here. Several review papers (Claverie and Diner 1976; de la Peña 1983; Brody 1983) are available to the interested reader.

The classical treatment of a charged particle deals with the closed system formed by it and any explicit charge distribution whose forces act on it. Stochastic electrodynamics (SED, for short) begins by the observation that in reality all charged particles move in a random background field due to the bremsstrahlung of all other charged particles in the universe; one cannot therefore use any closed-system theories; one has to have recourse, instead, to a stochastic process to describe the situation.

For a single particle relatively free from other matter, the random background is isotropic and homogeneous; its spectrum is Gaussian and characterized by

$$\langle E_i(r,t) \rangle = 0 = \langle B_i(r,t) \rangle , \tag{14.2}$$

$$\langle \tilde{E}_i(\omega)\tilde{E}_j(\omega') \rangle = \langle \tilde{B}_i(\omega)\tilde{B}_j(\omega') \rangle = \frac{4\hbar}{3\pi c^2}\omega^3 \delta(\omega - \omega')\delta_{ij'} . \tag{14.3}$$

Here $E_i(r,t)$, $B_i(r,t)$ are the field components, while $E_i(\omega)$, $B_i(\omega')$ are their Fourier transforms. When other matter is near, the distribution becomes more complex; in the simple case of a field describable by a non-zero temperature, the right-hand side of (14.3) is multiplied by the Planck factor, $\coth(\hbar\omega/kT)$, with T the temperature. This result is found by purely classical arguments, except for the assumption of the random background field; but it, as well as several other direct consequences of this assumption, coincides in form exactly with the quantum-electrodynamic results. But Eq. (14.2) and (14.3) occur in QED with a profoundly different significance, for the fluctuating field is taken to be virtual there. This coincidence in form and wide divergence in meaning appears repeatedly and implies the need for great caution in interpreting the results, be they discrepant or not from standard quantum mechanics. Note that Planck's constant h enters SED only through Eq. (14.3), as an amplitude parameter for the fluctuations of the random electromagnetic field.

The Braffort-Marshall equation (Braffort et al. 1954; Marshall 1963, 1965a) that describes the particle motion is a form of Langevin equation, namely

$$m\ddot{r}(t) = f(r,t) + m\tau\,\dddot{r}(t) + eE(t)\,, \qquad \tau = \frac{2e^2}{3mc^3}\,. \qquad (14.4)$$

This is an approximate equation: the "classical" force f and the random force eE are, a little artificially, separated, the particle's radiation reaction is described by the Abraham-Lorentz approximation, the space dependence of E is neglected, as are all relativistic effects and the entire magnetic field component. Several of these, in part fairly crude, approximations could be eliminated (though not the relativistic one, since we do not yet have any relativistic theory of stochastic processes); but the improved equations of motion are mathematically quite intractable. Even Eq. (14.4) presents the difficulty that it describes a significantly non-Markovian process, a type of process little is known about. So far only methods for dealing with these through Markovian approximations have been developed.

Despite these difficulties, Eq. (14.4) and the concept of the random electromagnetic field have scored a number of successes. Among these we may include (de la Peña 1983 and Brody 1983):

(1) The derivation of the Planck distribution. As no quantum hypothesis is needed for this, the photon concept has to be reconsidered.

(2) Modifications of the random background due to nearby matter give a quantitative explanation of the Casimir effect and of van der Waals forces.

(3) For the harmonic oscillator, explicit calculation yields the level spectrum, level widths and even transition probabilites; calculated to first order in τ, the quantum predictions are recovered exactly.

(4) Higher-order terms in τ provide values for the Lamb shift and the anomalous magnetic moment of electrons which reproduce the non-relativistic parts of these effect.

(5) For the square-well potential, one improvement on Eq. (14.4), concerning the Abraham-Lorentz term, gives solutions entirely compatible with the usual quantum ones.

For a number of effects it has not yet proved possible to carry out explicit calculations, mostly because of mathematical difficulties; but at the handwaving level adequate pictures emerge, covering the electron spin, some aspects of quantum statistics, and other points.

The handwaving level of SED also offers fruitful insight into puzzles such as the double slit. If we have probabilities $\Pr(x|A)$ and $\Pr(x|B)$ of hitting point x on the screen when only slit A or slit B is open, then we should have a probability of $\Pr(x|A) + \Pr(x|B)$ when both slits are open, because classically the probabilities of mutually exclusive events are additive; yet this sum reveals none of the quantum-mechanical interference effects. So goes the traditional argument. It contains, however, the tacit assumption that the probability $\Pr(x|A; B)$ of the particle going through A when B is also open is equal to $\Pr(x|A; \sim B)$, the probability of its going through A with B closed. SED shows this to be invalid: near a spatially periodic structure the random background field is modified, and for each frequency the radiation is enhanced in the direction of the Bragg angles. Since in SED a free electron responds mainly to frequencies near the de Broglie frequency of the random electromagnetic field, it is likely to suffer just those angular deviations that tend to form the interferences one expects. Thus it is not the particle that "knows" that the other slit is open; it is the random background, for which such "knowledge" is not surprising.

14.4 Problems and Solutions

If SED is a highly successful theory in the light of these achievements, it nevertheless has certain problems which have held up progress until quite recently. The two main difficulties are the lack of a suitable mathematical formalism and the question of the hydrogen atom's stability and level structure.

Quantum mechanics possesses an array of alternative formalisms, among which one can choose the most suitable one for a given problem; these formalisms are almost always equivalent. SED, on the other hand, has no general method, and for each of the cases mentioned above a specific way for solving it had to be developed. To a large extent this is due to the unexplored condition of the physics of open systems, and to our lack of knowledge about stochastic processes. The impressive body of mathematical theory does not tell us much about non-Markovian processes – yet Eq. (14.4) is of this kind; nothing is known about metastable states of stochastic processes – yet we need them for the problem of the hydrogen atom, as we shall now see.

On searching for those time-independent distributions that are Markov approximations to the solution of Eq. (14.4), one finds (Claverie and Soto 1982) that the ensemble average of the electron's binding energy is 0; the equilibrium state of the hydrogen atom is the auto-ionized one. This result, entirely at variance with both quantum theory and experiment, is nevertheless not unreasonable. In any potential well of bounded height, particles subjected to random fluctuations with a spectrum given by Eq. (14.3) will have a very small but non-zero probability of escaping. The auto-ionized state is then the only "equilibrium" state of the H atom, and the well known quantum states cannot but be metastable states which our present methods are unable to describe in stochastic processes.

This at first sight surprising conception becomes more acceptable if two points are considered. Firstly, a very crude estimate yields for the half life of the ground state in the H atom a value of $\sim 10^4$ years; this greatly exceeds anything at present observable, since it will be masked by thermal dissociation at any temperature greater than a very few degrees absolute. Secondly, it has proved possible to show that *if* metastable states exits, they will correspond to the energy levels of the H atom (de la Peña and Cetto 1984) in the following sense, the quasi-stable solution of Eq. (14.4) will be well approximated by a harmonic oscillator of frequency γ, say, which characterizes that level. Equation (14.4) then becomes

$$m\ddot{r} + m\gamma^2 r = m\tau\,\dddot{r} + eE , \qquad (14.5)$$

where the term $m\gamma^2 r$ can appear instead of $-f$, provided the two cancel on the average, in the sense that

$$\langle (f + m\gamma^2 r)r \rangle = 0 = \langle (f + m\gamma^2 r)\dot{r} \rangle . \qquad (14.6)$$

Now the expectation value of the electron's kinetic energy, K, in the harmonic oscillator is (de la Peña and Cetto 1979)

$$\langle K \rangle = \tfrac{1}{4}\hbar\omega \qquad (14.7)$$

per degree of freedom. But another relation between K and γ comes from Eq. (14.6); for the H atom; the force in a circular orbit of radius a is $f(r) = -(e^2/a^3)r$, so that $\gamma^2 = (e^2/ma^3)$, and therefore

$$\langle K \rangle = \frac{e^2}{2a} = \tfrac{1}{2}(me\gamma^2)^{1/2} . \qquad (14.8)$$

With two degrees of freedom, Eqs. (14.7) and (14.8) can both be satisfied provided $\gamma = me^4/h^2$, so that the orbital radius and the energy become

$$a = \frac{h^2}{me^2} , \quad \varepsilon = -\langle K \rangle = -\frac{me^4}{2h^2} . \qquad (14.9)$$

More elaborate calculations have been done; but this simple case exhibits the essential point that the notion of seeing the quantum levels of the H atom as metastable states of a stochastic process yields two conditions: Eq. (14.5) expresses the connection with the random background and is general, while Eq. (14.6) contains the specific dynamics. If the higher harmonics of γ are taken into account, we obtain the excited states. A number of further approximate results have already been obtained in similar ways, sufficient to show that it is indeed the metastable states of the stochastic motion that must be looked for, if we are to take the next major step ahead in SED.

There are numerous problems that still require further research. But so far from presenting conceptual questions, these are open questions of the kind that one may expect during the developmental phase of a theory.

One of these problems has further implications, namely that neutral particles should not, at first sight, possess quantum-like properties in SED. However, massive neutral particles appear to be composite; but their components are charged, and the random background fields acts on them. This action goes as e^2 and so does not depend on the sign of the charge. Massless neutral particles, on the other hand, are relativistic and cannot yet be treated by SED. More significant than this point is the suggestion made by Santos (1975, 1979) of a general stochastic theory. If the random electromagnetic field confers quantum behaviour on charged particles, so will the random backgrounds of other fields. There need nevertheless not be an equal number of different and possibly incompatible quantum theories, if we accept the notion that the abundance of particles which interact with two or more fields suffices to establish stochastic equilibrium between the random fluctuations of all fields; in particular, their amplitudes must be such that they carry the same average energy, so that Eq. (14.3) shows that $\hbar$ characterizes not just the electromagnetic fluctuations but is genuinely a universal constant. If this is the case, neutral particles acquire quantum properties from another field and so present no problem. The suggestion removes an important source of conflict between SED and theories based on other fields, such as gravitational stochastic ones. There remains, however, a methodological difference: SED is possible in a non-relativistic framework, and in fact there is much evidence for the reality of the random electromagnetic field; stochastic gravitation (or, as has also been proposed, a stochastic space-time structure) is necessarily relativistic and require, in view of our present ignorance of relativistic stochastic processes, numerous *ad hoc* assumptions.

References

Ballentine, L.E. (1970): Revs. Mod. Phys. **42**, 358

Braffort, P.B., Spighel, M., Tzara, C. (1954): C. R. Acad. Sci. (Paris) **239**, 157, 925

Braffort, P.B., Tzara, C. (1954): C. R. Acad. Sci. (Paris) **239**, 1779

Brody, T.A. (1975): Rev. Mex. Fís. **24**, 25

Brody, T.A. (1983): Rev. Mex. Fís. **29**, 461

Brody, T.A. (1989a): Rev. Mex. Fís. **35**, S80 (Chap. 15)

Brody, T.A. (1989b): Rev. Mex. Fís. **35**, S71 (Chap. 17)

Claverie, P., Diner, S. (1976): in O. Chalvet (ed.) *Localization and Delocalization in Quantum Chemistry* (Reidel, Dordrecht) pp. 395, 449

Claverie, P. Soto, F. (1982): J. Math. Phys. **23**, 753

Cohen, L. (1966): J. Math. Phys. **7**, 781

Cohen, L. (1973) in C. Hooker (ed.) *Contemporary Research in the Foundations and Philosophy of Quantum Mechanics* (Reidel, Dordrecht) p. 66; Cohen, L., Zaparovanny, Y.I. (1980): J. Math. Phys. **21**, 794

Einstein, A. (1949): in P.A. Schilpp (ed.) *Albert Einstein Philosopher-Scientist* (Library of Living Philosophers, Evanston, Ill.) p. 665

Einstein, A. (1953) in *Scientific Papers Presented to Max Born* (Oliver and Boyd, Edinburgh) p. 33

Einstein, A., Podolsky B., Rosen N. (1935): Phys. Rev. **47**, 777

Hopf, E. (1937): *Ergodentheorie* (Julius Springer, Berlin)

Marshall, T.W. (1963): Proc. Roy. Soc. **A276**, 475
Marshall, T.W. (1965a): Proc. Camb. Phil. Soc. **61**, 537
Marshall, T.W. (1965b): Nuovo Cimento **38**, 206
Nelson, E. (1966): Phys. Rev. **150**, 1076
Nelson, E. (1967): *Dynamical Theories of Brownian Motion* (Princeton University Press, Princeton, N.J.)
Neumann, J. von (1932): *Mathematische Grundlagen der Quantenmechanik* (Julius Springer, Berlin)
Paul, H. (1985): Amer. J. Phys. **53**, 318
Peña, L. de la (1983): in B. Gómez et. al. (eds.) *Stochastic Processes Applied to Physics and Other Related Fields* (World Scientific, Singapore) p 428
Peña, L. de la, Cetto, A.M. (1979): J. Math. Phys. **20**,
Peña, L. de la, Cetto, A.M. (1986): Nuovo Cim. **B92**, 189
Popper, K.R. (1980): *The Logic of Scientific Discovery* (Hutchinson, London)
Ross-Bonney, A.A. (1975): Nuovo Cimento **30B**, 55
Santos, E. (1975): preprint GIFT
Santos, E. (1979) in *Proceedings, Einstein Centennial Symposium* (Bogotá)
Shewell, J.R. (1959): Amer. J. Phys. **27**, 16
Slater, J.C. (1929): J. Franklin Inst. **207**, 449
Suppes, P., Zanotti, M. (1976): in P. Suppes (ed.) *Logic and Probability in Quantum Mechanics* (Reidel, Dordrecht), p. 715

15. Are Hidden Variables Possible? *

Abstract. The impossibility proof given by von Neumann for hidden variables to complete quantum mechanics is a direct consequence of the density-matrix formalism; since this stands or falls with quantum theory itself, "deterministic" or dispersionless hidden variables, but only these, must be ruled out. This is shown to be valid for all statistical theories in physics. That there exist theories with acceptable hidden variable structures then shows that dispersive hidden variables cannot be ruled out either classically or quantum-mechanically.

(i)

As soon as the essentially statistical nature of quantum mechanics was recognized, the apparently inevitable dispersions predicted by the theory and observed in experiment were thought by many authors to be explainable in terms of further dynamical variables, commonly called "hidden" in view of their unknown nature. By attributing different values to these variables for different systems in an experimental series where the wave function was the same for all systems, it was hoped that one could account for the fact that the choice of one among several possible eigenvalues upon measurement appeared to be random. The discussion was stilled for a time by von Neumann's proof (von Neumann, 1932) that such hidden variables could not consistently be introduced since "the present system of quantum mechanics would have to be objectively false, in order that another description of the elementary processes other than the statistical one be possible". We propose to show here, in part from von Neumann's own argument, that this conclusion is unduly pessimistic.

The problem was reopened by Bohm's successful construction (Bohm, 1952) of a model explicitly containing hidden variables that nevertheless exactly reduced to quantum mechanics. The model proved not to be a good physical theory in the sense either of inherent plausibility or of experimental verification; all the same it proved important simply because for the first time an unobjectionable counterexample to von Neumann's argument could be exhibited. This is all to the good: quantum mechanics is at the same time so successful and so much plagued by unresolved conceptual problems that a fundamental question like that of hidden variables should not simply be shelved. Yet the von Neumann proof led many authors (including von Neumann himself) to interpret the statistical nature of quantum mechanics as irreducible and in conflict with the causal conception underlying classical physics. The

---

* First published in Rev. Mex. Fís. **35** Supl. (1989) S80

unacceptability of this conclusion has led other to doubt the validity of von Neumann's proof; thus de Broglie (1953), Bell (1966) and, more recently, Bitsakis.

Now von Neumann's proof did contain certain weaknesses; but we shall argue here that rejecting it therefore is not possible without also rejecting quantum mechanics itself. However, at least some of the weaknesses in the proof have since been overcome (see Bell (1966) for a review and all references). We show here, furthermore, that the quantum case generalizes to all classical statistical theories. But such theories not only exist but possess well understood "hidden" variables. This apparent contradiction is resolved by correctly interpreting what is proved, in the sense that any hidden variables must themselves be statistical in nature and possess non-negligible dispersions, They may nevertheless be given fixed values, but then an entirely different theory results. This interpretation applies to von Neumann's argument; so far from constituting an impossibility proof, it should therefore be seen as giving a strong hint about the structure of hidden-variable theories to be developed for quantum mechanics.

(ii)

Von Neumann proves that one cannot add hidden variables so as to render quantum mechanics "deterministic" as a corollary to one of the main theorems in his book (von Neumann 1932). The theorem establishes the existence and principal property of density matrices. We quote:

Theorem DM: *To every quantum-mechanical ensemble there corresponds a linear semi-definite Hermitan matrix $\hat{\rho} = [\rho_{mn}]$, such that the expectation value of any operator $\hat{R}$ corresponding to a physical quantity is:*

$$\langle \hat{R} \rangle = \sum_{mn} \rho_{mn} R_{mn} = \mathrm{tr} \rho R \,.$$

(We note that "observable" is generally used now for von Neumann's "physical quantity".) From *Theorem DM* the required corollary is derived in a mathematically unexceptionable way. It states:

Corollary HV: *No quantum-mechanical ensemble described by a density matrix satisfying the requirements of Theorem DM is dispersion-free.*

Here an ensemble is termed dispersion-free if for all operators $\hat{R}$ we have that $\langle \hat{R}^2 \rangle = \langle \hat{R} \rangle^2$. On the other hand, pure or homogeneous ensembles of course exist, which cannot be split into subensembles of different expectation values; these correspond to vectors in the underlying Hilbert space. The details of the proof, with useful comments, are given by Albertson (1961); the theory of density matrices will be found e.g. in Fano (1957).

Now the proof von Neumann gives of *Theorem DM*, though mathematically correct, rests on postulates which may with good reason be doubted.

Thus the postulate that if $\hat{A}$ and $\hat{B}$ are the quantum operators that correspond to the classical observables a and b, the operator corresponding to $a+b$ will be $\hat{A}+\hat{B}$, even if $\hat{A}$ and $\hat{B}$ do not commute. As Bell discusses (Bell, 1966), this cannot always hold. For this and related reasons, the von Neumann proof is not really acceptable.

But one cannot conclude that the theorem, and hence its *Corollary HV*, is not valid, and this for two reasons:

(1) Gleason (1957) has given an alternative proof, at least for Hilbert spaces of dimensionality greater than 2, that escapes some of the criticisms made of von Neumann's proof, while Kochen and Specker (1967) provide yet a third proof which, while also not unobjectionable, resists yet other criticisms;

(2) It should have been obvious that the place occupied by *Theorem DM* in von Neumann's foundational structure renders it indispensable. Von Neumann (1932) offers two alternative postulational schemes for quantum mechanics: one, the Hilbert-space formalism, in which each vector in a finite or infinite-dimensional Hilbert space corresponds to a quantum state, i.e. a normalisable solution of the appropriate wave equation: and two, the density-matrix formalism based on *Theorem DM*. He shows, moreover, that these two formalisms are intimately related. By adding the plausible statistical postulate that for a mixture of states no interference terms between elements of the mixture should arise in any expectation values, one obtains from the Hilbert-space formalism a formalism exactly equivalent to the density-matrix one; and the subclass of homogeneous density matrices, i.e. those for which $\hat{\rho}^2 = \hat{\rho}$, turns out to be equivalent to the set of normalized vectors in the corresponding Hilbert space. If $\mathcal{D}$ represents the class of all density matrices, $\mathcal{I}$ the subclass of idempotent ones, r the ray space of normalized wave functions, and $\mathcal{SP}$ the statistical postulate, one may write, symbolically,

$$r \Leftrightarrow \mathcal{I} \subset \mathcal{D},$$

$$\mathcal{D} = r + \mathcal{SP}. \tag{15.1}$$

From *Theorem DM* the impossibility of non-dispersive ensembles follows at once. If this also makes hidden variables impossible, as von Neumann contends then indeed *Theorem DM* must be wrong before hidden variables can be accepted. But if we abandon *Theorem DM* as the basis for the density-matrix formalism for quantum mechanics, then the two relations (15.1) imply that the Hilbert-space formalism must also be abandoned. The introduction of any hidden variables is then nugatory.

The dilemma is inescapable so long as one forgets the fact that the impossibility of non-dispersive ensembles (which we may now accept as a well-established consequence of the quantum formalism) only rules out non-dispersive hidden variables. Indeed, von Neumann himself states clearly and repeatedly what is often overlooked, namely that the hidden variables he is concerned to prove inadmissible are the non-dispersive ones whose incorporation would eliminate the statistical element from quantum mechanics and reduce it to a classical theory of the type where the theoretical predictions correspond (to within experimental error limits) to the individual measured values. But this aim is meaningful only because, as Bitsakis points out, von Neumann confuses causality with determinism and hence does not even consider dispersive hidden variables; for if hidden variables are themselves of statistical character, they will not eliminate the statistical elements from quantum theory nor render it deterministic as for instance few-particle mechanics is; but they *will* make patent that quantum mechanics and causal description are compatible. But it is this that hidden-variable theorists wish to achieve, and not (as is often alleged) a return to classical theory. The statistical nature of quantum mechanics cannot be explained away without making it incompatible with experimental facts; but it can be explained.

We proceed to spell this out in further detail.

(iii)

That von Neumann's result is more general may be seen by deriving it from probability theory.

A physical theory must possess dynamical variables which satisfy equations of motion (when the independent variable is time) or other similar equations. Here we need only consider a single variable, say x. If the theory is statistical in nature, x will possess a cumulative probability-distribution function $P(x)$, and the equations of motion will describe the time evolution of $P(x)$. If now a further, "hidden", variable y is to be added to the theory without changing its earlier results, then firstly a joint distribution function $Q(x,y)$ must exist, and secondly its marginal must be $P(x)$:

$$P(x) = Q(x, \infty) . \tag{15.2}$$

The distribution function of y is given by

$$R(y) = Q(\infty, y) . \tag{15.3}$$

If y is to be "deterministic", it will take fixed value η, while $R(y)$ must be the Heaviside step function:

$$R(y) = H(y - \eta) , \qquad dR(y) = \delta(y - \eta)dy . \tag{15.4}$$

From the definition of a conditional probability we have

$$Q(x, y) = \int_{-\infty}^{y} P_c(x|y')dR(y') , \tag{15.5}$$

where $P_c(x|y)$ is the probability that the dynamical variable take the value x or less when the added variable has the value y. Then

$$Q(x, y) = \begin{cases} 0 & y < \eta \\ P_c(x|\eta) & y > \eta , \end{cases} \tag{15.6}$$

and the value at $y = \eta$ may be taken either as 0 or as $P_c(x|\eta)$. Using Eq. (15.2) yields

$$P(x) = P_c(x|\eta) , \tag{15.7}$$

and combining (15.6) with (15.7) one has

$$Q(x,y) = \begin{cases} 0 & y < \eta \\ P(x) & y > \eta . \end{cases} \tag{15.8}$$

With (15.4) this gives

$$Q(x,y) = P(x)R(y) . \tag{15.9}$$

Equation (15.9) establishes that a deterministic added variable, in the sense stated above, is statistically independent of the original variable x. This is compatible with its being linked to x through an equation of motion or its equivalent, and it cannot therefore be integrated into the structure of the theory for x. Thus Eq. (15.9) is the generalization to any statistical theory of von Neumann's proof that deterministic hidden variables cannot be introduced into the structure of quantum mechanics.

Three comments should avoid some misunderstanding.

i) If the distribution of x is also a Heaviside function, Eq. (15.9) is valid but irrelevant. For in this case the theory that contains x is no longer a statistical theory.

ii) The arguments leading to Eq. (15.9) do not affect any parameters in the theory or in any model built on it; whether already in the original theory or introduced along with y, their distribution is always the Heaviside step function.

iii) The argument has been stated in terms of a single fixed value for y but can easily be extended to the case of y non-zero in an interval. Instead of Eq. (15.4) one has

$$R(y) = \begin{cases} 0 & y < \eta \\ 1 & y > \eta' , \end{cases} \tag{15.4'}$$

while $R(y)$ increases monotonically in the interval (η, η'). Equation (15.8) now becomes

$$Q(x,y) = \begin{cases} 0 & y < \eta \\ P(x) & y > \eta' . \end{cases} \tag{15.8'}$$

In the interval (η, η') some different expression holds. Equation (15.9) is to be replaced by

$$Q(x,y) = P(x)R(y) \quad \forall y \notin (\eta, \eta') . \tag{15.9'}$$

If the interval (η, η') is small, say of the order of the experimental error limits on y, the incompatability of y with the existing theoretical structure is established to within that error limit. But if all physically relevant values of y fall inside (η, η'), then y is not dispersionless. Equation (15.9') does not hold in the region of interest, and the present argument does not impede the introduction of y into the theory. Intermediate cases must be examined on their merits.

(iv)

The conclusions drawn from Eq. (15.9) or (15.9') clearly hold for a classical theory such as statistical mechanics. This is in apparent contradiction with the fact that here the hidden variables exist and are described by classical Newtonian mechanics and would generally be called deterministic. The difficulty is due to a conceptual confusion fostered by the unclear way in which the foundations of statistical mechanics commonly are presented in physics textbooks.

A description of an N-particle system such as a volume of gas is perfectly possible, at least in principle, within the framework of classical mechanics. The problem of finding the initial positions and velocities (which is often, but with little justification, cited as the reason for the statistical approach) can in many cases be solved, for instance in the case of medium-sized stellar clusters or in that of a computational model. It would depend entirely on the initial conditions and therefore differ greatly for two identical volumes of gas at the same temperature and pressure, or for the same volume in thermal equilibrium at different times. What is sought, even in computer models, is a description of such many-body systems that is independent of the initial conditions; this is not possible in classical mechanics but is achieved by averaging over all possible initial conditions (or a statistically adequate subset of them). It is not the size of N but this need to eliminate the initial conditions that justifies the approach of statistical mechanics; indeed, statistical mechanics provides a powerful tool for dealing with systems of small N , well within our present powers, if not of analytical solution then of a computational one. But in the ensemble of possible trajectories over phase space the hidden variables do have the dispersions needed according to the conclusions of the preceding section.

The large-N justification is not always wrong; but it springs from a different consideration, that of the relative size of the fluctuations about the ensemble averages. When these fluctuations are small compared to the relevant observational errors, the ensemble theory offers much greater physical insight as well as an easier approach to applications. When they are much larger, the classical theory is often preferable. Now in many cases the relative size of the fluctuations is of the order of $N^{-1/2}$; hence a seeming justification of the statistical approach when N is large.

The case of statistical mechanics also offers a good illustration of the further point that while the hidden variables for a statistical theory must themselves be statistical, they can also be "deterministic" in the sense of having zero dispersions; but the theory based on them has then a wholly different character, as incompatible in conception with the statistical theory as in fact classical and statistical mechanics are. The ensemble concept and the process of averaging over the ensemble establish the passage from the classical to the statistical theory; the inverse passage is not possible.

Ignoring the distinction between the two theory types leads to such errors as stipulating that the classical hidden-variable theory should reproduce the algebraic structure of the statistical theory. Von Neumann's additivity postulate is of this kind, as Bell (1966) shows by means of an illuminating example. A similar problem arises in the impossibility proof of Kochen and Specker (1967): they consider the function f_A that, for a given operator $\hat{A}$, leads from the hidden-variable space Ω to the predictions

$$\langle \hat{A} \rangle = \int_\Omega f_A(\omega) d\mu_\psi(\omega) .$$

They then require that

$$f_{g(A)} = g(f_A) .$$

The authors' justification (Kochen and Specker 1967) is that "In any theory, one way of measuring A^2 consists in measuring A and squaring the resulting value. In fact, this may be used as the *definition* of a function of an observable." This confuses the two levels. The theory predicts $\langle \hat{A} \rangle$, whose square is $\langle \hat{A} \rangle^2$. Experimentally, one finds individual observations, which are not predicted by the theory (only their average is), one squares them and then averages. If there is dispersion the two results differ.

(v)

The classical impossibility proof for hidden variables, consequent upon Eqs. (15.9) and (15.9′), may then be interpreted in the following way:

If hidden variables are to be introduced in a statistical theory in order to obtain a more detailed description, they must themselves be dispersive. The same variable may also figure in non-statistical form in another theory, which however will be of different type.

$$(15.10)$$

Formulating the conclusion in this way underlines its positive value instead of exhibiting only what it forbids.

Conclusion (15.10) is straightforwardly applicable to quantum mechanics. As was seen in Sect. 2, the von Neumann proof leads to the same conclusion for quantum mechanics; it may therefore be seen as a special case of (15.10).

Since quantum mechanics is dynamical theory, (15.10) strongly suggest that any hidden-variable formulation must be a stochastic one in order to be successful. That this hint is in the right direction is confirmed by the fact that not only Bohm's (Bohm, 1952) original hidden-variable theory but all later ones (Bohm and Bub, 1966 or Wiener and Siegel, 1953, 1955; Siegel and Wiener, 1956 are good examples) are of statistical type and involve stochastic variables. The same is true of Nelson's (Nelson 1966) much more complete formulation. Whatever the value as physical theories of these extensions of the quantum formalism, they exhibit the capability of hidden-variable theories to provide new interpretative elements of much significance.

One cannot, however, conclude that adding hidden variables to the quantum formalism is the way out of the conceptual confusion that surrounds quantum mechanics. As we have seen dispersive variables could indeed be added; but they are undesirable, because rather than patching up an existing theory one should resolve the difficulties by erecting a conceptually more coherent theory on a new and physically more satisfactory basis. It may have been some such consideration that kept Einstein, otherwise a vigorous critic of the deficiencies of the standard view, from giving his approval to hidden-variable theories. And it is certainly new theorization that has in recent years led to the development of the idea that the amplitudes of the random background in the electromagnetic field could play the role of hidden variables. The quantum formalism is only a very good approximation to the exact ensemble averages of the motions of charged particles in such random fields, and so a theory on this basis is not simply a hidden-variable extension of quantum mechanics. The hidden variables may also take non-fluctuating values: we then recover classical electromagnetic theory, in which quantum behaviour appears. For a recent review and references see Brody (1983).

Acknowledgements. I would like to thank L. de la Peña for several stimulating discussions, and A.K. Theophilou, Head of the Department of Theoretical Physics at the "Democritos", both for extended hospitality and for many helpful comments.

References

Albertson, J. (1961): Amer. J. Phys. **29**, 478
Bell, J.S. (1966): Rev. Mod. Phys. **38**, 441
Bohm, D. (1952): Phys. Rev. **85** 166, 180
Bohm, D., Bub, J. (1966): Rev. Mod. Phys. **38**, 453
Brody, T.A. (1983): Rev. Mex. Fís. **29**, 461
Broglie, L. de (1953): *La physique quantique restera-t-elle indéterministe?* (Gauthier-Villars, Paris)
Fano, U. (1957): Rev. Mod. Phys. **29**, 74
Gleason, A.M. (1957): J. Math. Mech. **6**, 885
Kochen, S., Specker, E.P. (1967): J. Math. Mech. **17**, 59
Nelson, E. (1966): Phys. Rev. **150**, 1079
Nelson, E. (1967): *Dynamical Theories of Brownian Motion* (Princeton University Press, Princeton, N.J.)
Neumann, J. von (1932): *Mathematische Grundlagen der Quantenmechanik* (Springer Verlag Berlin [Engl. tr. Princeton University Press, Princeton, N.J. 1955]
Siegel, A., Wiener, N. (1956): Phys. Rev. **101**, 429
Wiener, N., Siegel, A. (1953): Phys. Rev. **91**, 1951
Wiener, N., Siegel A. (1955): Nuovo Cim. Suppl. **2**, 304

16. The Bell Inequality I: Joint Measurability[*]

Abstract. An examination of the derivations of the Bell inequality shows that neither hidden-variable nor locality presupposes that two spin projections along different directions for the same particle are jointly measurable, or equivalently, that they have a joint probability distribution. The failure of joint measurability in the quantum case and in the non-quantum ones that are presented is traced to an unsatisfactory choice of system made for describing the physical situation. It is then not possible to draw any philosophical conclusions from the Bell inequality; in particular, Einsteinian realism is not affected by it.

16.1 Introduction

The debate initiated by Einstein, Podolsky and Rosen (1935) and Bohr (1935) has in the last few years centred largely on the inequality derived by Bell (1965). His derivation made two explicit assumptions, that a hidden-variable explanation be feasible, and that it be local; that quantum mechanics does not satisfy the inequality is then taken to mean that, in Rohrlich's words, "local hidden-variable theory is dead" (Rohrlich 1983). The presumption of non-locality has variously been used to cast doubt on Einstein's view of reality (Fry 1984; d'Espagnat 1984), or to justify the notion of superluminal information transport (and its implications) (Costa de Beauregard 1979), or even to bolster up parapsychological speculations (Walker 1975; Capra 1980). Since it is this side of the debate which has spilled over into the popular press, a most undesirable mispresentation has been created of what research in physics is about.

I propose to show here that such extrapolations are baseless, because the real assumptions underlying the Bell inequality are neither the hidden-variable nor the locality one, but the assumption that the spin projection of a particle can be measured in more than one direction without any mutual interference; this will below be called the joint-measurability assumption, JMA, either directly or in the equivalent form of the assumption that a joint probability distribution exists for two or more spin projections on the same particle. From the JMA there follows also the possibility of reordering the sets of experimental data so as to exhibit the fact that they satisfy the Bell inequality.

In Sect. 16.2 five derivations are presented: Bell's derivation (Bell 1965), a quantum-mechanical one, Santos' quantum-logical one (Santos 1985), and

* First published in: Revista Mexicana de Física **35** Suplemento (1989) 52–70.

two probability-theoretic derivations due to Wigner (1970) and Holt (1973), and Suppes and Zanotti (1981), respectively; it will be seen that only the first uses any hidden variables, and only the first two make any locality assumptions. Section 16.3, using another derivation (Eberhard 1977, 1978; Stapp 1977; Peres 1982) based on the treatment of experimental data, shows the need for the JMA if the data are to satisfy the inequality. Section 16.4 examines how failure of the JMA springs from an inadequate selection of the physical system; making the system selection explicit allows a resolution of the conceptual difficulty. None of the conclusions of these three sections are restricted to quantum mechanics; some classical systems that also violate the Bell inequality are therefore presented in Sect. 16.5. The last section sums up and has some general comments. Some of the weaknesses of the locality concept as used in discussions of the Bell inequality are discussed elsewhere (Brody, in this volume Chap. 17).

The upshot of the argument is that in the Einstein-Bohr debate the Bell inequality does not lead unequivocally to a victory for Bohr's side, as is commonly said, and even less so for any of its more recent distortions. To the contrary, reality remains obstinately Einsteinian.

16.2 The Derivation of the Bell Inequality

16.2.1 Bell's Hidden-Variable Derivation

Bell (1965) considered the set-up

$$D_A \longleftarrow F_A(\alpha) \longleftarrow S \longrightarrow F_B(\beta) \longrightarrow D_B \tag{16.1}$$

with the source S emitting pairs of spin $\frac{1}{2}$ particles of total spin 0; the particles A and B go through filters oriented at angles α and β in the plane perpendicular to their motion; these could be Stern-Gerlach magnets; the detector D_A and D_B register $+1$ for spin up and -1 for spin down. Actual experiments have mostly used cascade photons from atomic deexcitations (see Bruno, d'Agostino, Maroni 1977 or Clauser, Horne, Shimony, Holt 1969) for reviews); this differs from the spin case only in details that are inessential for the present purposes.

Denoting the individual results registered by D_A and D_B by a and b, respectively,

$$a = \pm 1, \qquad b = \pm 1 \tag{16.2}$$

an experimental correlation coefficient $r_{\alpha\beta}$ is obtained from n repetitions of a measurement pair. A theoretical prediction $\rho_{\alpha\beta}$ for this quantity is found as follows. Assume that there are hidden variables λ with a probability $\mu(\lambda)$; then the locality assumption implies that $a = a(\alpha, \lambda)$ and $b = b(\beta, \lambda)$. The rotational symmetry of the set-up (16.1) now gives that

$$\langle a \rangle = \int a(\alpha, \lambda) d\mu(\lambda) = \langle b \rangle = 0$$

$$\langle a^2 \rangle = \int a^2(\alpha, \lambda) d\mu(\lambda) = \langle b^2 \rangle = 1 \qquad (16.3)$$

and therefore we have

$$\rho_{\alpha\beta} = \int a(\alpha, \lambda) b(\beta, \lambda) d\mu(\lambda) . \qquad (16.4)$$

If the filters in (16.1) had been oriented along directions α' and β', two new measurements a' and b' (suppressing the arguments and transferring the primes for simplicity) are obtained; three further correlation coefficients $\rho_{\alpha'\beta}, \rho_{\alpha\beta'}, \rho_{\alpha'\beta'}$ may then be written, and one has

$$V_{\pm} \equiv \rho_{\alpha\beta} - \rho_{\alpha'\beta} \pm (\rho_{\alpha\beta'} + \rho_{\alpha'\beta'}) = \int [ab - a'b \pm (ab' + a'b')] d\mu . \qquad (16.5)$$

Now

$$ab - a'b \pm (ab' + a'b') = \pm 2 , \qquad (16.6)$$

since either $a - a' = 0$, $a + a' = \pm 2$ or $a - a' = \pm 2$, $a + a' = 0$. Hence

$$V = \max(|V_+|, |V_-|) = |\rho_{\alpha\beta} - \rho_{\alpha'\beta}| + |\rho_{\alpha\beta'} + \rho_{\alpha'\beta'}| \leq 2 . \qquad (16.7)$$

This is the Bell inequality in the form given by Clauser et al. (1969). Many other inequalities of this kind are known (Roy, Singh 1978; Garg, Mermin 1984), but since the discussion below applies with small changes to them, these variants will be ignored.

As is well known (Bell 1965; Bruno, d'Agostino, Maroni 1977; Clauser, Shimony 1978), Eq. (16.7) is not compatible with the prediction

$$\rho_{\alpha\beta} = \langle (\hat{\sigma}'_A \cdot \alpha)(\hat{\sigma}'_B \cdot \beta) \rangle = -\cos(\alpha - \beta) \qquad (16.8)$$

of quantum mechanics. In (16.8), α and β are unit vectors in the directions α and β, while $\hat{\sigma}'_A$ and $\hat{\sigma}'_B$ are the spin operators for particles A and B, the apostrophe indicating that they are normalized to give eigenvalues ± 1.

The experimental evidence on the whole favours the quantum-mechanical result (16.8), though there is more room for doubts than is often conceded (Santos 1984; Selleri 1984, Garuccio, Selleri 1984; Marshall 1983, 1984; Marshall, Santos, Selleri 1983).

This derivation of Eq. (16.7) assumes (1) the existence of hidden variables λ that account for the values a and b, (2) the locality assumption that a contains no β dependence, (3) that for a given λ we know all four values a, b, a', b' and (4) that μ depends neither on α nor on β. The four assumptions are not independent of each other, of course. The third is equivalent to the JMA; it is needed for the second member of (16.5).

16.2.2 A Quantum Derivation

Consider a system composed of two subsystems A and B and described by a density matrix $\hat{W}_{AB}$. Filter procedures, with operators $\hat{F}(i, \vartheta)$, $i = A, B$ can be carried out on the two subsystems, where ϑ is an angular parameter. To agree with Eq. (16.2) the $\hat{F}$ are taken to have eigenvalues ± 1, and to satisfy

$$\operatorname{tr} \hat{F}(i, \vartheta) \hat{W}_{AB} = 0$$
$$\operatorname{tr} \hat{F}^2(i, \vartheta) \hat{W}_{AB} = 1$$

$\hat{F}(A, \vartheta)$ and $\hat{F}(B, \phi)$ commute for all ϑ, ϕ. Then

$$\rho_{\alpha\beta} = \operatorname{tr} \hat{F}(A, \alpha) \hat{F}(B, \beta) \hat{W}_{AB}$$

and thus

$$V = |\operatorname{tr} \hat{F}(A, \alpha) \hat{F}(B, \beta) \hat{W}_{AB} - \operatorname{tr} \hat{F}(A, \alpha') \hat{F}(B, \beta) \hat{W}_{AB}|$$
$$+ |\operatorname{tr} \hat{F}(A, \alpha) \hat{F}(B, \beta') \hat{W}_{AB} - \operatorname{tr} \hat{F}(A, \alpha') \hat{F}(B, \beta') \hat{W}_{AB}| .$$

It is tempting to carry out the trace over the B variables, to find ($\hat{W}_A = \operatorname{tr}_B \hat{W}_{AB}$)

$$V \le |\operatorname{tr}_A [\hat{F}(A, \alpha) - \hat{F}(A, \alpha')] \hat{W}_A| + |\operatorname{tr} [\hat{F}(A, \alpha) + \hat{F}(A, \alpha')] \hat{W}_A| \le 2 , \quad (16.9)$$

where use has been made of Eq. (16.6). But this is not valid since, in general,

$$\operatorname{tr} \hat{F}(B, \vartheta) \hat{W}_{AB} \neq \operatorname{tr} \hat{F}(B, \phi) \hat{W}_{AB} .$$

Now when the operators $\hat{F}(B, \vartheta)$ and $\hat{F}(A, \vartheta + \pi)$ act on states $\hat{W}_{AB}$ with total spin 0, they behave exactly alike (not, of course, acting on $\hat{F}(i, \phi) \hat{W}_{AB}$). Hence by writing V in terms of operators on A only, the trace over the B coordinates is immediate, and we find

$$V = |\operatorname{tr}_A \hat{F}(A, \alpha) \hat{F}(A, \beta + \pi) \hat{W}_A - \operatorname{tr}_A \hat{F}(A, \alpha') \hat{F}(A, \beta + \pi) \hat{W}_A|$$
$$+ |\operatorname{tr}_A \hat{F}(A, \alpha) \hat{F}(A, \beta' + \pi) \hat{W}_A + \operatorname{tr}_A \hat{F}(A, \alpha') \hat{F}(A, \beta' + \pi) \hat{W}_A| .$$

Therefore, provided $\hat{F}(A, \vartheta)$ and $\hat{F}(A, \phi)$ commute for $\vartheta \neq \pi$, Eq. (16.9) follows and the Bell inequality (16.7) holds; but if they do not commute, the inequality cannot be derived.

This formalism is local, for through $\operatorname{tr}_B \hat{F}(B, \vartheta) \hat{W}_{AB}$ depends on ϑ, $\operatorname{tr}_A \operatorname{tr}_B \hat{F}(B, \vartheta) \hat{W}_{AB}$ does not. This derivation assumes (1) locality, (2) the JMA. The latter takes the form of the commutation requirement for two operators on the same particle, since only then does the inequality follow. The quantum nature of the proof excludes hidden variables.

16.2.3 A Quantum-Logical Derivation

This derivation is due to Santos (1985). He defines a separation between two propositions x and y on the quantum lattice as

$$s(x, y) = p((x \cap y') \cup (x' \cap y)), \qquad (16.10)$$

where x' is the orthocomplement of x and $p(x)$ is the probability measure that defines the quantum state of the system. Now the quantum separation (16.10) satisfies the triangle inequality

$$s(x, y) + s(y, z) \geq s(x, z)$$

provided x, y, z all belong to one Boolean sublattice. Since this cannot be the case for spin measurements, two of which would have to be done on the same particle, Santos derives the inequality

$$s(x_1, y_1) + s(x_2, y_1) + s(x_2, y_2) \geq s(x_1, y_2) \qquad (16.11)$$

valid when all four propositions belong to one Boolean sublattice. Since this is not generally the case, Santos adds a "locality condition": We have no reason to identify the proposition x_1 when measured with y_1, written (x_1/y_1), with the proposition x_1 when measured with y_2; since these propositions concern incompatible spin components, a complete measurement has in fact 8 propositions rather than 4. He then defines as *local* any theory in which, for a fixed i, the propositions (x_i/y_k) are identified for all k whenever the region in which the y_k are measured is spatially separated from that of x_i. In such a theory, he concludes, (16.11) will hold, and from it the Bell inequality follows quite straightforwardly.

The quantum-logical derivation goes through, therefore in two cases. Either the four propositions belong to a Boolean sublattice and are therefore jointly measurable; this is the JMA. Or the theory is "local"; but if (x_1/y_1) and (x_1/y_2) can be identified, that is to say, if their truth values are identical, then x_1 can be measured with y_1 and also with y_2, and so y_1 and y_2 can be measured jointly. The second case reduces to the first. To make this clear, consider the $8n$ propositions $(x_i/y_k)_m$, $m = 1 \ldots n, i = 1, 2$; the truth value of $(x_i/y_k)_m$ is not in general that of $(x_i/y_k)_{m'}$. It is, however, possible (under not very restrictive conditions) to reorder one of the proposition sets so that they are equal in most cases. What is not possible is to do this simultaneously for all $8n$ propositions, unless the JMA holds. This will be shown in Sect. 16.3.

16.2.4 The Wigner-Holt Probability-Theoretic Derivation

We give this derivation, due to Wigner (1970), in a revised form of Holt's (1973). Although Wigner begins his discussion in terms of hidden variables, both his notation and the nature of his argument make them superfluous; we do not therefore introduce them. A probability space in which the four possible results a, a', b, b' form the four axes is considered. To the 2^4 points in this space probabilities $p_{aa'bb'}$ are assigned, which satisfy

$$\sum_{aa'bb'} p_{aa'bb'} = 1 .$$

If we assume that they also satisfy Eq. (16.3), the correlation coefficients are

$$\rho_{\alpha\beta} = \sum_{aa'bb'} abp_{aa'bb'} \tag{16.12}$$

and so on. One has immediately that

$$\rho_{\alpha\beta} - \rho_{\alpha'\beta} \pm (\rho_{\alpha\beta'} + \rho_{\alpha'\beta'}) = \sum_{aa'bb'} [ab - a'b \pm (ab' + a'b')]p_{aa'bb'} .$$

Equation (16.6) then yields the Bell inequality in the form of Eq. (16.7).

Locality, in the sense that the value found for a in one measurement does not depend on whether b or b' is determined with it, cannot be expressed here since only probabilities are given. Not even weak locality, in the sense that the probability of α is independent of what is measured with it, can be stated since the probabilities depend on all four variables. Only related conditions, such as

$$\Pr(a|b = 1) = \Pr(a|b = -1)$$

or

$$\Pr(a|b = 1) = \Pr(a|b' = 1)$$

may be stipulated, in the form

$$\sum_{a'b'} p_{aa'1b'} = \sum_{a'b'} p_{aa'-1b'}$$

or

$$\sum_{a'b'} p_{aa'1b'} = \sum_{a'b} p_{aa'b1} .$$

But the Bell inequality is derived whether such conditions hold or not. Thus no locality condition is involved in this derivation. On the other hand, the existence of a joint probability distribution is expressly required; we show in Sect. 16.4.1 below that this is equivalent to the JMA.

16.2.5 A Pure Probability-Theoretic Derivation

Suppes and Zanotti (1981) have given what seems to be the most general derivation of the Bell inequality, based only on probability theory. They establish the following theorem:

Let x, y, z be three random variables taking ± 1 as values and having 0 expectations; their correlation coefficients $\rho(x, y), \rho(x, z), \rho(y, z)$ are given. Then a necessary and sufficient condition for a joint probability distribution of the three variables to exist is that the inequalities

$$-1 \leq \rho(x, y) + \rho(x, z) + \rho(y, z) \leq 1 + 2 \min[\rho(x, y), \rho(x, z), \rho(y, z)] \tag{16.13}$$

be satisfied. The proof is based on the fact that of the eight discrete values forming the joint distribution, seven are determined by the known means and correlations and by their sum being 1. Equation (16.13) is then the condition that the eighth probability lie between 0 and 1.

This theorem is now applied to the triplets $(a, a'b), (a, a', b'), (a', b, b')$ to give, e.g.

$$-1 \leq \rho_{\alpha\alpha'} + \rho_{\alpha\beta'} + \rho_{\alpha'\beta'}$$
$$-1 \leq -\rho_{\alpha\alpha'} - \rho_{\alpha'\beta} + \rho_{\alpha\beta}$$

which sum to

$$-2 \leq \rho_{\alpha\beta} - \rho_{\alpha'\beta} + \rho_{\alpha\beta'} + \rho_{\alpha'\beta'} \; .$$

Combining this with similar relations yields the Bell inequality (16.7).

In this derivation no hidden variables occur, and a locality assumption would be meaningless. The only assumption required is the JMA, which here takes the form of the assumption that the four three-dimensional distributions exist (equivalent to the existence of the four-dimensional distribution, since they are discrete-valued). We note that explicit use is made of the correlations $\rho_{\alpha\alpha'}$ and $\rho_{\beta\beta'}$ which are not well defined in quantum theory.

16.2.6 Summary

Only the first of these derivations effectively uses hidden variables, while in the second they are impossible. Only the first two make any locality assumption; Santos' locality condition will be seen below to have a different origin. But in all of them either the JMA is assumed to hold or a joint probability distribution is taken to exist. Since quantum mechanics, which does not satisfy the JMA, violates the Bell inequality, it is the JMA that appears as the key assumption in deriving the inequality.

The irrelevance of the hidden-variable hypothesis has previously been noted by Eberhard (1982) and Stapp (1982). The nature of the locality assumptions made in these derivations is further discussed in this volume, Chap. 17)

16.3 The Reordering Problem

Another derivation of the Bell inequality is due to Eberhard (1977, 1978) and Stapp (1977); it was given a simpler form by Peres (1982). In the derivations of Sect. 16.2, the values of a, a', b, b' are seen as "possessed" by the particles so that they can figure in the same equations, even though they are not jointly measurable. To sidestep the resulting conceptual problems (to which we return below), here the experimental observations, a_i and b_i, $i = 1, \ldots n$, are considered. They yield an experimental correlation coefficient

$$r_{\alpha\beta} = \frac{1}{n} \sum_{i=1}^{n} a_i b_i \; , \tag{16.14}$$

in terms of which we have

$$\rho_{\alpha\beta} \dot{=} r_{\alpha\beta} \, .$$

(Here and below, $\dot{=}$ is to be read "equal in the limit of n large enough", i.e. when any difference is of the order of the expected statistical fluctuations.)

If measurement b'_i along the direction β' had been made, a correlation coefficient

$$\rho_{\alpha\beta} \dot{=} \frac{1}{n} \sum_{i=1}^{n} a_i b'_i \tag{16.15}$$

would be obtained, and similarly for the other two. Here it is argued that locality implies that for a given i, the a_i in Eq. (16.14) is equal to that in Eq. (16.15), so that the four correlation coefficients may be combined as

$$\rho_{\alpha\beta} - \rho_{\alpha'\beta} \pm (\rho_{\alpha\beta'} + \rho_{\alpha'\beta'}) \dot{=} \frac{1}{n} \sum_{i=1}^{n} [a_i b'_i - a'_i b_i \pm (a_i b'_i + a'_i b'_i)] \leq 2$$

where Eq. (16.6) has been applied to each summand in the second member so as to yield the Bell inequality (16.7).

The "locality" argument used here, essentially the same as that employed by Santos (see Sect. 16.2.3), stipulates a condition such as

$$a_i(\alpha, \beta) = a_i(\alpha) \, ,$$

or perhaps better

$$a_i(\alpha, \beta) = a_i(\alpha, \beta') \, . \tag{16.16}$$

But a functional dependence of this kind can be attributed to a theoretical quantity, not to an experimental result, for which we can at most say that it was found together with this or that other result. In Santos' notation the condition becomes

$$(a_i/b_i) = (a_i/b'_i) \, .$$

If the JMA holds, then in fact a_i, b_i and b'_i can be measured together and (16.16) holds. If the JMA does not hold, then (a_i/b_i) and (a_i/b'_i) belong to independent experimental runs, and (16.16) cannot be true as it stands. It may be made true by reordering the data from one of the runs, provided that

$$\frac{1}{n} \sum_{i=1}^{n} a_i(\alpha, \beta) \dot{=} \frac{1}{n} \sum_{j=1}^{n} a_j(\alpha, \beta') \, . \tag{16.17}$$

(It can be shown (see Chap. 17) that (16.17) is meaningful whereas (16.16) is not). This condition is given by Stapp (1982), but he did not see that it is not sufficient. This becomes clear in an explicit description of the reordering process for all four sets of n data pairs:

1. In each series from one run, place the pairs with positive a at the top, and within each part of a given sign, place the pairs with positive b at the top. Pairs must not, of course, be disassociated. This is always possible. Then

2. Provided Eq. (16.17) and its equivalent for α' hold, the a values in columns 1 and 2 are paired, as are those in colums 3 and 4.
3. Let i_+, j_+, k_+, l_+ be the numbers of pairs in the four columns, respectively, where a and b are positive, and i_-, j_-, k_-, l_- the corresponding numbers of negative-negative pairs. The two pairings of b values are then achieved if the number of cases in which the b differs in sign between columns 1 and 2 is the same as that between columns 3 and 4; that is to say

$$|i_+ - j_+| \doteq |k_+ - l_+|$$
$$|i_- - j_-| \doteq |k_- - l_-| \tag{16.18}$$

for the positive-a part and the negative-a part, respectively.

Conditions (16.18) imply the Bell inequality; we show this for the simpler case where on account of rotational invariance each column has altogether p positive a values and q positive b values. Then the product sum in the first column is

$$\langle ab \rangle = 2(i_+ + i_- - p) - n$$

and so

$$\rho_{\alpha\beta} \doteq \frac{2n(i_+ + i_- - p) - n^2 - (2p - n)(2q - n)}{[n^2 - (2p-n)^2]^{\frac{1}{2}}[n^2 - (2q-n)^2]^{\frac{1}{2}}} .$$

The sum of the two Equations (16.18) then takes the form

$$|\rho_{\alpha\beta} - \rho_{\alpha\beta'}| = |\rho_{\alpha'\beta} - \rho_{\alpha'\beta'}| . \tag{16.19}$$

For only two correlation coefficients we always have

$$|\rho_{\alpha'\beta} - \rho_{\alpha'\beta'}| + |\rho_{\alpha'\beta} + \rho_{\alpha'\beta'}| \leq 2 \tag{16.20}$$

Combining (16.19) and (16.20) yields the Bell inequality (16.7).

That there are non-reorderable sets of data, and that in fact their number is much larger than that of the reorderable ones, is evident because the total number of different sets is, for even n,

$$N_{\text{tot}} = \frac{1}{n!} \binom{n}{\frac{1}{2}n}^8$$

while that of the reorderable sets is only

$$N_{\text{reord}} = \frac{1}{n!} \binom{n}{\frac{1}{2}n}^4 .$$

Two examples of non-reorderable sets are

$\alpha\beta$	$\alpha\beta'$	$\alpha'\beta$	$\alpha'\beta'$		$\alpha\beta$	$\alpha\beta'$	$\alpha'\beta$	$\alpha'\beta'$
++	++	+−	+−		++	++	++	+−
−−	−−	−+	−+		−−	−−	−−	−+
+−	+−	+−	++		++	+−	++	++
−+	−+	−+	−−		−−	−+	−−	−−
++	+−	+−	++		++	+−	++	++
−−	−+	−+	−−		−−	−+	−−	−−

The first satisfies the Bell inequality, the second does not.

As regards the Eberhard-Stapp-Peres derivation of the beginning of this section, it is now clear that it needs the property of reorderability in order to go through; the data must thus have the structure they would have if they came from a jointly measurable experiment.

16.4 The Joint-Measurability Assumption

16.4.1 Joint Measurability and Joint Probability Distribution

The joint-measurability assumption refers to the possibility of measuring two (or more) physical quantities without mutual interference; this last expression is to be understood in the sense that neither measurement affects the value obtained by the other, not necessarily in the sense that they are carried out simultaneously. In the case of spin projections, the JMA is violated, but it is still possible to pass a particle through a second Stern-Gerlach magnet after its spin projection has been measured a first time (assuming that detection does not absorb it). It is of course possible to calculate the quantum-mechanical correlation coefficients for this case, and they satisfy, as expected the Bell inequality (De Baere 1984).

The JMA characterizes the experimental set-up but may also be expressed theoretically. This occurs in the Bell derivation of Sect. 16.2.1, where the use of Eq. (16.6) requires that e.g. $ab - a'b = (a - a')b$, so that for every λ in the range of integration $a = a(\alpha, \lambda)$ and $a' = a(\alpha', \lambda)$ must be simultaneously known; this is not possible except coincidentally unless they refer to the same particle and are thus jointly measurable.[1] Bell's original derivation, instead of Eq. (16.6), used integrals such as $\int aa'bb' d\mu(\lambda)$, to which this point also applies. The quantum derivation also requires the JMA, so that the operators $\hat{F}(A, \alpha)$ and $\hat{F}(A, \alpha')$ commute.

In the two probability-theoretic derivation of Sect. 16.2 the Bell inequality follows when a joint probability distribution for the four measured quantities is assumed. If we assume that the theoretical model on which this distribution is based provides an adequate account of the data from the experiment, then this asssumption is equivalent to the JMA: if the JMA holds, a sequence of joint measurements can be carried out of the four variables, and from them experimental estimates can be made of one fourth-order correlation coefficient, four third-order ones, and six second-order ones; four of the last enter into the Bell inequality. To account for all these, a joint probability distribution must

---

[1] If we take a and a' to be found from separate runs in an experimental series, then it must be assumed that it is possible to reproduce the λ exactly, or at least to determine sufficiently well to decide which a goes with which a'; as de Baere (1984) has noted, this is an impossible task. He deduces, quite correctly, that the violation of Bell's inequality does not have anything to do with locality. The present point of view seems preferable, however, in that it involves no presupposition about hidden variables.

exist to provide theoretical values for them. Inversely, given a joint probability distribution, the JMA is needed so that the corresponding experimental correlations can be measured.

The two coefficients that do not enter into the usual form of the Bell inequality are $\rho_{\alpha\alpha'}$ and $\rho_{\beta\beta'}$. If the JMA holds, these should be experimentally measurable and theoretically accounted for. Neither in the spin-projection case nor in the cascade-photon one can they be measured or derived from quantum theory. On the other hand, the models of Sects. 16.2 and 16.3 all furnish unambiguous predictions for these coefficients:

Section 16.2.1 $\rho_{\alpha\alpha'} = \int a(\alpha, \lambda)a(\alpha', \lambda)d\mu(\lambda)$

Section 16.2.2 $\mathrm{tr}\hat{F}(A, \alpha)\hat{F}(A, \alpha')\hat{W}_{AB}$

Section 16.2.3 $s(a_1, a_2')$

Section 16.2.4 $\sum_{aa'bb'} p_{aa'bb'}$

Section 16.3 $\frac{1}{n}\sum a_i a_i'$

and the Suppes-Zanotti derivation of Sect. 16.2.5 uses these correlations explicitly. This makes it plain that the inability of these models to predict correctly the situation in quantum theory lies in their acceptance of the JMA.[2]

The equivalence of the existence of a joint probability distribution and the Bell inequality has been proved earlier (Suppes and Zanotti 1981; De Muynck and Abu-Zeid 1984; Fine 1982). This equivalence does not conflict with our conclusions concerning reorderability, the condition for which, Eqs. (16.18), is not equivalent either to the Bell inequality or to the joint probability distribution, i.e. the JMA.

16.4.2 The JMA and Physical Reality

The JMA could be thought to raise the same doubts about physical reality that were dissipated when locality was shown not to be involved in the deduction of the Bell inequality. For if two quantities cannot be measured together, it is not clear that they can be said actually to exist; nevertheless, spin projections satisfy the EPR criterion (Einstein, Podolsky and Rosen 1983) or physical reality. But this is to ignore that properties do not exist independently and are not "possessed" in the same sense as objects are possessed by their owner, who can sell them but cannot sell his age, say. Hence properties come into being and disappear with the physical systems with which they are associated.

Thus there arises the possibility of using too limited a model to describe their relation in the system. If a model is used which exceeds the relevant system, no problems develop; but if a subsystem is modelled, difficulties can appear. In going from an adequate model to a submodel, a property may be affected in one of three ways: (i) it may remain unchanged, as for instance the temperature of a (not too small) fraction of a macroscopic system in thermal

---

[2] The relevance of these correlations to an adequate interpretation of the Bell inequality was first stressed by Lochak (1975, 1977). The point was somewhat cavalierly dismissed by d'Espagnat (1984), simply because in many derivations they do not appear explicitly.

equilibrium; (ii) it may change in value, as for instance the mass of the same fraction; or (iii) it may cease to exist, as for instance the binding energy of a hydrogen atom, which is not a property of either component particle.

Too small a model can be extended by embedding it in a larger one. The following situations can then arise:

1. The extension may be unique, as far as the property of interest is concerned. In this case the "small" model provides all necessary information, and the property may even remain constant, as in case (i) above. The extension may give conceptual clarity.
2. When the extension is not unique, parameters that specify the choice are needed; the choice is usually dictated by experimental reasons, not theoretical ones.
   2.1 The choices are mutually exclusive, e.g. through involving incompatible experimental set-ups.
   2.2 If the choices do not exclude each other, one must distinguish. Let the property of interest be a function $q = q(\phi, \xi)$ of the choice parameter(s) ϕ and any needed small-model parameters ξ. Then
   2.2.1 If q is a 1:1 function of ϕ, then for another value ϕ' but the same ξ we have

$$q' = f(q, \phi', \xi)$$

a well-defined function, and the correlation $\rho(q, q')$ must exist.
   2.2.2 The function q has no inverse, f does not exist, nor does the correlation between q and q'.

Case 2.2.1 is the case in which inequalities of the Bell type are expected to exist. In case 1 they cannot be defined, because the relevant parameters do not exist, e.g. the angle parameters in the spin-projection case. In case 2.1 the JMA does not hold, and the inequalities cannot be derived. Case 2.2.2 is marginal: though the different quantities are jointly measurable, they are so each in a different set-up, and the Bell inequality may not be satisfied. The measurement of total spin (actually of the multiplicity $2s + 1$) corresponds to case 1, since the orientation of the inhomogeneous magentic field is irrelevant. Spin projections correspond to case 2.1: the JMA does not hold for them. An example of case 2.2.2 will be discussed in the next section.

This then means that a spin projection is not a property of the particle by itself, as it is just after emission from the source, but of the joint system (particle + Stern-Gerlach magnet); this is made more comprehensible by the quantum picture, of a wave function with cylindrical symmetry before the interaction with the inhomogeneous field, of planar symmetry in the density matrix after it. But in the absence of a deeper understanding of the physical nature of spin it does not appear possible to furnish a more precise description.

The above account in no way conflicts with the philosophical picture of an independently existing physical reality. The process of creation of a property is a real and physical process; it does not (normally) occur in the process of detection, and even less when measured results are apprehended by the physicist. All that has been done is to complete the Einsteinian picture where it was not sufficiently detailed.

The classification given above is not absolute. A non-linear transformation of the physical quantity that expresses the property of interest may change it from category to category. This is trivially obvious when q can take a continuous range of values. Dichotomic variables (such as the projections of a total spin $\frac{1}{2}$) are exceptional only in that two of the four possible transformations are linear, while the other two eliminate the dependence on ϕ and so translate it to case 1.

16.5 Some Classical Examples

The concept of system extension, and the resulting possiblity of the JMA not holding, is not restricted to quantum systems. Cases 1 and 2.2.1 hardly need illustrating; accordingly we present examples for the other two.

16.5.1. Classical violations of the Bell inequality

That classical systems which violate the Bell inequality should exist seems first to have been suggested by Suppes and Zanotti (1981); the first detailed description is given by Scalera (1983, 1984) and consists of two ribbons issuing helicoidally, with a common angular velocity, from a source; whenever their inclinations are along preassigned directions, they are marked, and two detectors measure the energy on each ribbon between two successive markers. The correlation between these energies is then, under suitable conditions, shown to violate the Bell inequality.

Such a model could not easily be realized in the laboratory; this is remedied by Notarrigo (1984), whose model uses two rows of equal masses linked by equal springs and moving only in the planes transverse to the rows. In each row, one end initiates the motion under random conditions which are the same for the two rows; the other masses start at rest. The other ends of the rows are fixed and reflect the excitation wave. In each row one mass is constrained to move along a certain direction, the "easy" direction; this is different for the two rows. Beyond the filter so established the motion is along the easy direction only; for one of the masses there, a reference point is chosen and an event is noted if, during a time window chosen to give a predetermined event probability, the mass passes the reference point moving outward. The quantity of interest is then the correlation coefficient between events on the two rows. It is obtained from a computer simulation, and found to approximate the quantum dependence, Eq. (16.8), except for the sign, when the number of masses in each row is not too small (≥ 20); the Bell inequality is then violated.

This model, like the Scalera one, is local: there is no interaction of any kind between the two rows, beyond the identity of the starting impulse which here corresponds to the condition of total spin 0 in the set-up of Eq. (16.1). The only "hidden" variables are those involved in the randomization of the initial conditions. Since it is not possible to set more than one easy direction on each row, the model belongs to category 2.1 of the preceding section: accordingly the JMA does not hold for it, and the Bell inequality cannot be expected to hold either.

16.5.2 The Harmonic Oscillator

Further insight is gained by considering an analytically soluble model. Consider two harmonic oscillators of the same frequency but with different phase angles:

$$a = a_0 + a_1 \cos(\omega t + \alpha)$$
$$b = b_0 + b_1 \cos(\omega t + \beta) .$$

(16.21)

Then

$$\langle a \rangle = a_0 , \quad \langle a^2 \rangle = a_0^2 + \frac{1}{2} a_1^2$$

$$\langle b \rangle = b_0 , \quad \langle b^2 \rangle = b_0^2 + \frac{1}{2} b_1^2$$

(16.22)

$$\langle ab \rangle = a_0 b_0 + \frac{1}{2} a_1 b_1 \cos(\alpha - \beta) .$$

The averages in (16.22) are taken either over a complete cycle $T = 2\pi$, or over an ensemble of uniformly distributed starting times. The harmonic oscillator is known to be ergodic and so these two averages coincide.

The correlation coefficient between the two harmonic oscillators of Eq. (16.21) is then

$$\rho_{\alpha\beta} = \cos(\alpha - \beta) .$$

(16.23)

Except for the sign, this is the quantum correlation (16.8), and, like it, it violates the Bell inequality.

Since, as noted in Sect. 16.4, by a simple non-linear transformation the violation may be eliminated, it is of interest to discretise the problem by redefining the variables of Eq. (16.21) as

$$a = \begin{cases} +1 & \text{if } a_0 + a_1 \cos(\omega t + \alpha) > 0 \\ -1 & \text{otherwise} \end{cases}$$

(16.24)

and similarly for b. In the case most favourable for violation, the correlation coefficient takes the limiting form found by Selleri (1978)

$$\rho_{\alpha\beta} = 1 - \frac{2|\alpha - \beta|}{\pi} .$$

The discriminant V of Eq. (16.7) now reaches the value 2 but does not exceed it. A small higher-harmonic term in (16.24) will make it exceed 2:

$$a = \begin{cases} +1 & \text{if } a_0 + a_1 [\cos(\omega t + \alpha) + \varepsilon \cos 2(\omega t + \alpha)] > 0 \\ -1 & \text{otherwise} \end{cases}$$

and similarly for b. Even for $\varepsilon = 0.1$, a computer simulation has shown that $V > 2$ over a considerable region, starting at $\alpha - \beta = 1.8°$, $\alpha - \beta' = 30.6°$, $\alpha - \alpha' = 16.2°$; for small $\varepsilon, V - 2 = \varepsilon^2/2$.

Other small non-linearities have similar effects. A quartic term added to the harmonic-oscillator potential or a term in t^2 in the phase angles of Eq. (16.21) lead in analogous fashion to a violation of the Bell inequality.

A procedural point requires a comment. Some authors (Bergia and Cannata 1982; Baracca, Bergia, Livi and Restignoli 1975) have argued for the use of the covariance $\mathrm{cov}(a, b) = \langle ab \rangle - \langle a \rangle \langle b \rangle$ in calculating V instead of the correlation coefficient $\rho(a, b) = \mathrm{cov}(a, b)/[\mathrm{var}(a) \cdot \mathrm{var}(b)]^{\frac{1}{2}}$, where $\mathrm{var}(a) = \mathrm{cov}(a, a)$. The covariance, however, has dimensions and therefore changes with scale changes; thus using the real values $\pm\frac{1}{2}$ for the spins would either make even the quantum covariance satisfy the Bell inequality or require a right-hand side $2[\mathrm{var}(a) \cdot \mathrm{var}(b)]^{\frac{1}{2}} = \frac{1}{2}$ instead of 2 in the inequality, Eqs. (16.7) and (16.20). The dimensionless correlation coefficient is therefore to be preferred. A related problem appears in a paper by Barut and Meystre (1984). They treat the classical three-dimensional harmonic oscillator in a very "anschaulich" way, and find that it violates the Bell inequality even in a discretised version; they attribute this to a difference between the quantum and the classical normalizations. In fact their correlation coefficient covers the range $(-1, 1)$, as it should, and their normalization factor is similar in origin to the one mentioned above for the spin case, which is eliminated by the artifice of using eigenvalues ± 1 for the spin of a spin $-\frac{1}{2}$ particle.

The harmonic oscillator, seen as a one-dimensional system, clearly falls in category 2.1 of the classification above: an oscillator cannot simultaneously have two phase angles α and α'. But a two-dimensional rotor may have an unlimited number of projections into harmonic oscillators, all of them simultaneously measurable: a laboratory set-up that realises such a scheme is easily imagined. However, the function q of ϕ (here the angle α) has no unique inverse, for to each q there correspond two values of α, with

$$\alpha_1 + \alpha_2 = 2\pi - 2\omega t \,.$$

Thus this is a case in category 2.2.2. It illustrates, moreover, that restructuring the model describing the system may move a case from category to category.

16.6 Concluding Remarks

Three general comments may be made before concluding.

1. As was shown in Sect. 16.3, reordering a set of experimental data in such a way that quadruplets satisfying (16.6) can unambiguously be extracted and the Bell inequality satisfied is possible only under restricted conditions; it cannot in general be done for the data from four separate runs at

different angle settings. Consequently, the Bell inequality places a restriction on the correlation coefficients, and coefficients that vary unrestrictedly may violate it. Once this is realized, it is obvious that it is not the violation but the satisfaction of the inequality that could be thought to need transmitting further information between the A and B measurements of the set-up (16.1). This is precisely the opposite of what is claimed in order to argue for the superluminal transmission of information (Costa de Beauregard 1979); such arguments are therefore suspect from the outset.

2. That two different spin projections cannot be determined on the same particle without mutual interference is, of course, well known. By taking (sometimes only implicitly) the JMA as starting point, the deductions of the Bell inequality systematically ignore this fact, however. This is not in itself to be considered a mistake. When building a theoretical model for a physical situation, many even essential characteristics must of needs be ignored, and the success of the model justifies these oversimplifications a posteriori. But if the model misdescribes a significant aspect, such omission must be revised. This has not so far been done as regards the interpretation of the Bell inequality.

3. The concept of a physical system as a finite part of the universe, circumscribed by the experimental techniques for its preparation and isolation, and represented theoretically by a model, has been seen to play a significant rôle in the perception of what the JMA implies. In particular, the use of more than one system at the same time, with their corresponding models, the transitions between which require rather careful description, is exemplified here. It is the author's belief that this situation is not unique. More than one conceptual difficulty bedevilling the foundations of physics would probably be resolved if due attention were paid to stating just what systems are being used and how transitions between them are to be handled. It is unfortunate that the many cases where the system definition is obvious have led to a tradition in the physics literature of ignoring the entire question; together with the obscurity due to the lack of any philosophical account of the system concept, this has created much needless confusion.

In conlcusion, it has been shown that all the derivations here examined of the Bell inequality use the JMA, either directly or by assuming a joint probability distribution. Only a few, on the other hand, need hidden variables or a locality assumption. Thus the JMA stands out as the key assumption. Now for many variable pairs, classical as well as quantum-mechanical, the JMA is not valid. This raises, as we have seen, interesting questions concerning the system concept and that of the corresponding theoretical model; but in no way could it imply that we must either doubt the separability of physical systems when no possible interaction between them can exist, or else abandon any realist position, such as that defended by Einstein.

Thus a final caveat against drawing overhasty philosophical conclusions before the underlying physics is well understood may not be out of place.

Acknowledgements. I have had stimulating discussions and exchanges of letters with numerous colleagues, above all with E. Bitsakis, G. Fehér, M.A. Fisher, N. Hadjisavvas, T.E. Küchler, P. Savignon, F. Selleri, J. Stachel, P. Suppes and C. Theobald. I am also grateful to T.W. Marshall and E. Santos, who offered me both hospitality and much useful comment. P.E. Hodgson provided a much appreciated stay at Oxford, together with numberless comments and some searching questions; he also gave a careful reading to a first draft of this paper. The work was done during a sabbatical stay at the Democritos Research Centre, to whose stimulating atmohpere it owes much. A large part of the paper has benefitted from the patience and helpful remarks of A.K. Theophilou. But my greatest debt is to A.M. Cetto and L. de la Peña, in collaboration with whom most of the ideas expounded here were conceived.

References

Baere, W. de (1984): Lett. Nuovo Cim. **39**, 234
Baere, W. de (1984): Lett. Nuovo Cim. **40** (1984), 488
Barut, A.O., Meystre, P. (1984a): Acta Phys. Austr. **56**, 13
Barut, A.O., Meystre, P. (1984b): Phys. Lett. **105A**, 458
Beauregard, O. Costa de (1979): Nuovo Cim. **51B**, 267
Bell, J.S. (1965): Physics 1, 195
Bergia, S., Cannata F. (1982): Found Phys. **12**, 843; see also Baracca, A., Bergia, S. Livi, R., Restignoli, M. (1975): Int. J. Theor. Phys. **15**, 41
Bohr, N. (1935): Phys. Rev. **48**, 696
Brody, T.A., Peña, L. de la (1979): Nuovo Cim. **58B** (1979), 455
Brody, T.A. (1989): "The Bell Inequality II: Locality" chap. 17, in this volume
Bruno, M, d'Agostino, M., Maroni, C. (1977): Nuovo Cim. **40B**, 143
Capra, F. (1980): "Le Tao de la Physique" in *Science et conscience* (Stock éditeurs, Paris), p. 43
Clauser, J.F., Shimony, A., Horne, M.A., Shimony, A., Holt, R.A. (1969): Phys. Rev. Lett. **23**, 880
Clauser, J.F., Shimony, A. (1978): Reps. Prog. Phys. **41** , 1881
Eberhard, P.H. (1977): Nuovo Cim. **38B**, 75
Eberhard, P.H. (1978): Nuovo Cim. **46B**, 392
Eberhard, P. (1982): Phys. Rev. Lett. **49**, 1474
Einstein, A., Podolsky, B., Rosen, N. (1935): Phys. Rev. **47**, 777
d'Espagnat, B. (1984): Phys. Reports **110**, 201
Fry, E. (1984): Physics Today **37(1)**, S26
Fine, A. (1982): Phys. Rev. Lett. **48**, 291
Fine, A. (1982): J. Math. Phys. **23**, 1306
Garg, A., Mermin, N.D. (1984): Found. Phys. **14**, 1
Holt, R.A. (1973): *Ph.D. thesis* (Harvard University)
Küchler, T.E. (private communication)
Lochak, G. (1975): Epistem. Lett. **6**, 41
Lochak, G. (1976): Found Phys. Lett. **6**, 173
Lochak, G. (1977) in J. Leite Lopes, M. Paty (eds.) *Quantum Mechanics Half a Century Later* (Reidel, Dordrecht)
Marshall, T.W. (1983): Phys. Lett. **99A**, 163
Marshall, T.W. (1984): **100A**, 225
Marshall, T.W., Santos, E., Selleri, F. (1983): Phys. Lett. **98A**, 5
Marshall, T.W., Santos, E., Selleri, F. (1983): Lett. Nuovo Cim. **38**, 417

Muynck, W. de, Abu-Zeid, O. (1984): Phys. Lett. **100A**, 485
Notarrigo, S. (1984): Nuovo Cim. **83B**, 173
Peres, A. (1982): Amer. J. Phys. **46**, 1470
Peña L. de la, Cetto, A.M., Brody, T.A. (1972): Lett. Nuovo Cim. **5**, 177
Rohrlich, F. (1983): Science **221**, 1251
Roy, S.M., Singh, V. (1978): J. Phys. **A11** (1978), L167
Santos, E. (1984): Phys. Lett. **101A**, 379; and private communication
Santos, E. (1985): (preprint) (Universidad de Santander)
Scalera, G.C. (1983): Lett. Nuovo Cim. **38**, 16
Scalera, G.C. (1984): Lett. Nuovo Cim. **40**, 353 (1984)
Selleri, F. (1984): Lett. Nuovo Cim. **39**, 252
    See also Garuccio, A., Selleri, F. (1984): Phys. Lett. **103A**, 99
Selleri, F. (1978): Found. Phys. **8**, 103
Selleri, F., Tarozzi, G. (1981): Riv. Nuovo Cim. **4(2)**, 1
Stapp, H.P. (1977): Nuovo Cim. **40B**, 191
Stapp, H.P. (1982): Phys. Rev. Lett **49**, 1470
Suppes, P., Zanotti, M. (1981): Synthèse **48**, 191 (preprint 17 August 1981)
Walker, E.H. (1975) in L. Oteri (ed.) *Quantum Physics and Parapsychology* (Parapsychology Foundation)
Wigner, E.P. (1970): Amer. J. Phys. **38**, 1005

17. The Bell Inequality II: Locality[*]

Abstract. Three forms of the locality condition used in connection with the Bell inequality are examined and compared to the causality rule that causal effects cannot propagate at a speed greater than that of light. It is shown that they differ substantially among each other and from the causality rule; they cannot therefore serve to derive conclusions implying the failure of causality from the violations of the Bell inequality.

(i)

The various forms of the Bell inequality (Bell 1965; Clauser, Shimony 1978) are generally thought to involve a locality condition, which ensures that the two particles or photons emitted in a correlated state do not subsequently influence each other. That certain quantum systems violate the inequalities is considered to create a serious conceptual problem for the understanding of quantum mechanics: if locality cannot any longer be assumed, one cannot give adequate theoretical descriptions of any physical systems without explicit reference to all other systems it might have interacted with in the past. Alternatively, the problem can be seen as lying in our conceptions of causality, or of physical reality (Selleri 1983).

In the preceding chapter it was shown that these violations of the Bell inequality involve the failure, not of locality but of another assumption, that of joint measurability. This assumes that it is possible to measure the quantity under study (spin projection or photon polarisation) of a single particle in more than one direction without mutual interference. But the locality condition presupposed in the discussions of the Bell inequality also merits some attention, and here we examine the three main forms it has taken in the literature.

The only locality condition generally accepted is the causality rule, derived from relatively theory, that causal connections cannot be propagated at speeds greater than that of light. For the present purpose this condition may be formulated as follows: Two space-time regions R_A and R_B are local to each other (or separable, in another terminology) if, for any two points x_A, x_B

$$(x_A - X_B)_\mu (x_A - x_B)^\mu < 0 , \qquad x_A \in R_A , \quad x_B \in R_B , \qquad (17.1)$$

---

[*] First published in: Revista Mexicana de Física **35** Suplemento (1989) S71–S79

where the Einstein summation convention is used and a signature $(+--)(-)$ for the metric is assumed. The following points about this condition may be noted:

1) The condition is negative. It does not imply that if two events x_A and x_B do not satisfy (17.1) then there must exist a causal connection between them.

2) The condition is useful only if the physical systems A and B described as lying in R_A and R_B have finite life times, in view of the fact that the past light cones of any two points whatsoever intersect. Relativistic field theory restricts its application to specific events rather than systems perduring any length of time.

3) In the limit $c \to \infty$ all point pairs violate (17.1), so that no non-relativistic equivalent to (17.1) can be written.

Point (2) implies that one must distinguish between direct interactions (transmitted from one system to the other during their existence) and parallel features in their behaviour due to common causes lying in their joint past (this may be called "inherited correlation", using correlation in its widest sense). This distinction remains relevant in the non-relativistic case, even though no equivalent to (17.1) can be given provided one accepts that physical systems exist only for finite times.

That the distinction is in fact relevant can be seen by considering how direct interactions between the systems A and B could be detected experimentally. If one or more of the properties describing B changes value whenever an external interference with A (occuring after A and B have separated) alters those of A, evidence for such an interaction has been obtained; on the other hand, any such interference tends to destroy an inherited correlation.

This method for distinguishing between direct interaction and inherited correlation is available only if the measurement procedures do not themselves significantly alter the systems, for in that case we cannot discriminate between the effect of the direct interaction and that of the previous measurement. If the measurements do affect the systems, one must resort to preparing several pairs of systems, their common origin (and any causes associated with it) being held constant or systematically varied, according to experimental requirement. Constancy here will allow the experimenter to observe the results of posterior interference with some of the pairs, while variation without later interference will identify inherited correlations. Note that the latter may be observed only in this way, whatever the measurements may do to the systems.

If the common cause is not subject to experimental control, or if factors outside the system[1] also act in determining the measured property, its values will fluctuate and individual pairs of observations will show up neither direct interaction nor inherited correlation. Only a statistically adequate set of observation pairs will do so, and correspondingly a probabilistic model is required.

---

[1] I.e. outside the theoretical description and experimental control; topographically such factors may belong to the system's surroundings or be internal to it but not included in the theoretical model. Thus no assumption of determinism is needed.

In this respect no distinction between classical and quantum mechanics need be drawn.

There is a common factor in all these cases: it is that of variation. No single pair of observations on A and B could establish a conclusion either of interdependence or of its absence. This is so even when each measurement pair results from a long series of individual observations: two such series are needed for any conclusion. For the particular case of the Bell inequality, at least three series are necessary, one for each correlation coefficient.

The case of Bell's inequality combines both the points. The measurements cannot be repeated, and they are statistical in nature. Because of the first point, the discussion is often carried out in terms of counterfactual expressions ("if we had measured at different angle settings, the outcome would have been") and their use has sometimes been doubted on philosophical grounds, hence the title of one paper (Peres 1982): 'Unperformed Experiments Have No Results'. Such discussions are, however, both inconclusive and irrelevant, since they miss the essential distinction: a theoretical model will correctly predict unperformed measurements, but it will not predict unperformable ones. Hence the results of unperformed but feasible measurements can be combined with experimental results, while combining measurements the theory prohibits in this way can lead to absurd conclusions. To put it more abstractly, counterfactuals are admissible relative to the underlying physical theory (or its model) but must be rejected if they violate its stipulations.

The causal condition (17.1), besides being plausible, has found useful and uncontroversial applications in relativistic quantum field theory and related theories; it will here be taken, therefore, as the standard of comparison for the locality conditions formulated in connection with the Bell inequality.

(ii)

Bell (1965) defined two quantities, $a(\alpha, \lambda)$ and $b(\beta, \lambda)$, taking the values $+1$ and -1, to be the results predicted for spin-projection measurements of two spin $1/2$ systems A and B issuing from the dissociation of a system of total spin 0; the spin projections are measured along directions α and β respectively, and λ represents the (set of) hidden variables supposedly sufficient to determine a and b. His locality condition was that a should not depend on β nor any other variable characterising system B, and similarly that b should not depend on α. In particular, a should not depend on b, and vice-versa.

This condition is by no means satisfactory as a condition of locality:

1) It can be satisfied even when (17.1) is violated, since two quantities need have no functionally expressible interrelation even though they are attached to the same space-time region. This is why Home and Sengupta (1984) can derive a Bell inequality for the spin and the orbital angular momentum of one electron bound in an atomic orbit.

2) It can be violated when (17.1) is satisfied. In the extreme case when $b(\beta, \lambda)$ possesses an inverse, $\lambda = \lambda(b, \beta)$, we have

$$a(\alpha, \lambda) = a(\alpha, \lambda(b, \beta)) = a(\alpha, b, \beta) . \qquad (17.2)$$

But even if, as for the spin case, a does not have an inverse, the space Λ over which the λ vary can be divided into subspaces $\Lambda_b(\lambda)$, and the value of b determines in which of them the values of the λ must lie; though this does not suffice to determine a, the probability of either value of a will now depend on b. The trouble is that Bell's condition does not discriminate between direct interaction and inherited correlation; this is inevitable, since no time specifications can be attached to λ in a meaningful way.

3) The condition has some undesirable consequences; these will be discussed below, in connection with thesave Clauser-Horne factorisability condition which includes it.

Thus, though there is considerable overlap between them, the Bell condition does not coincide sated with the causality rule stated above, nor can it discriminate between direct interaction and inherited correlation as the causality rule does. But unless the violations of the Bell inequality could unambiguously be attributed to failure of the causality rule, the far-reaching interpretations usually offered of this situation would not stand up. Moreover, as shown in Sect. i, the disagreement between Bell's inequality and quantum mechanics can be explained rather simply; it can, indeed, appear while the Bell condition is fully satisfied.

(iii)

Another form of locality condition has been developed by Stapp (1977, 1985a,b) and used by Eberhard (1977, 1978) and, less evidently, by Peres (1982); in a somewhat different context, it is also used by Santos (1985). It is in some sense the 'experimental' form of Bell's condition, in that it refers explicitly to the actual observations. Thus Stapp (1985) first defines "The results or values appearing in R_A and R_B" as $a = a(c_A, c_B, d)$ and $b = b(c_A, c_B, d)$, where c_A, c_B characterise the measuring set-up for systems A and B, respectively while d indicates all other relevant parameters; he then states that his locality condition "is represented by the conditions $a(c_A, c_B, d) = a(c_A, d)$ and $b = b(c_B, d)$: the result produced in R_A is independent of the choice of experiment made in R_B, and vice versa."[2] Stapp, in fact, uses an explicitly counterfactual version (1985), in the following sense: We have measured a and b; had we measured a and $b' = b(c_B', d)$ instead, we would have obtained the same value for a. But in the spin case these possibilities are mutually exclusive. Had a and b' been measured, the a from the alternative measurement would not have been available to make the comparison. It is possible to suppose otherwise; but at the price of conflict with quantum mechanics. That such a conflict arises here is made evident by the fact that Stapp's formalism admits correlation coefficients between a and a' or b and b'. These correlations are not accessible by experiment, nor are they well defined in quantum theory; yet Stapp (Eqs. 17.6) and (17.7) in Stapp (1985) uses them to deduce a contradiction with quantum mechanics.

---

[2] The notation has been slightly adapted from Stapp's.

The approach is thus invalid. What has been overlooked is the statistical nature of the quantities that are being measured. Using the notation of Sect. i, Eq. (17.16), Stapp's locality conditions may be written

$$a_i(\alpha, \beta) = a_i(\alpha, \beta') , \qquad a_i(\alpha', \beta) = a_i(\alpha', \beta')$$
$$b_i(\alpha, \beta) = b_i(\alpha', \beta) , \qquad b_i(\alpha, \beta') = b_i(\alpha', \beta') \tag{17.3}$$

But what is both experimentally accessible and actually satisfied are the conditions

$$\langle a(\alpha, \beta) \rangle = \langle a(\alpha, \beta') \rangle \text{ etc.} \tag{17.4}$$

where

$$\langle a(\alpha, \beta) \rangle = \frac{1}{n} \sum_{i=1}^{n} a_i(\alpha, \beta) \text{ etc.} \tag{17.5}$$

are experimental averages. Conditions corresponding to (17.4) for the theoretical expectations also hold. As discussed above, these conditions establish for statistically fluctuating quantities that there is no direct interaction; they are thus locality conditions. They agree with the meaning of (17.1), for their space-time requirements are implicitly guaranteed because each pair of observations must be obtained from particles coming from a specific dissociation.

But the Bell inequality will only follow from conditions such as (17.3); starting with (17.3) one easily finds, for instance, that

$$a_i(\alpha, \beta)b_i(\alpha, \beta) - a_i(\alpha, \beta')b_i(\alpha, \beta') \pm [a_i(\alpha', \beta)b_i(\alpha', \beta) + a_i(\alpha', \beta')b_i(\alpha'\beta')] = \pm 2$$

and hence, summing over $i = 1 \ldots n$, dividing by n and going to the limit of large n,

$$-2 \leq P_{\alpha\beta} - P_{\alpha\beta'} \pm (P_{\alpha'\beta} + P_{\alpha'\beta'}) \leq 2 \tag{17.6}$$

which is the desired inequality. However, as shown in Sect. i, the locality conditions (17.4) do not suffice to derive (17.5). A further conditions must also hold. It is needed because in any actual data only a fraction satisfies (17.3), or can be made to satisfy it by rearranging the order of the data (which may be done since the chronological order should be irrelevant). This further condition is essentially equivalent to the joint-measurability assumption. But under this assumption, (17.3) either is automatically satisfied if all four measurements a, a', b, b' are carried out on each pair of particles, or it can be satisfied by rearrangement.

It may be concluded that Stapp's form of locality condition goes considerably beyond the physically justifiable locality condition (17.4).

(iv)

The third form of locality condition to be examined here is the so-called factorisability requirement (Clauser, Horne, 1974): the joint probability density for the results a, b should be written as[3]

---

[3] The notation adopted here is intended to make clear the distinction between (1) the statistical varieties a, b; (2) the statistical variable λ on which the probability densities q and r are conditioned; and (3) the parameters α and β which do not have statistical character.

$$p(a, b; \alpha, \beta) = \int q(a|\lambda; \alpha) r(b|\lambda; \beta) d\mu(\lambda) \qquad (17.7)$$

where μ is the distribution of the "hidden" variables λ over their space Λ, with

$$\int_{\Lambda} d\mu(\lambda) = 1$$

while q and r are the conditional probability densities of a and b, with α and β being the corresponding angular parameters.

The use of (17.7) in the sense of a locality condition creates several problems. It has been shown by Theophilou and Brody that given any bivariate distribution $p(x, y)$ there exists an infinite family of conditional distributions $q(x|z)r(y|z)$ such that

$$p(x, y) = \int q(x|z) r(y|z) d\mu(z) . \qquad (17.8)$$

But if q and r depend on parameters α and β, respectively, so in general will $\mu(z)$. Now because (17.8) is always possible, it cannot be factorisability in itself that constitutes a locality condition; it cannot discriminate between local and non-local situations.

There remains only one possible stipulation that could be seen as a locality condition. In writing (17.7) with no B dependence in a (*i.e.* neither b nor β) and no A dependence in b, the Bell condition has already been incorporated. But (17.6) cannot be used to derive a Bell-type inequality unless μ is independent of α or β, for in any derivation bivariate densities $p(a, b; \alpha, \beta)$ for different angle settings must be combined in a single integral with a common $\mu(\lambda)$, so that the analogue of (17.6) will apply to the integrand. This is not possible here, as will be shown in two ways.

Firstly, following a suggestion due to Varma we consider only system A. Then the probability of finding a value a should be independent of α because of the cylindrical symmetry of the set-up:

$$p(a) = \int_{\Lambda} q(a|\lambda; \alpha) d\mu(\lambda) . \qquad (17.9)$$

Hence

$$\int_{\Lambda} d\mu(\lambda)[q(a|\lambda) - q(a|\lambda; \alpha')] = 0 . \qquad (17.10)$$

There should then exist an operator F that transforms the q for one value of α into that for another. Equation (17.10) then shows that this operator is a functional of the hidden-variable distribution μ. Taking $\alpha' = 0$, we then have

$$q(a|\lambda; a) = F_{\mu, \alpha} q(a|\lambda; 0) . \qquad (17.11)$$

If $F_{\mu, \alpha}$ possesses an inverse, it is possible to absorb the α-dependence into the hidden-variable distribution as $\tilde{\mu}(\lambda; \alpha)$:

$$p(a) = \int_{\Lambda} \tilde{q}(a|\lambda) d\tilde{\mu}(\lambda; \alpha) , \qquad (17.12)$$

where $\tilde{q}(a|\lambda) = q(a|\lambda; 0)$. Intermediate cases also exist. Equations (17.11) and (17.12) show that the Clauser-Horne ansatz (17.7), contrary to what is commonly stated, does not separate cleanly between the description of the systems A and B on leaving the source and the effect of the analyser-detector set-up.

Secondly, we consider the physical significance of the λ variables. Some of them will represent constants of the motion of A and B and thus have no time dependence; where a conservation law holds, $\lambda_A = -\lambda_B$. But there will be others that do depend on time, for instance in theories such as stochastic electrodynamics, in which quantum particles move under the influence of random background field. But it is then evident that the values of the λ that figure in q and r must be those that prevail when the particle enters into interaction with the analyser and not those that existed at the source; it is then unsurprising that $\mu(\lambda)$ should depend on certain analyser characteristics such as the angles α and β.

Thus the absence of an angle dependence in $\mu(\lambda)$ can hardly be qualified as a locality condition. If systems A and B do not interact, this can be correctly expressed by stipulating that μ can be factorised in a form such as

$$\mu(\lambda; \lambda, \beta) = \mu_A(\lambda_A, \lambda_0, \lambda_1; \alpha)\mu_B(\lambda_B, \lambda_0, \lambda_1; \beta), \qquad (17.13)$$

where λ_0 are the constants of the motion A and B share, λ_1 are those for which conservation laws hold, while λ_A, λ_B are not conserved. No further factorisation is possible: if there were no variables common to the two factors, all correlation coefficients would be 0, and if these common variables could figure in a separate factor by themselves, the situation of Eq. (17.7) would be reproduced.

But a form like (17.13) evidently does not permit the derivation of a Bell inequality. To sum up, Bell's locality condition comes closest to agreement with the relativistic condition but still differs from it in significant ways. The Stapp condition is composed of two parts; one is a genuine locality condition, namely Eq. (17.4), but is not by itself sufficient to give the Bell inequality; the other is the joint-measurability assumption (see Sect. i) and has nothing to do with locality. The factorisability condition is a locality condition only insofar as it includes the Bell condition; what it adds to this – the demand the $\mu(\lambda)$ should not depend on the angles α and β – is not fully compatible with other conditions the model must satisfy and has little physical justification.

It may be concluded that the three locality conditions here (1) differ among each other significantly, and (2) do not agree with what locality means in other parts of physics. Therefore (3) they do not bear the weight of the interpretation commonly placed on them; even if it had not been shown in Sect. i that the relevant factor in the derivation of the Bell inequality is the joint-measurability assumption, the violation of the inequality by quantum systems could not be used to conclude anything concerning locality in its proper meaning.

Acknowledgements. I have greatly benefitted from discussions with H.P. Stapp, R.K. Varma, F. Selleri and many other participants in the Urbino conference, in which I could participate through Prof. Selleri's kindness. I have also received decisive help from A.K. Theophilou, as well as the extended hospitality of his group at the Democritos Centre.

References

Bell, J.S. (1965): Physics **1**, 195
Brody, T.A. (1989): Rev. Mex. Fís. **35** Supl. S52 and Chap. 16 of this volume
Brody, T.A. (1983): Rev. Mex. Fís. **29**, 461; see also Chap 14
Clauser, J.F., Horne, M.A. (1974): Phys. Rev. **D10**, 526
Clauser, J.F., Shimony, A. (1978): Reps. Prog. Phys. **41**, 1881
Eberhard, P. (1977): Nuovo Cim. **38B**, 191
Eberhard, P. (1978): Nuovo Cim. **46B**, 392
Home, D., Sengupta, S. (1984): Phys. Lett. **102A**, 159
Peres, A. (1982): Amer. J. Phys. **46**, 1470
Santos, E. (1985) (Preprint) (Universidad de Santander)
Selleri, F. (1983): *Die Debatte um die Quantentheorie* (F. Vieweg, Braunschweig)
Stapp, H.P. (1977): Nuovo Cim. **40B**, 191
Stapp, H.P. (1985a): Amer. J. Phys. **53**, 306
Stapp, H.P. (1985b): (Preprint LBL 20094)

18. The Irrelevance of the Bell Inequality[*]

18.1 Problem and Conclusion

It is almost half a century since the great quantum debate began, sparked off by Einstein, Podolsky and Rosen (1935) on one side and Bohr (1953) on the other. In that long time only one new element has been introduced, the inequality derived by Bell (1965). That this inequality is violated by quantum mechanics both theoretically and experimentally (though see the papers by Santos (1985) and Selleri (1985)) has been held by many to presage the definitive defeat of the Einsteinian realist position in this debate. The argument is this: The Bell inequality rests on nothing but the two assumptions that hidden-variable extensions of quantum mechanics are possible and that two systems that have been in contact can move far enough apart for all interactions between them to be negligible; this last is the locality assumption. If the inequality is violated by quantum mechanics, then one or both assumptions must be false (unless we wish to reject quantum mechanics). Ruling out the hidden variables goes a long way to making it impossible to develop theories that go beyond quantum mechanics. Eliminating locality is even more serious: if the universe is too strongly non-local, we could not isolate a part of it and study it in relative independence of the remainder, so that essentially the scientific endeavour has been shown to be senseless. Even without this drastic conclusion, we might have to accept some at present unknown interaction, presumably propagated at speeds exceeding that of light, in order to explain the abnormally high correlations exhibited by quantum mechanics for systems separated by a space-like interval. This idea has been the source of much facile speculation, even spilling over into the popular press, concerning the supposed parapsychological consequences of modern physics; but even among the more responsible members of the physics community doubts have arisen. We are in fact living a situation that Einstein had foreseen many years ago (Einstein 1948).

But I hope to show that the outlook is not so bleak. I shall argue, firstly, that there are derivations of the Bell inequality that do not depend on the

---

[*] First published in: E.I. Bitsakis and N. Tambakis (eds.): *Determinism in Physics.* Gutenberg Publishing Company (Athens, 1985)

notions of hidden variables or locality; thus its violation has little to do with them and in particular does not imply that we have to abandon them. But then why does quantum mechanics violate it? I propose to show that all the derivations use a further assumption which has not so far been spelt out, namely that two different variables (e.g. the spin projections in two different directions) can be measured on the same particle without mutual interference. This is, of course, not the case for many quantum variables, and that their correlations should not satisfy the Bell inequality is thus not particularly mysterious. Now if this sort of explanation holds true, similar situations should arise in classical physics; I therefore exhibit some classical models that violate the Bell inequality.

The conclusion from these arguments is then that Bell's inequality, so far from representing the final victory of Bohr's side in the debate, turns out to be irrelevant to the problem.

18.2 Four Derivations of Bell's Inequality

What is Bell's inequality? Consider a system of total spin 0 which breaks up into two particles A and B of spin $\frac{1}{2}$. We measure A's spin projection along a direction α (perpendicular to the direction of motion) and B's along β; denote the results, which may be $+1$ or -1, by a and b. Quantum theory predicts that the correlation between a and b will be

$$\varrho_{ab} = -\cos(\alpha - \beta) \, . \tag{18.1}$$

Consider now another direction α' for A, giving results $a' = a(\alpha', \lambda)$, and similarly β' for B, giving $b' = b(\beta', \lambda)$. Here we are supposing the existence of a set of variables λ, which we call "hidden" because we do not know anything about them, but which between them determine what a will be for any measurement given the angle α, and similarly for the other angles. A typical Bell inequality (there are in fact many such; this one is given by Shimony (Clauser & Shimony 1978)) among the correlations we can find here is then

$$V = |\varrho_{ab} - \varrho_{a'b}| + |\varrho'_{ab} + \varrho_{a'b'}| \leq 2 \, . \tag{18.2}$$

It is now not difficult to see that the quantum-mechanical result (18.1) violates (18.2) quite significantly; thus for $\alpha = 0°$, $\beta = 45°$, $\alpha' = 90°$, $\beta' = 135°$, the value of V in (18.2) is $2\sqrt{2}$! To understand the origin of this contradiction, we examine how (18.2) may be obtained.

(a) Bell (this is a slight adaption of Bell's original proof).

We assume locality, i.e. that a, for instance, does not depend on β but only on α and the hidden variables λ. Then we can write

$$\varrho_{ab} = \int a(\alpha, \lambda) b(\beta, \lambda) \phi(\lambda) d\lambda \, , \tag{18.3}$$

where $\phi(\lambda)$ is the statistical distribution of the λ's. We then have

$$\varrho_{ab} - \varrho_{a'b} = \int \phi(\lambda)d\lambda[a(\alpha,\lambda)b(\beta,\lambda) - a(\alpha',\lambda)b(\beta,\lambda)] , \qquad (18.4)$$

and from this and similar relations simple algebra leads to Eq. (18.2) and its violation by quantum mechanics.

(b) Another derivation is due to Wigner (1970) and (in more systematic form) Holt (Holt & Pipkin 1974). Since each of the four values a, b, a', b' can only be $+1$ or -1, there are 16 possible combinations, to which they assign probabilities $p_0, p_1, \cdots, p_{15}$, say; we can write the correlations in terms of them, for instance

$$\begin{aligned}\varrho_{ab} = p_0 &+ p_1 - p_2 - p_3 + p_4 + p_5 - p_6 - p_7 - p_8 - p_9 + \\ &+ p_{10} + p_{11} - p_{12} - p_{13} + p_{14} + p_{15}\end{aligned} \qquad (18.5)$$

The rule is simple: use all probabilities, giving them a $+$ sign when a and b are equal, a $-$ sign when they differ. None of the probabilities can be negative, and their sum must of course equal 1. Quite straight-forward algebra, with no further assumption, then leads once again to Eq. (18.2).

(c) A third derivation, due to Eberhard (1977) and Stapp (1971), is here presented as given by Peres (1978). He takes the four values $a, b, a'b'$ to refer to specific experimental observations. Now

$$ab - a'b + ab' + a'b = (a - a')b + (a + a')b' = \pm 2 \qquad (18.6)$$

since one of the parentheses must be 0 and the other $+2$, the b or b' at most contributing a sign change. If we have a set of n such quadruples, then for n large enough

$$|\frac{1}{n}\sum(ab - a'b + ab' + a'b')| \simeq |\varrho_{ab} - \varrho_{a'b} + \varrho_{ab'} + \varrho_{a'b'}| \leq 2 . \qquad (18.7)$$

An analogous argument leads to

$$|\varrho_{ab} - \varrho_{a'b} - \varrho_{ab'} - \varrho_{a'b'}| \leq 2 , \qquad (18.8)$$

(18.7) and (18.8) together yield (18.2) again.

(d) The last derivation considered here was given by Suppes & Zanotti (1981). They first demonstrate a simple theorem in probability theory, as follows. Let x, y, z be three stochastic variables taking only the values $+1$ and -1 and having means 0. Then there exists a joint probability distribution $\Phi(x, y, z)$ if and only if the pairwise correlations satisfy

$$-1 \leq \varrho_{xy} + \varrho_{yz} + \varrho_{zx} \leq 1 \pm 2\min(\varrho_{xy}, \varrho_{yz}, \varrho_{zx}) . \qquad (18.9)$$

This theorem is now applied to the four triples $\{a, a', b\}$, $\{a, a', b'\}$, $\{a, b, b'\}$, $\{a', b, b'\}$, to give (for the first inequality in (18.9) used on the first triple)

$$-1 \leq \varrho_{aa'} + \varrho_{a'b} + \varrho_{ab'} , \qquad (18.10)$$

and (for the second inequality and the second triple)

$$-1 \leq -\varrho_{aa'} - \varrho_{ab'} + \varrho_{a'b'} .$$ (18.11)

Summing (18.10) and (18.11), we find

$$-2 \leq \varrho_{ab} - \varrho_{ab'} + \varrho_{a'b} + \varrho_{a'b'} .$$

Combining several such inequalities, we recover Eq. (18.2).

18.3 What Do the Derivations Presuppose?

These four derivations represent the range of what has appeared in the literature; other proofs have been published, of course, but they are either variants
of combinations of these four, and we need not examine them in detail.

Now it is easily seen that – apart from very general suppositions concerning
the validity of mathematics and so on – the four derivations assume that

(1) We have four variables, a, b, a', b' that take the values $+1$ or -1.
(2) These variables associate pairwise to form physically meaningful relations
ϱ_{ab}, etc.

Beyond that only the first derivation requires a locality condition, and only
this one so much as mentions hidden variables. That no locality is involved is
particularly clear in Wigner's derivation, where any condition that for instance
the probability that $a = 1$ is the same whether $b = 1$ or $b = -1$ would take a
form such as

$$p_0 + p_1 + p_2 + p_3 = p_4 + p_5 + p_6 + p_7$$

Nothing of the sort is required; once the probabilities are assumed to exist and
the correlations can therefore be given in terms of them, the Bell inequality
follows. Analogous arguments lead to the same conclusion for the other two
demonstrations. The last is in fact purely probability-theoretic, and it is not
even clear how a locality condition could be formulated for it; and in the Peres
case, the sixteen possible quadruplets of values could occur with any frequency
whatever – even 0 – without affecting the validity of the argument, so that
again locality is not needed.

Here then we have four derivations that independently arrive at a certain
result; one of the four requires two stipulations (hidden variables and locality)
that the others do not need. The result the derivations furnish is not acceptable, because it conflicts with another, well established, theory and because
there is some experimental evidence against it. Can we conclude that the two
extra stipulations of the first derivation must be rejected? Of course not; for
the other three derivations would still be valid. We could only do it, in fact, if
these derivations were in some sense doubtful – which they are as little as the
first one is. That many physicists, including the author of the second derivation, have jumped to such conclusions seems to me to explicable at best if for
them "the wish was father to the thought".

But then, how is the contradiction between (18.1) and (18.2) to be explained? Why does quantum mechanics not satisfy Bell's inequality? The reason is quite simple: our two assumptions above are not complete. This is particularly clear in Peres' derivation. On the l.h.s of Eq. (18.7), all four variables occur twice; for the equation to hold, both occurrences must take the same value. This is trivial, but has implications when we consider Eq. (18.8). Here, if sums over products like ab are to yield physically meaningful correlations, the values a and b must come from the same pair; and so must a and b'. Now if we want to use (18.7) and (18.8), the occurrences of a in each summand must systematically coincide for all n quadruples, however large n may be. The simplest, indeed the only, way to achieve this is if both are the value given by one and the same A particle; but this is paired with only one B particle, which must therefore yield both b and b'. In a similar way we see that a and a' must both come from a single A particle in each summand of (18.8). Moreover, these measurements must be carried out in such a way that they do not interfere with each other: it should not matter if we measure b or b' first, because we must be able to pair either with the corresponding a value as if the pair a, b or the pair a, b' had alone been determined. And the same goes for a and a'. If the physical system is such that all this is possible we shall say that we have joint measuribility.

The Suppes-Zanotti derivation offers us the corresponding theoretical requirement: the existence of a joint probability distribution, which is the explicit protasis of their theorem. It is not difficult to see that this is the theoretical counterpart: from an appropriate joint distribution we can derive correlations between the variables; if the distribution is to apply to the system we study, we must be able to determine experimental values for the correlations, and to do this we must be able to measure associated pairs of values (or triplets or quadruplets, for higher-order correlations). If we can measure one value only on a given system and have to measure the other one on a different system, the association is arbitrary and the product sum is physically meaningless. Thus we arrive at the requirement of joint measurability. Inversely, given joint measurability, we can find experimental values for the correlation, and the theory must be able to compute theoretical ones to explain the experimental values; but it can only do this if a joint probability distribution exists for the variables being measured. Thus joint measurability and joint probability distribution are experimental and theoretical counterparts of each other, under the condition that the theory adequately fits the physical system.

Now the Suppes-Zanotti derivation explicitly involves joint distributions for all the four triples of our variables; the Wigner-Holt one has the slightly stronger requirement that the joint quadruple distribution exists, since the sixteen probabilities constituting it are taken to be well defined; in the Peres derivation we have explicit joint measurability, since the four variables are supposed to have been experimentally obtained (if they exist jointly but are not jointly measurable it becomes impossible to explain why theoretical predictions and experimental outcomes agree, as they seem to do); and in the Bell case it is stipulated that specifying λ permits determining all four values (at least in principle) from the angles $\alpha, \beta, \alpha'\beta'$ through Eq. (18.3), and their coexistence then allows a joint probability distribution to be written for them. This is necessary: note that in Eq. (18.4) three of the four variables occur within the same integral, associated with the same λ; and it is not possible to derive Eq. (18.2) if the integral is broken up and rewritten with different integration variables.

The conclusion we have formed is strengthened by observing that in all four methods of deriving Eq. (18.2) the rather remarkable correlations $\varrho_{aa'}$ and $\varrho_{bb'}$ are well defined, can be written on the same footing as the others, and occur in Bell-type inequalities; indeed, in one of the derivations they are explicitly used, see Eqs. (18.10) and (18.11).

It is clear, then, that the assumption (2) made above must be amended to read

(2') the dichometic variables possess a joint probability distribution and can therefore be measured without disturbing each other.

(The Suppes-Zanotti derivation relaxes this condition slightly, at the price of a condition on the means of the variables not needed for the other derivations.)

The violation of the inequality (18.2) by quantum mechanics is now trivially explained, for of course neither in the spin case originally envisaged by Bell nor in the photon-cascade experiments studied later do the variables satisfy (2'): we may measure a and b on a particle pair; we can even deduce a value b' provided the angle β' coincided with α, and similarly for a'; but these angles never violate the inequality, and we cannot determine all four projections (or polarisations) for arbitrary angles.

18.4 Classical Violations

It might be thought that only quantum mechanics can exhibit violations of the Bell inequality, because only quantum system ever lack joint measurability. But this is not so; it is a common situation, and most often occurs for properties which are affected in a fundamental way by being measured. The life time of a light bulb, for instance, which depends on the voltage applied in a well studied way, obviously cannot be measured on the same bulb for two different voltages. Therefore, no joint probability distribution of the life times at two different voltages can exist. If there were a bulb-making machine that produces them in closely similar pairs but with sizeable fluctuations from pair to pair, a Bell-type situation might arise. But to make this line of argument conclusive, examples that allow quantitative evaluation are needed. Several such are now known.

Consider, for instance, two points moving uniformly around two identical circles with angular velocity ω and in phase. If we measure their projections onto two straight lines inclined at angles α and β, we can write the results at time t as

$$a = \cos(\omega t + \alpha) \qquad b = \cos(\omega t + \beta)\,, \tag{18.12}$$

where we have taken the radius of the circle as 1. Then the correlation between the two projections (calculated as the time average over a complete cycle) is

$$\varrho_{ab} = \cos(\alpha - \beta), \qquad (18.13)$$

and so, if we again consider the same angles $\alpha = 0°$, $\beta = 45°$, $\alpha' = 90°$, $\beta' = 135°$ as before, the maximum of V in Eq. (18.2) is $2\sqrt{2}$, for except for the different sign, this is the quantum-mechanical correlation.

However, these variables a and b are continuous, and it might be objected that a non-linear transformation will generate new variables with a different correlation. Of course this argument could be turned around: for any system with continuous variables, we can find a non-linear transformation that produces new variables which violate the Bell inequality. But let us dichotomise our a and b by defining

$$\xi = \begin{cases} +1 & \text{if } a > 0 \\ -1 & \leq \end{cases} \qquad \eta = \begin{cases} +1 & \text{if } b > 0 \\ -1 & \leq \end{cases} \qquad (18.14)$$

The correlation $\varrho_{\xi\eta}$ is now a linear function of $\alpha - \beta$ and V never exceeds 2: the inequality is not violated.

We therefore add a small second-harmonic term to (18.12) before dichotomising according to (18.14):

$$\begin{aligned} a &= \cos(\omega t + \alpha) + \varepsilon \cos 2(\omega t + \alpha) \\ b &= \cos(\omega t + \beta) + \varepsilon \cos 2(\omega t + \beta). \end{aligned} \qquad (18.15)$$

Now we do violate the inequality. The calculation is not difficult though rather tedious, and it yields that

$$V \simeq 2 + 2\varepsilon$$

for small ε, and somewhat less as ε increases.

There are also stochastic models that exhibit a violation of the Bell inequality but otherwise are perfectly classical. The first one to be developed was given by Notarrigo (1983) and models the photon-cascade situation. The second one, due to Scalera (1983, 1984) is rather more abstract.

As final remark, it should be noted that the condition of non-joint measurability, though necessary, is not by itself sufficient to ensure that the Bell inequality should be violated. Complete necessary and sufficient conditions are not known.

References

Bell, J.S. (1965): Physics 1, 195
Bohr, N. (1953): Phys. Rev. 48, 696
Clauser, J.F., Shimony, A. (1978): Rep. Prog. Phys. 41, 1881
Eberhard, P.H. (1977): Nuovo Cimento 38B, 75
Einstein, A. (1948): Dialectica 2, 320 (I owe this reference to Dr. T.W. Marshall)
Einstein, A., Podolsky, B., Rosen, N. (1935): Phys. Rev. 47, 777

Holt, R.A., Pipkin, F.M. (1974): (Preprint) (Harvard University)
Notarrigo, S. (1983): (Preprint) (Univ. of Catania, Italy)
Peres, A. (1978): Amer. J. Phys. **46**, 745
Santos, E. (1985): in Bitsakis, E. and Tambakis, N.: *Determinism in Physics*, (Gutemberg Publ. Co., Athens)
Scalera, G.C. (1983): Lett. Nuovo Cim. **38**, 16
Scalera, G.C. (1984): Lett. Nuovo Cim. **40**, 353
Selleri, F. (1985) in Bitsakis, E. and Tambakis, N.: *Determinism in Physics*, (Gutemberg Publ. Co., Athens)
Stapp, H.P. (1971): Phys. Rev. **D3**, 1303
Suppes, P., Zanotti, M. (1981): (Preprint) Stanford University)
Wigner, E. (1970): Amer. J. Phys. **38**, 1005

19. Measurement and State Representation

Experimental physics is a highly developed and central part of physics. In spite of this undeniable fact there remains a great deal of confusion concerning the meaning of an experimental measurement, mostly in connection with problems in the interpretation of quantum mechanics. There appear to be two opposing points of view, which might, in a slightly caricatural way, be termed the Bohrian and the automatically realist view. To Bohr (Bohr 1935) we owe the conception that measuring device and measured system form an unanalysable whole; hence it is not possible unambigously to assign properties (and the values they take) to the system, nor – as a corollary – to understand the physics of the measuring device unless it is seen in interaction with a specific system to be measured. The opposite view, accepted by many as the only possible realist one, is that all measurement procedures, properly carried out, yield values of properties directly assignable to (i.e. *possessed by)* the system the measurement is carried out on. In both views, the problem of what the measurement tells us about the real world is decided once and for all, and seems therefore to require no further analysis.

Yet neither view is able to give a satisfactory account of problems such as the Bell inequality (Bell 1964, 1966) or the conclusions drawn from it (Clauser & Shimony 1978; Bruno et al. 1977; De Baere 1986; Brody 1986, 1989b). The Suppes-Zanotti theorem (Suppes and Zanotti 1981; Fine 1982a,b) shows that the violation of the Bell inequality by certain quantum situations arises because the four relevant variables do not have a joint probability distribution or, this being the experimental equivalent, because spin projections (or polarisations) cannot be measured in two different directions on the same particle (or photon) (Brody 1985); another way of stating this is that an unhappy choice of variables for describing the experimental situation raises an unjustified expectation that a joint distribution should exist (Brody 1989a). But this still leaves the fact that only one spin can be determined on each particle unaccounted for. The Bohrian view does not provide an explanation; it only suggests that we are wrong in looking for an explanation. The automatic realist view, on the other hand, implies not only that each particle carries information concerning the spin projection in all possible directions along which it might have been measured, though only one is realised; since all this information is intrinsic to the particle and therefore *exists* before ever the particle enters any measuring device, this view also implies that the Bell inequality should be obeyed.

The difficulty is resolved by adopting and developing further the extremely natural point first made by Margenau (1937), that both state preparation and measurement are needed for any experiment, yet are distinct. Among the wealth of different situations that may – and do – arise, there are some that have Bohrian characteristics and others that have automatic realist ones; yet a detailed analysis in which measuring device and measuring system are clearly distinguished still remains possible (in contradistinction to Bohr's view) and moreover necessary (as against the automatic realist view); on the other hand, the measured value will sometimes be seen to *belong* to the measured system, on other occasions to a larger system. In Sect. (i) state preparations and measurements, and the distinction between them, will be discussed and their relationship to the projection postulate (v. Neumann 1955) exhibited, along the lines established by Margenau (1937). Section (ii) studies how state-preparation procedures may be adapted to measurement. Section (iii) considers the application of this concept to the Bohm-Bell experiment (Bell 1964, 1966), while Sect. (iv) extends the analysis to successive measurements. Finally, Sect. (v) draws some general conclusions.

It should be noted at the outset that the discussion given here represents only a first step. It will be necessary to carry the argument further in at least two directions: (1) a number of concrete experimental situations should be analysed from the point of view presented here, in order to confirm and extend the categories given here; and (2) the assumption is here made that the compound system of *device + measured system* is adequately described by a deterministic model. Cases such as non-ergodic systems, where the measured value may depend critically on the measuring time, still remain to be studied; as e.g. Palmer (1982) discusses, qualitatively different phenomena are observed for different time scales.

(i)

Margenau (1937) observed what every experimentalist knows, what is in fact the basis of experimentation, but what philosophy of science – following in this the empiricist tradition – only rarely acknowledges: in order to obtain meaningful information, a measurement (or observation) is not enough. It is necessary to prepare the physical system to be studied in an at least partly known state, then to let it interact with known forces (due to appropriately determined fields, or to other systems in a correspondingly prepared state), and finally to carry out one or more measurements; and the result of the experiment is not the measured value but its correlation (in a very general sense) with the state preparation. Empiricist philosophy, in ignoring this fact and looking only at observation, created a false image of the process of knowledge acquisition as essentially passive, as a mere receiving of sense information, from which the outer world has somehow to be built up. Corresponding to this we have the view of scientific theory as an interpretative structure only; of its real function, fully one half – its role in guiding us through the selection, design, building and operation of the state-preparation devices – is thus swept under the carpet. In comparison with this, the question whether theory is merely a convenient calculating device or somehow pictures reality seems wholly besides the point.

The basic structure of an experiment is thus the preparation of a known state, its interaction with other systems or devices, themselves appropriately prepared, and the final measurement. The symbolic abbreviation $E = PIM$ is convenient. Some general comments on the three phases are necessary.

I, the interaction phase, may be interaction with an agency external to the system considered, in which case conservation laws for energy and momentum will not in general hold for the system alone; it may be the time evolution of a closed system, generally described by an evolution operator, some feature of which the experiment is expected to reveal; or it may be the identity operator, when the measurement phase M is immediately contiguous to the state preparation phase P. We shall not further analyse I here.

P may consist in an elaborate set-up which guarantees the production of physical systems in the required state; thus particle accelerators yield a beam of one type of particles with a certain energy, sometimes also in a determined polarisation state. P may be several independent devices, e.g. in collision experiments. At the other extreme, P may simply be an appropriate selection procedure among already existing systems; in the so-called observational sciences such as astronomy this is the predominant or the only state-preparation procedure. Commonly both types of state-preparation methods are required, either through an experimental arrangement such as anticoincidence circuits that select the desired systems among those coming from another device of P type, or by the selection among observed events of those that satisfy some criterion such as a transverse-momentum transfer within certain limits. It will be clear from these examples that state preparation, while logically prior to measurement, need not occur before it in time; the selection form, in particular, is often carried out after the measurement, e.g. when a computer does data-reduction calculations on previously registered experimental results.

Measurement, M, may be multiple; either because several properties are being measured simultanously on the same system; or because the interaction has resulted in several component systems now moving apart independently; or, finally, because the desired information is an ensemble property and has to be obtained from many repetitions of a measurement, as for an angular distribution or an excitation function in nuclear physics.

Measurements and state preparations are in one respect inverses of each other. When preparing a system in a state previously decided upon, then (to the extent the theoretical model on which the preparation procedure is based on works satisfactorily) the state of the system after preparation is known; its state beforehand, however, is not only not known, or only partly specifiable, it is also irrelevant. Measurements, on the other hand, yield information about the state of the system as it enters the measuring device; the system's state after measurement is, in general, of no interest. The only exception is a measurement that is also a state preparation for a following measurement; this special case will be discussed below.

The fact that the post-measurement state of a physical system is irrelevant is borne out by the observation that in very many cases the measurement process destroys the system, while in others it profoundly alters its state; only very exceptionally is the state unchanged by the measurement process. This is the case both in classical and in quantum mechanics; yet the distinction between measurements that do not alter the system's state and those that do, i.e. between those of the first and those of the second kind (v. Neumann 1955) has not proved useful in application.

But if the ulterior fate of a measured system is of such little significance, why has it received so much attention? Why, in particular, did v. Neumann propose the projection postulate precisely to describe it? I believe the reason to be simply the confusion between state preparation and measurement, between P and M. For P we do indeed have to know the system's state after the procedure; for M it is irrelevant; but if the two are confused, the requirement on P appears to be necessary for M, with the added complication (which now is insoluble) that M may have several outcomes while P has only one, chosen by the appropriate arrangement of slits or other selection devices. If the distinction between P and M is clearly kept in mind (and the experimenter must do so) the projection *postulate* is not required; instead there appears the need so to design the P devices that the systems produced by it are in the state corresponding to the preset (not measured) eigenvalue of the appropriate operator; i.e. what might be called a projection desideratum.

(ii)

While the *functional* distinction between P and M is clearcut, so that a given device may be either P or M in a given experimental set-up, the device that is M in one set-up could be used as P in another, or inversely; we now examine briefly the circumstances under which this is possible.

For a measuring device to be used as P, two conditions must be satisfied: (1) the measured system must leave the device with the measured property either maintaining the value found, or changing it in a foreseeable way; and (2) among the systems that leave the device it must be possible to select those possessing a predetermined value of the measured property.

Condition (1) eliminates all measuring devices that destroy the system or make it unrecognisable (e.g. the photographic emulsion for low-energy electrons) or alter unpredictably the measured value (e.g. the position variable a relatively long time after scattering) for the purposes of preparing systems in a suitable state before interaction and measurement. If P is done by a selection procedure after interaction, condition (1) does not necessarily apply, of course; but then two separate measurements must be done (one is the relevant M, the other supplies the data for P), and instead of (1) we have a compatibility condition.

A measuring device must be able to give at least two different responses. Condition (2) is needed to select those systems that possess the desired value of the measured quantity; it is therefore necessary (but not sufficient) to modify the M device by adding to it a selection device, if it is to be used for P.

In the opposite direction, the use of P as M, the P device may be employed as a binary M device if it is non-generating, memory-incomplete, precise and accurate. A device is *imprecise* if the state of the systems it produces differs in the mean from the nominal one; it is *inaccurate* to the extent that the state produced fluctuates statistically about the mean; these two characteristics, though important for practical applications, are only mentioned here for completeness' sake. A P device is generating if it generates the system type it produces from systems of other types, or if it builds them; it is *non-generating* if it accepts systems of the desired type and merely transforms their state into the required one. Thus a positron source, whether a positron-emitting radioisotope or an accelerator using pair creation, is a generating P device, and it is evident that there is no way in which it could be adapted to carrying out any measurements whatever on an incoming positron beam. Lastly, a non-generating P device is *incomplete* if only a certain fraction of the incoming systems is changed into the desired state; such a device, to be used for P, must contain some arrangement for discrimination between systems in the desired state and those in other states. Of course all real P devices are to a greater or lesser degree incomplete; but this incompleteness may be merely of statistical type, so that it carries little or no memory of the system's state before entry into the device; in a *memory-incomplete* device, however, the systems produced carry some information concerning their earlier state. It is these devices, and only these, that may be adapted to become M devices by extracting this information.

The Stern-Gerlach magnet is precisely of this type. It produces two beams of spinning particles (of spin $\frac{1}{2}$, one with its spin parallel to the field in homogeneity, the other antiparallel, from an incoming beam of particles with any spin orientation. The discrimination mentioned above is achieved by the directional separation between the two beams, allowing the absorption or deflection of the unwanted one. Note, however, that the information concerning the previous state is purely statistical: from the fact that a particle is in the antiparallel beam nothing except not previously perpendicular can be concluded about its spin orientation before entering the Stern-Gerlach field, while the complete incoming spin density matrix can be derived only from the set of relative frequencies of particles in the parallel and antiparallel beams for all orientations of Stern-Gerlach field. Thus a single measurement result furnishes at most the probability estimates to be derived from a sample of poor statistical quality.

The modification required to convert the Stern-Gerlach magnet from a P to an M one is the addition of a particle counter in each outgoing beam, so as to determine these frequencies; this conversion is possible only because of the presence of the second beam, of particles unwanted, if the device is used for P. That the incompleteness is essential if the modification is to be possible is seen if the device is made complete: place a spin-flip field into the antiparallel beam and unite the two beams again: now every particle has the same spin orientation, determined by the orientation of the Stern-Gerlach field, and no information concerning their earlier spin state can be obtained from any measurement on them.

The Stern-Gerlach field is a P device; when adapted for M, the value obtained is that of the particle's spin projection *after* it has been altered by the field; hence an individual observation cannot yield certain information about the spin state before entering the field, which is what an M procedure should provide; however, the relative frequency in the two channels does carry information of statistical character.

Many other M procedures are converted P procedures; the values measured with them must then be understood to apply to the systems after interaction with the device. In particular, polarisation measurements on electromagnetic waves may be analysed along the same lines as the spin-projection measurements discussed above. But other cases arise; e.g. the localisation of particles by means of a slit is a P procedure that can be converted to M, yet the interaction with the slit alters the momentum (significantly so for quantum systems) but not the position; hence the measured position is also that of the particle as it enters the device. There are also numerous M procedures that are not converted P procedures, and for them the measured value corresponds to the state of the system as it enters the device.

(iii)

These considerations are also valid for the simultaneous determination of two related spin projections, as in the Bohm-Bell experiment (Bell 1964, 1966), where a spin-zero system disintegrated into two spin $\frac{1}{2}$ particles the spin projections of which are measured along directions α and β, say. The resulting conclusions may be stated as follows:

1) The individual measurement result, e.g. $(+\ +)$ in an obvious notation, gives the spin projections *after* they have been altered by interaction with the Stern-Gerlach fields.
2) The determination of the correlation coefficients from a sufficiently large set of individual samples taken at all angles α and β yields statistical information about the spin state describing the incoming particle beam. Each correlation coefficient is, individually, a function of the instrument settings α and β, and represents the correlation of the outgoing particle beams.

In other words, because we measure spins (and also polarisations) by means of converted P devices that alter the spin projections, the spin correlations concern the spin projections as they come out, and not those that enter the devices. The latter – the spins possessed by the particles generated when the spin-0 systems split into two – may or may not satisfy the Bell inequality; that the post-measurement spins do not satisfy it has little bearing on the behaviour of the pre-measurement values.

The same conclusion follows also from the usual quantum formalism for the correlations. If the spin operators for the two particles (normalised to have eigenvalues ± 1) are σ_1 and σ_2, while the two Stern-Gerlach fields are oriented along unit vectors α and β respectively, then the correlation is

$$C_{\alpha\beta} = \langle \psi | (\sigma_1 \cdot \alpha)(\sigma_2 \cdot \beta) | \psi \rangle \,, \qquad (19.1)$$

where

$$\psi = \tfrac{1}{\sqrt{2}}(u_+ v_- - u_- v_+) u_1(\pm) u_2(\pm) \,, \qquad (19.2)$$

with $u_i(\pm)$ the spin states of the two particles along the reference direction with respect to which the angles α and β are measured.

Yet $\sigma_1 \cdot \alpha$, for instance, is the spin projection the particle will have after it has passed through the inhomogeneous field along α; this is not only confirmed by subsequent measurement, it also corresponds to the classical situation of Newtonian spinning particle: if its spin vector is $\mathbf{s}$, then $\mathbf{s} \cdot \alpha$ is not the component along α in the ordinary sense, because $\mathbf{s}$ is an axial vector; it is what the spin of the particle will be if we force it to rotate along α. If we wish to estimate the spin correlation *before* measurement, we must compute

$$C' = \langle \psi | (\sigma_1 \cdot \sigma_2) | \psi \rangle = -1 \,, \qquad (19.3)$$

which may be interpreted as the spin projection of one particle along the spin projection of the other. Here there cannot, of course, exist any angle dependence, and no Bell-type inequality is conceivable.

It is to be noted that the expectation (19.3) is obtained not only with wavefunctions of the form (19.2) but with any state of total spin 0; hence the conclusion given does not depend on any assumption of rotational symmetry about the axis of propagation.

This analysis of the Bell problem is dependent on the type of measuring device used in the experiments; if at any time a better understanding of particle spin allows us to build an M device for spins that is not a converted P device, quite different results will be obtained. Only a detailed analysis of how such a new instrument works could decide whether its results should be described by Eq. (19.3), or by another one; and therefore whether the observations made with it should satisfy the Bell inequality or not.

In previous papers (Brody 1985, 1989b) the author showed that all derivations of the Bell inequality tacitly or explicitly made the assumption that the four measurement results a, a', b, b' (obtained in directions α, α' for particle 1 in directions β, β' for particle 2) possess a joint probability distribution; or, equivalently, that a and a' are jointly measureable (i.e. without disturbing each other's value), and likewise b and b'. On the other hand, no locality assumptions need be made. The present analysis makes it possible to understand why in the Bell situation no joint measurability can exist, so that the Bell inequality need not to be satisfied.

References

Baere, W. de (1986): Adv. Electronics and Elec. Phys. **68**, 245
Bell, J. (1964): Physics (N.Y.) **1**, 195
Bell, J. (1966): Rev. Mod. Phys. **38**, 447
Bohr, N. (1935): Phys. Rev. **48**, 696
Brody, T. (1985) in E.I. Bitsakis, N. Tambakis (eds): *Determinism in Physics* (Gutenberg, Athens)
Brody, T. (1989a): Rev. Mex. Fis. Supl. xxx (PL)
Brody, T. (1989b): Rev. Mex. Fis. Supl. xxx (I: Joint)
Bruno, M., d'Agostini, M., Maroni, C. (1977): Nuovo Cim. **40B**, 143
Clauser, J.F., Shimony, A. (1978): Rep. Prog. Phys. **41**, 1881
Fine, A. (1982a): Phys. Rev. Lett. **48**, 291
Fine, A. (1982b): J. Math. Phys. **23**, 1306
Margenau, H. (1937): Phil. of Sci. **4**, 337
Neumann, J. von (1955): *Mathematical Foundations of Quantum Mechanics* (Princeton Univ. Press) (Original German version published in 1932)
Palmer, R.G. (1982): Adv. Phys. **31**, 669
Suppes, P., Zanotti, M. (1981): Synthese **48**, 191

20. On Quantum Logic[*]

Abstract. The status and justification of quantum logic are reviewed. On the basis of several independent arguments it is concluded that it cannot be a logic in the philosophical sense of a general theory concerning the structure of valid inferences. Taken as a calculus for combining quantum-mechanical propositions, it leaves a number of significant aspects of quantum physics unaccounted for. It is shown, moreover, that quantum logic, so far from being more general than Boolean logic, forms a subset of a slight and natural extension of Boolean logic, a subset which corresponds to incomplete statements. The philosophical background of this unsatisfactory state of affairs is briefly explored.

20.1 Introduction

The conceptual difficulties which accompanied the birth of quantum mechanics and still provoke endless discussion have given rise to many different attempts at resolution. Not the least extraordinary of these is what has become known as quantum logic.

The first fully worked-out proposal for an alternative logic as a solution to these problems is to be found in a paper by Birkhoff and von Neumann (Birkhoff, v. Neumann 1936). Their starting point is that the propositions expressing experimental observations on quantum variables do not satisfy the distribution laws of Boolean logic. For instance[1], if p is the proposition "the x-projection of the spin of an electron is $\frac{1}{2}$" and q is "the z-projection of the same electron's spin is $\frac{1}{2}$," then p' is the "x-projection of the spin is $-\frac{1}{2}$" because it must be either $\frac{1}{2}$ or $-\frac{1}{2}$. Hence

$$q \cap (p \cup p') = q \cap 1 = q , \qquad (20.1)$$

while

$$(q \cap p) \cup (q \cap p') = 0 \cup 0 = 0 , \qquad (20.2)$$

because neither the pair q and p nor the pair q and p' can be simultaneously true: the corresponding quantum operators do not commute. Thus the logic of quantum mechanics cannot be Boolean, and the authors conclude that it

---

[*] First published in: *Foundations of Physics*, Vol. 14, No. 5 (1984)
[1] The example actually used by Birkhoff and von Neumann was a little misleading and so gave rise to a confusing controversey (Popper 1968); to avoid this we use a somewhat simpler case.

must form an orthocomplemented modular lattice of propositions; that is to say, the set $\{p, q, r, \ldots\}$ of quantum propositions has the following properties:

(1) there exists a relation $\subseteq$ that establishes a partial order on the set, where $p \subseteq q$ means that whenever p is true then so is q;

(2) for any p, q there exists in the set an element $p \cup q$ (which is least with respect to $\subseteq$) such that $p \subseteq (p \cup q)$, $q \subseteq (p \cup q)$, and also a greatest element $p \cap q$ such that $(p \cap q) \subseteq p$, $(p \cap q) \subseteq q$;

(3) there exists a least element 0 (the false or absurd proposition) such that for all p, $0 \subseteq p$, and a greatest element 1 (the true or trivial proposition) such that for all p, $p \subseteq 1$;

(4) there exists an orthocomplementation $'$ which satisfies: $(p')' = p$, $p \subseteq q$ implies $q' \subseteq p'$, $p \cap p' = 0$ and $p \cup p' = 1$;

(5) the modular property holds, that is to say, for all p, q, r, if $p \subseteq q$, then $p \cup (r \cap q) = (p \cup r) \cap q$.

The challenge this paper presented was not taken up until the beginning of the sixties. Mackey (1963) based his approach on two types of primitive notions, propositions expressing observational results and states defined as functionals from these propositions to $\{0, 1\}$, i.e., the set of probabilities for the propositions to be true in a given case constitute its quantum-mechanical state description. This new concept was taken over by Jauch and his collaborators (Jauch 1968; Piron 1976); for the propositions, however, they postulate an atomic orthocomplemented lattice which is not necessarily modular. An atom here is a proposition a such that $0 \subseteq x \subseteq a$ implies either $x = 0$ or $x = a$; and a lattice is atomic if every proposition can be written as a union of atoms. Later authors (Gudder 1967, 1970, 1979) have weakened the structure from a lattice to an orthocomplemented partially ordered set (a phrase we shall abbreviate to oposet). The atomic axiom has, however, generally been accepted, though it raises problems we mention below. During these last decades, numerous philosophers, for instance Putnam (1969) or Suppes (1966), have adopted the "quantum-logical" point of view.

But what is remarkable in all this postwar work is the changed perspective. Birkhoff and von Neumann had pursued their work with the clearly set-out aim of resolving all the doubts and puzzles about the fundamental concepts of quantum mechanics by establishing that the only valid logic underlying this theory made questions as to the "real" values of position and momentum possessed by the particles quite meaningless and rendered inoperative any hidden variables behind the probabilistic structure; their formalism was in fact intended to underpin what is now, somewhat misleadingly, known as the Copenhagen school. Yet now we have Piron (1976, p. 2), for instance, writing

"The starting point was to take seriously Einstein's criticism of the usual interpretation of quantum mechanics and thus to describe a physical system in terms of elements of reality. The Hilbert space formalism was then obtained from a set of essentially nonstatistical axioms as the appropriate description of a physical system," while at the same time Gudder (1970, 1980) develops quantum-logical structures that admit hidden variables in a quite reasonable way. Thus the original intention has been reversed among the physicists who now work in this field; indeed, for many of them it is merely an alternative scheme of axiomatization that they build up, and no definite aim seems to underlie many of the papers.

This remarkable switch has not yet penetrated the philosophical literature, where a main aim is linked to something most physicists appear to have abandoned: the view that quantum logic really constitutes a logic in the full sense of the word and is to be interpreted as a viable alternative to Boolean logic.

The situation leaves numberless questions unanswered, and the present tendency would seem to be leave them unanswered. We comment on this point at the end of the paper; for the rest, our purpose will be to answer a few of these open questions. In Sect. 20.2 we ask whether quantum logic is indeed a logic and conclude that it is not, on the basis of five different points. In Sect. 20.3 we examine how far quantum logic, taken now simply as calculus of quantum propositions, satisfies the neeeds of quantum physics; it turns out to leave significant gaps, both on the experimental and on the theoretical side. Section 20.4 looks at the relation of quantum to Boolean logic, with the result that Boolean logic is not simply a special case, a sublogic, of quantum logic: rather, a straightforward and plausible extension of Boolean logic contains quantum logic. Finally, Sect. 20.5 is concerned with certain more general and philosophical points which have not so far received the attention they merit; among them is the matter of the atomic postulate.

20.2 Is Quantum Logic a Logic?

This question has been extensively debated; but much of the discussion is not relevant to our present purposes, being centered about the problem whether an alternative logic is, philosophically speaking, conceivable. "Logic" here has the meaning of a general theory of valid inference; its generality is proportional to its independence of the semantic contents of the inferences, and – though too frequently without a full realization of it – the debate has been largely about how far such independence can be achieved. The traditional viewpoint, that it can and must be an absolute independence, has been challenged, first by the construction of alternative logical theories (e.g., multivalued logics), and then by a more profound questioning of its basis through intuitionist logic; more recently the extreme view of total dependence has been proposed, leading to the incommensurability of scientific theories. A related question is whether the choice of appropriate logic can be made on the basis of experimental evidence; to put it more briefly (and more misleadingly) the question is, does logic have an empirical basis?

Here we need not enter into these matters, except to state the point of view from which the following arguments will be presented: while logic is certainly not an empirical science in the sense of responding directly to experimental or observational data, it is equally certainly an abstraction from experience and is subject to indirect confrontation with the real world, through the mediation of all the experimental sciences that make use of it; this mediation may itself be indirect, via the mathematical techniques employed. Thus alternative logics are entirely conceivable, though not likely to be restricted to particular sciences and even less to one specific theory. The real question is then whether quantum logic fills such a role, and it is this we shall examine, leaving arguments such as Putnam's (1969) aside. Five points arise.

1. The coexistence, for a given scientific theory, of two logics (in the full meaning indicated above) is inconceivable. This may be seen as follows: Let D_1 be the set of conclusions arrived at under logic L_1, and D_2 those for logic L_2. In view of the specific and restricted aims of scientific theories, we take D_1 and D_2 not as the sets of all possible conclusions, but as the sets of those which (up to now) have been drawn and found to be relevant to the theory. Three cases arise:

(a) $D_1 = D_2$: the distinction between the logics is irrelevant and it is plausible that a common logic can be found of which L_1 and L_2 are versions or sublogics;

(b) $D_1 \subset D_2$ (or vice versa): then L_1 is expected to be a sublogic of L_2 (or vice versa), and use of the latter is entirely adequate;

(c) the difference set $D_1 - D_2$ is nonempty, and L_1 and L_2 are incompatible.

With quantum logic we are evidently in case (c), case (a) being obviously inapplicable, while case (b) can be excluded by the fact that certain quantum propositions do not hold in Boolean logic (as they should if the latter were a sublogic of the former, i.e., if D_1 were a subset of D_2). Thus the choice between the two logics for a given derivation matters, yet no consistent criterion for choosing a logic can exist. For by means of such a criterion we could build a logic L' containing both L_1 and L_2, so that the situation reduces to a single logic.

If this is so, then Boolean logic would have to be eliminated from the foundations of quantum mechanics; but this would require the reformulation of the very considerable arsenal of mathematical techniques that quantum theory uses, a task that is not practicable and is conceivably impossible, in view of the restrictive nature of quantum logic.

This argument implies that two incompatible logics cannot be combined within a single theoretical framework, but does not eliminate the possibility of two independent theories in the same field using different logics; their eclectic combination, often needed in applications to concrete problems, might be awkward yet not impossible, since the exigencies of each specific problem determine, with little margin for ambiguity, the precise mix of theoretical concepts needed.[2]

---

[2] The "logical reconstruction" of semiclassical models of, e.g., nuclear structure on the basis of quantum logic on the one hand, and Boolean logic on the other, has not so far been attempted; it might indeed be thought to be impossible. Yet in view

2. If quantum logic is the only acceptable logic for quantum mechanics (indeed, if any but Boolean logic is required), the fact that the entire development of quantum theory could occur without it needs explaining; but to the author's knowledge no such explanation is yet forthcoming. Moreover, the Boolean-based derivation of quantum results should sometimes be wrong, yet no such misleading derivations appear to be known.

3. A logic in the full sense is intended to structure inferences and must therefore have a least one rule of inference that allows one from the truth of propositions of the form $\alpha, \beta, \ldots$ to conclude that a proposition of the form γ holds. In Boolean logic we have the *modus ponendo ponens*

$$\frac{\begin{array}{c} p \to q \\ p \end{array}}{q} \tag{20.3}$$

Other rules of inference can be derived from this (Hilbert & Ackermann 1949). But such a rule is necessarily a relation among propositions, i.e., among elements of the logic. In quantum logic this is not possible, even though a connective $\subseteq$ with properties analogous to the Boolean implication $\to$ exists; for with p, q in the lattice, $p \subseteq q$ is not necessarily an element of the lattice (or oposet, in the Gudder approach). The problem is not solved by introducing such rules as

$$\frac{\begin{array}{c} p \cup q \\ q' \end{array}}{p} \tag{20.4}$$

because with this a commutative sublattice would not be a proper Boolean sublogic. It is this lack of inference rules that has convinced many authors (e.g., Jauch (1968), Piron (1976), and Gudder (1971) that quantum logic is not a logic in the philosophical sense.

The so-called Sasaki hook, $p > q = p' \cup (p \cap q)$, has on occasion been suggested as a basis for inference rules. This does belong to the lattice, can enter into the *modus ponendo ponens*, and in a Boolean sublattice reduces to the material implication $\to$. But it has been shown to have significant weaknesses, under its alternative name of Stalnaker counterfactual conditional (Hardegree 1976): it is only weakly transitive so that many important argument chains cannot validly be formed with it; the equivalence $p \equiv q = (p > q) \cap (q > p)$ functions only when $p, p'q, q'$ all belong to the same Boolean sublattice; and there many physically interesting lattices in which the significant combinations $p \cap q$ (or even all of them) are 0, so that the Sasaki hook becomes useless. These restrictions on its applicability make it useless for practical purposes, and the problem of inference rules remains[3]

---

of the many highly succesful models of this kind, it would seem to be important to have more than just an intuitive account of their logical nature.

[3] Hardegree (1976) has described a kind of modal inference rule which can be constructed within quantum logic, and Mittelstaedt (1960, 1961, 1979) gives a description of inference in the sense of Lorenzen's operative methods. Since such proceedings require an even profounder restructuring of physics (and not merely of quantum theory), without affecting the validity of the other points made in this section, we shall not pursue these ideas here.

4. A further point also arises from the fact that the quantum-logical connective → is not always in the lattice. Together with it, the two-sided connective, i.e., the equivalence, and therefore all the statements of the essential properties of the lattice, covered in points (2) to (5) of the description in Sect. 20.1, will usually not belong to the lattice (or oposet). They form part of the metalogic for the lattice, and there is no evidence that this metalogic is anything but Boolean; in particular, if it were nondistributive, the use made of the quantifiers ∀, ∃ in deriving the three rules of inference for the first-order predicate calculus (Kleene 1952) would create obvious difficulties.

Now it may plausibly be argued that a necessary (though not sufficient) condition for a formal structure to be interpretable as a logic in the philosophical sense is that one interpretation should be the universe of its own statements; it can then serve as its own metalogic (questions of consistency and completeness apart). The heuristic behind this condition is evident: if it does not hold, the formal structure has only limited interpretations, the metalogic must be appealed to in applications so as to that the interpretation is valid, and thus this metalogic has a better claim to be regarded as a philosophical logic.[4] But quantum logic does not satisfy this condition, and therefore, if there is a logic in the philosophical sense, it will presumably be the Boolean one, which does.

5. Quantum mechanics achieves much of its generality as a physical theory because it reproduces, as a limiting case, the results of classical mechanics. If quantum logic is its underlying logic, it should provide for this passage to the limit. But it does not. At the simplest level, it does not contain Planck's constant h, and so the limit operation $h \to 0$ is not possible. More to the point (since the classical limit of quantum mechanics is not reducible to the – rather meaningless – vanishing of a universal constant), there is no way to develop the incompatibility relation, say, into a series the higher terms of which vanish in some approximation; indeed, the whole conception of limit laws for logics lacks precedents.

Thus there are considerable difficulties in the view that quantum logic is a logic in the full sense of the word. It would be preferable to regard it as the propositional calculus to be used for quantum propositions. However, this alternative presents its own difficulties, which are of two kinds. As we shall see in the next section, quantum logic has significant shortcomings as a representation of quantum physics; but also the view that it is not logic involves an apparently unacceptable problem. Quantum-logical statements possess truth values, and in view of their nature as experimental results, these values are of

---

[4] An uninterpreted formalism requires no metalogic; but when it is interpreted, the counterparts of its fundamental elements in the interpreting structure form the basis of a theory whose underlying structure is precisely the metalogic considered here. Thus the condition may be seen as a test of sufficient universality. It should be noted that even though a logic satisfies the condition (and this may turn out to be undecidable), this does not mean that it is the unique philosophical logic: there could well be other logics that also satisfy it. Nor does it mean that there must necessarily be at least one such logic.

the same kind as those of Boolean propositions. Indeed, as Maczynski (1971) has shown, the quantum-logical structure forms a Lindenbaum-Tarski algebra, as does standard logic, and in its Boolean sublogics the relations $\subseteq$ goes over into the ordinary material implication $\rightarrow$. Thus the same demarcation problem arises as was discussed in point 1 above: there can be no (logically formulable) criterion which tells us when to use quantum logic and when to use Boolean. Jauch (1968) tries to avoid the problem by accepting the universal validity of Boolean logic; he writes (p. 77) that "the calculus of formal logic, on the other hand, is obtained by making an analysis of the meaning of propositions. It is true under all circumstances and even tautologically so. Thus, ordinary logic is used even in quantum mechanics of systems with a propositional calculus vastly different from that of formal logic"; but on this view both the propositional calculus and ordinary formal logic should apply to the same propositions, giving rise in some cases to incompatible results.

In fact, Jauch (like most writers on quantum logic) does not carry out such a program but switches from one logic to the other on an intuitive basis. No attempt is made to formulate an explicit criterion. Another instance of this switch is the Putnam-Heelan debate excellently described by Jammer (1974, pp. 406–407). Now formally inconsistent theorizing of this kind is usual enough in physics, as has been discussed elsewhere (Brody 1981), and much of the strength of the science derives from it; but it is incompatible with the construction of the logical framework of a theory.

20.3 Quantum Logic as a Propositional Calculus

The main difficulty that a quantum-logical calculus has to face is its inadequacy in providing foundations for all of quantum physics. We shall consider this under the three headings of its description of relevant experimental data, of its relation to standard Schrödinger theory, and of the connections to the rest of quantum physics.

1. Quantum mechanics has an extensive experimental background; in quantum logic this is to be represented by the propositions that form the elements of the lattice (or the oposet). These propositions variously called questions (Mackey 1963; Piron 1976), yes-no experiments (Jauch 1968) or filters (Shimony 1971), express the results of dichotomic experiments, i.e., whether a certain condition is satisfied or not: the particle either goes or does not go through a slit, a measurement result falls or does not fall within a given interval, and so on. More complex results are taken to be built up from such propositions, each of which carries a single bit of information. The quantum state, defined as a probability measure over the lattice, assigns appropriate probabilities to all possible experimental outcomes to which these individual bits refer. The inevitable margin of error and uncertainty in all experimentation is thus not allowed for; but even if we accept this idealization, the experimental information available for formulating theoretical models is seen to be insufficient in two respects:

(a) Writers on quantum logic either state categorically or tacitly assume that experimental results can be expressed by yes-no experiments, be it directly or as equivalence classes formed from them (Piron 1976); but this leaves out several important types of measurement. One case is that of a correlation: it is obtained by averaging over pairs of experimental results, each expressible as a yes-no experiment (or set of them), but each result must be associated with one specific result of the other kind making up the pairs. If the logic is constructed by considering the n individual systems as one single "big" system, the correct pairing of yes-no value sets cannot be expressed in it (they are all propositions valid of the "big" system), and since the averaging can be described in the logic, any value between -1 and 1 can be obtained for the correlation coefficient. If, on the other hand, we consider n repetitions of the individual systems, the pairing may be correctly described, but now the pairs belong to different, though identically structured, logics, and the averaging procedure takes us outside them. Yet both descriptions are needed. A similar difficulty arises whenever one result is derived from two separate measurements and we must account both for the separate measurements and for the conclusion drawn from them: such is the case of $E - \Delta E$ measurements to determine both the nature and the energy of a particle. The difficulty is compounded when many data are needed to derive a single value, say when the momentum of a particle is derived from many succesive position measurements which yield the curvature in a magnetic field. A similar but more extended problem is that of determining a resonance energy by adapting a curve to a segment of the excitation function (each point of which is found from many data) and deriving its peak value.

(b) An actual experiment involves very much more than just the final observation. At the very least, the system requires a preparation procedure to ensure the right initial conditions (including the identity of the system components), an interaction that corresponds to the desired state evolution, and only then the measurement or measurements. Quantum logic has room only for the last; for though the state function does in some sense represent the preparation procedure (or at least its outcome), no way is specified for its experimental determination, and we cannot therefore decide whether a proposed procedure does in fact prepare the desired state. Yet the fact that a particle passed a certain slit is of no interest unless we also know enough about the experimental setup to decide if this is a theoretically expected or explainable outcome.

Thus the experimental information available within the quantum-logical structure is seriously impoverished.

2. Quantum logic as a propositional calculus should extend to all the propositions in the theory, or else another demarcation problem arises: if there are parts of the theory that require different deductive tools, on what basis is one to decide which to use? And without a criterion for this purpose, the theory breaks up into disconnected fragments. If therefore there are significant

aspects of quantum theory not covered by quantum logic, doubt must arise as to its validity as an underpinning for the axiomatization of quantum physics.

One such aspect has already been mentioned: the inability of quantum logic to account for the classical limits of quantum mechanics.

A second one springs from the quantum state concept. Quantum states are defined, following Mackey (1963), as a measure functions from the proposition lattice (or the oposet) to the interval $[0, 1]$ and therefore correspond to unit vectors in one specific Hilbert space. This creates a difficulty when linking scattering theory to quantum logic. Its "naive" version, due to Born (1926), requires nonnormalizable wave functions which are solutions of the Schrödinger equation but do not belong to any Hilbert space; more modern versions use a multiplicity of Hilbert spaces, usually one per scattering channel. It is possible to see these as subspaces of a single Hilbert space, but at the price of creating nontrivial problems concerning the existence of the scattering states (Amrein 1974; Pearson 1974; Reed, Simon 1979); sometimes a Banach space is made use of. The relation of such structures to quantum-logical descriptions does not seem to have been studied; but it seems likely that scattering theory could fit in only at the cost of quite profound reformulations. Similar situations arise whenever originally separate systems enter into interaction, or interacting systems separate: a single Hilbert space must factorize into the direct product of two Hilbert spaces, a state of affairs for which quantum logic does not appear to offer a model.

A third problem is that quantum logic offers only a static description of physical systems, as is evidenced by the nonexistence of a transition probability (Gudder 1978); Maczyñski's attempt (Maczyñski 1981) to define a transition amplitude is not satisfactory since it admits no transitions between eigenstates of the same operator, and in the classical limit all transition probabilities vanish; much the same problem arises with the definition proposed by Deliyannis (1973, 1976) and Cantoni (1975), as is already clear from the use made of it to establish the identity of states. In fact, these quantities are either identical with the overlap integral or else closely related to it, and contain no time dependence, as a genuine transition probability should. In order to describe the time evolution of quantum systems, one therefore has to introduce an evolution operator through a quite separate postulate, which is difficult to formulate except in Hilbert space, as in Jauch's (1968) approach. Here, then, quantum logic is incomplete, and one has to go back to traditional methods. Only the state vector is represented in quantum logic, and it does not completely determine the evolution operator, or, equivalently, the Hamiltonian. To see this, suppose a state vector Ψ satisfies

$$H_0 \Psi = i\hbar \partial_t \Psi ,$$

where H_0 is a suitable Hamiltonian. It will then also satisfy

$$(H_0 + H_1)\Psi = i\hbar \partial_t \Psi ,$$

provided Ψ is an eigenfunction of H_1 with eigenvalue 0; yet the future evolution of Ψ is affected, for if H_0 and H_1 do not commute,

$$e^{iH_0 t}\Psi \neq e^{i(H_0 + H_1)t}\Psi .$$

This problem is linked to that of scattering theory: The observable time evolution of a physical system implies that in some sense it is open, connected to the rest of the world. The time evolution of quantum systems therefore involves states that are not unit vectors in $L^2(C)$, the usual Hilbert space of quantum mechanics, for these satisfy the appropriate conservation laws (including those for energy and matter) and thus do not describe open systems.

3. Another indadequacy in quantum logic as a basis for a physical theory is that it has no loophole left for any extensions. In "ordinary" quantum mechanics the Schrödinger equation is only the first wave equation to have been formulated; extending it, first to cover spin, then by way of the Dirac equation and others to the relativistic case, held many difficult problems but was not impossible as a matter of principle. The difference is this: In ordinary quantum theory the central concept is that of a wave equation, and generalizing from Schrödinger's to other equations whose solutions require wider spaces than L^2 is quite natural. But in quantum logic an oposet of propositions is an object complete in itself, with no direct link to any underlying space-time structure, so that the attempt to extend it to something which is mappable to $L^2(C) \times L^2(C)$ rather than $L^2(C)$ can be expected to be a different theory; and in particular it is not obvious that it would contain present-day quantum logic as a special or limiting case.

Many of these problems have been excellently stated by Gudder (1978) and Mielnik (1981). Other difficulties of a more internal nature have also appeared, such as the fact that it is possible to formulate propositions that are meaningful in ordinary quantum mechanics but have no representation in quantum-logical lattices (Mielnik 1976; Cooke, Hilgevoord, 1981).

Lastly, a peculiar point may be mentioned: Certain observables in quantum mechanics, e.g., mass and charge, do not have fluctuating values; accordingly, theory treats them as (classical) parameters. Their place in quantum-logical schemes, however, is an open question.

We conclude from this discussion that quantum logic does not provide an adequate basis for the axiomatization of quantum theory. Certain of the deficiencies we have mentioned could well yield to future research; but their number and extent, together with their profound roots in the inadequacies of the quantum-logical account of experimental reality, raise this suspicion: any reworking of quantum logic may have to be so profound and thoroughgoing as to create a wholly new theory. And whether such a reworking is even possible one cannot, of course, foretell.

20.4 The Relationship of Quantum to Boolean Logic

Defenders of quantum logic in the philosophical sense commonly see it not only as a substitute for Boolean logic but as a generalization of it. We have seen above (Sect. 20.2) that the formalism of quantum logic is not a substitute, and is in fact not even complete since it will not accommodate an implication relation of a form suitable for inference rules. We propose to show here that quantum logic, so far from being a generalization of Boolean logic, is an incomplete form of a slight extension of it, the incompleteness consisting in its being a logic of incomplete statements.

Intuitively, a certain incompleteness is obvious. Thus for the spin example discussed in Sect. 20.1 we note that the explicitations given there,, e.g., that p is the proposition "the x-projection of an electron's spin is $+\frac{1}{2}$" or, more briefly, "x is up", are not complete. Quantum theory predicts not only that "x is up" but also that other spin projections are not determined.[5] But then, restricting ourselves to the only relevant spin directions, the three propositions become

$$p: \ x \text{ is up, } z \text{ is not determined;} \quad p = U(x) \cap N(z)$$
$$\tilde{p}: \ x \text{ is down, } z \text{ is not determined;} \quad \tilde{p} = D(x) \cap N(z)$$
$$q: \ z \text{ is up, } x \text{ is not determined;} \quad q = U(z) \cap N(x)$$

where we have used $\tilde{p}$ rather than p' because it is not the complement of p, while $N(x)$ is used for the proposition that the x-projection is not determined. This is clearly a stronger statement than $U(x) \cup D(x)$ since it expresses the relevant aspect of the Heisenberg inequality; one could in fact write

$$N(x) = \{\sim U(x) \cap \sim D(x)\} \ .$$

It is trivial to verify that now

$$q \cap (p \cup \tilde{p}) = 0 = (q \cap p) \cup (q \cap \tilde{p})$$
$$q \cup (p \cap \tilde{p}) = q = (q \cup p) \cap (q \cup \tilde{p}) ,$$

so that the logic is distributive and Boolean; the negation p' is seen to be

$$p' = \tilde{p} \cup q \cup \tilde{q} ,$$

where

$$\tilde{q} = D(z) \cap N(x) \ .$$

The same conclusion results if Boolean logic is restricted by an additional axiom

$$\{\sim U(x) \cap \sim D(x)\} \cup \{\sim U(z) \cap \sim D(z)\} ,$$

---

[5] We need not enter here into the thorny interpretation problems involved in deciding just why and in what sense the other spin projections are not determined; the fact that quantum theory does not determine them suffices in the present connection.

which merely states that either the x-projection or the z-projection of the spin is not determined; the alternation is of the two predicates $N(x)$ and $N(z)$ defined above.

These two formalisms are somewhat *ad hoc*. They cannot be extended to other, more complex situations; but they do suggest that in some sense quantum logic should appear as an incomplete Boolean logic, and not the other way around. The new feature which appears (most clearly stated in the first formalism) is that different propositions entering into the same discussion belong to somewhat different universes of discourse; since here this can be made explicit through the N predicate, completing the propositions to form elements of a Boolean logic is possible. But we must also introduce the general stipulation that every proposition in the logic must be of the form

$$P(x) \cap N(\Omega - x) \, ,$$

where Ω designates the space of possible arguments for the propositional function P.

But in general this will not work. Neither the special form, N, of the validity restriction nor the connective $\cap$ used to link it to P are of sufficient generality; and the completion will not always yield a simple Boolean algebra. Once the overly specific features are eliminated, we arrive at the general concept of a *molecular logic,* which may be described by the following (deliberately informal) rule set:

(1) *Elementary propositions* are those which are not further analyzed within the logic but for which additional qualifying circumstances have to be specified; they are thus either universal propositions or logical constants, and their relations are governed by conventional Boolean logic. Note that the elementary character of such propositions must be understood as relative to the particular deductive chain in which they occur.

(2) *Molecular* (or compound) *propositions* are formed by adding to an elementary proposition one or more propositions stating the required modifications of its meaning; these propositions could in their turn be molecular, but we shall only consider the simpler case when they are elementary.

(3) The truth value of a molecular proposition is a multicomponent structure; one component is the truth value of the basic elementary proposition, while the other propositions in the molecule furnish the other components.

(4) If a molecule c has a binary logical relation R to two others, a and b, then each component of c is a standard n-ary logical relation to the components a and b; this includes the basic elementary proposition c_0. A unary relation is defined analogously.

These four stipulations leave far too wide a range of possible structures open, and for each particular application further rules would need to be developed; such restrictions are also needed if a molecular logic is to have a suitable implication relation. Note that these logics are molecular extensions of standard Boolean logic, because the components of the molecules assume only the values "true" or "false".

It is convenient to impose the restriction that all molecules should have the same construction and in particular the same number of components; then any molecule can figure in any of the three positions of the relation $c = R(a, b)$. If some or all of the modifying components are now used to specify compatibility conditions, then one can very simply define both the set theoretical complement as a Boolean negation and the orthocomplement as a choice negation. This strongly suggests that the quantum logics be modelled in suitable molecular logics. That this is acutally the case can be seen below.

The structure of the various quantum logics is determined by the associated Heinemann diagram. In such a diagram, a downward-going line links two nodes which are connected by $\subseteq$); for an example, see Fig. 20.1. Evidently, the topmost node represents the trivial proposition 1 and the lowest one represents 0. From the point of view of a node, however, only the sequence of implications leading, in either direction, to 0 and 1 is relevant. If all nodes are numbered (in an arbitrary order), then the sequence of node numbers, with the node in question suitably distinguished, provides the

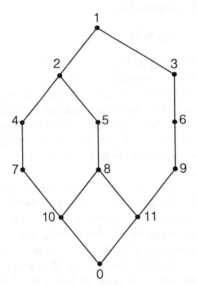

Fig. 20.1. A typical Heinemann diagram

necessary information. It is convenient to order the sequence by ascending and descending distance from the node considered; thus in Fig. 20.1 the sequence for the node marked 5 will be

$$S_5 = ((1)\,(2)\,5\,(8)\,(10\ 11)\,(0))$$

where nodes 10 and 11 are in the same pair of parentheses because they are at the same distance from node 5; the parenthesis sets on the left of the nonparenthesized number 5 are higher up, in order, while those on the right are lower down. Similarly,

$$S_3 = ((1)\ 3\ (6)\ (9)\ (11)\ (0))\,.$$

We then have that the meet or ∩ of nodes 3 and 5 is node 11, the first node number common to the right-hand halves of the two sequences. Its sequence is formed from the union of node sets at corresponding levels; node 10 is eliminated, however, because from a given node we cannot reach another one at the same level. Hence

$$S_{11} = ((1)\ (2\ 3)\ (5\ 6)\ (8\ 9)\ 11\ (0))\,,$$

and a glance at Fig. 20.1 will confirm that this is the correct sequence. For other operations the sequence are built up analoguously; e.g., the join ∪ of nodes 3 and 5 is clearly 1. The orthocomplement of a node is not explicit in the topology of the Heinemann diagram and accordingly is best indicated separately. Thus if node 5 has 4 as its orthocomplement, we may write for its complete environment the expression

$$E_5 = (4\ S_5) = (4\ ((1)\ (2)\ 5\ (8)\ (10\ 11)\ (0)))\,.$$

Because they can be handled in the computer by known techniques, we have written these expressions as so-called list structures (Brody 1968). Such structures can straightforwardly and in one-to-one fashion be translated into binary numbers, and their rules of combination can be expressed as (rather complicated) Boolean functions of the bits composing these numbers; since the details are not needed here, a formal proof of this – fairly obvious – result will be left for publication elsewhere, together with an account of the computer realizations of quantum logics.

The development of such structures is not intended as a serious working tool for quantum logic; though molecular logic could well find useful applications in other directions, the purpose of introducing it here was to exhibit in what sense quantum logic, so far from being a genuine generalization of Boolean logic, is an incomplete form of it. Admittedly molecular logic is in its turn a generalization of Boolean logic; if so, it is a slight and rather natural one, and it does exhibit quantum logic as the logic of incomplete statements about quantum systems: It is the logic we obtain from the appropriate molecular logic by simply ignoring all but the basic elementary propositions in the molecules while maintaining the relationships among the propositions given by the molecular logic.

20.5 Philosophical Aspects

As already mentioned, present-day formulations of quantum logic include the postulate that the structure (lattice or oposet) is atomic. A proposition a is atomic if no other proposition can be interpolated in the implication $0 \rightarrow a$, and the structure is atomic if all its elements may be written as unions of atomic propositions.

The atomic postulate is needed in order to achieve the restriction of what is otherwise a far too general logic to the quantum case. This need makes itself felt differently in different axiomatizations, and no proof of its universality is so far known; but it appears, in one form or another, in all cases. Thus in the work of Deliyannis (1973, 1976) the atomic postulate makes it possible to map quantum-logical states into Hilbert space. With Guz (1978) atomicity establishes a one-to-one correspondence between pure states and the atoms of the logic. This case is particularly interesting because here atomicity is not a separate postulate but derives from the two basic axioms that establish both the logic and the measure which defines the pure states. A third example is found in Piron (1976), where the atomic postulate plays a fundamental role in the description of the measuring process.

Thus, as Gudder (1978) has underlined, atomicity is an indispensable part of the structure. Yet it is physically unsatisfactory, on two counts:

(a) As already pointed out by Birkhoff and von Neumann (1936), we could not accept "x is a rational number" as a satisfactory statement of an experimental result, in view of the unavoidable experimental imprecision. They therefore propose to use only the Lebesgue-measurable subsets of the real line as possible experimental outcomes; yet a lattice formed from them in nonatomic and thus cannot serve as basis for the more recent versions of quantum logic, because now sets differing by sets of measure zero will belong to the same equivalence class.

(b) If the statement of an experimental outcome is atomic, so that it cannot be restated in terms of more elementary propositions within the framework of the quantum logic, then no further analysis is possible. (Combining several quantum-logical structures into one framework does not seem possible, as already mentioned above.) This conflicts with the constant need in physics to provide a full account of how experimental data are arrived at, so as to render possible their critique and review.

Birkhoff and von Neumann are clearly right in proposing Lebesgue-measurable sets as the appropriate starting point; the difficulties with present-day quantum logic crop up when, following Mackey (1963), one introduces a measure over the lattice or oposet for constructing the states, and so were not apparent to them. It might be argued that it is not sensible to impose this restriction to Lebesgue-measurable sets on quantum theory alone, for it is quite evidently valid also in classical physics. This would require a thorough reconstruction of all the underlying mathematics, beginning with analysis on the real line, and this is clearly beyond the bounds of practical possibility. It might therefore be prudent to accept the Borel-set structure as the starting point, even though it introduces a spurious element of unattainable precision – but only provisionally, and remembering that undesirable consequences may spring from it. And in view of point (b) above, the atomic assumption could well be seen as such an undesirable consequence. But unfortunately these conditions are often forgotten by those who work on quantum logic.

Point (b) must be added to the list of difficulties for linking quantum logic to an experimental basis. Moreover, since all propositions in an atomic lattice are built up from atoms, quantum logic now becomes an exact replica of the kind of physical theory adumbrated by the logical positivists in their earlier phases, e.g., by Carnap (1929) the atoms are what he called Protokollsätze-statements whose validity is in some fashion directly discernible and which are therefore not questioned, at least not within the ambit of the theory. The well-known weaknesses of such a philosophical position[6] create two problems for quantum logic:

One is the conceptual one implied by point (b) above: if the atomic propositions escape analysis within the theory, how can we establish that the set of such propositons we have chosen is the right one, neither omitting relevant ones nor including incorrect or extraneous ones? Any attempt to formulate criteria will either be based on quantum logic and must therefore suppose the propositions to be nonatomic and hence analyzable, or else will use another, conflicting, logic.

The other problem is that of verificationism. Carnap intended his concept of *Protokollsätze* with a justification outside the theory's scope to verify the theory and so to give it meaning. For quantum logic such questions have not been raised, presumably because it attempts to restructure an existing physical theory that has received ample experimental confirmation in its original version. But it should also be possible to ask how far experimental evidence upholds the quantum-logical scheme. Yet accepting the atomic postulate implies that any prediction made by quantum logic is reducible to a union of atoms and thus in no way goes beyond the original experimental evidence on which the theory is based. We must therefore either reject the atomic postulate or else suppose that only a proper subset of the atoms actually has an experimental backing. The first alternative creates the difficulties outlined above, while the second leaves unsolved the problem of how to analyze the experimenter's statements concerning the remaining atoms, how (in other words) to decide on their proper place in the theory.

Another aspect of the latent positivism behind quantum logic is its tendency to gloss over the distinction between theory and experiment.[7] It has,

---

[6] For a critical review from the standpoint of logical positivism itself, see Hempel (1965); other perspectives are discussed in Achinstein and Barker (1969), a collection of papers of which that by D. Shapere ("Notes towards a post-positivistic interpretation of science", p. 115) is particularly relevant to the present purposes. For a general criticism of logical positivism see, e.g. Cornforth (1946).

[7] This distinction is not as rigid as many traditional philosophies would view it: Indeed, though quite definite in any given case, it often has to be drawn differently in two superficially similar situations. It may be ignored where the theory is well understood and well established: This is, of course, a chief aim of scientific research. But in all those cases where either the validity or the precise meaning of a theory is in question it has to be drawn rather exactly, and that this may not be an easy task is no excuse for ignoring the distinction.

to take an important example, helped in perpetuating the confusion over the concept of incompatibility. It is an experimental fact that we cannot in general measure both the x- and the z-projection of a particle's spin; we expect quantum mechanics to account for this fact; but why should we expect the two relevant propositions to be incompatible, as quantum-logical systems have it? In a purely classical situation, NaCl may be either a solid crystal or dissolved in water; the two experimental cases cannot coexist; but the two propositions stating the facts are not therefore *logically* incompatible. Indeed, logical incompatibility, whether classical or quantum-mechanical, keeps us from discovering a physical account of the incompatibilty and is therefore undesirable. If an example is needed, it will be found in the usual discussion of the double-slit experiment:

We begin with the sentences

a_1: electrons in the beam pass through slit 1
a_2: electrons in the beam pass through slit 2
  b: the distribution of electron hits on the screen is the arithmetic sum of those obtained with only a single slit open

The argument now runs:

$$\frac{\begin{array}{c}(a_1 \cup a_2) \to b \\ \sim b\end{array}}{\sim (a_1 \cup a_2)}$$

Now the minor is experimentally quite well established; so if we find that the conclusion is unacceptable (as many people do), then either *the modus tollendo tollens* used here must be rejected (this has not so far been proposed) or the major premise must be impugned – on logical grounds.[8] But posing this alternative ignores the possibility that the major premise, though logically well formed, is factually incorrect; and a little further analysis shows that this possibility cannot be neglected. We can derive b from the following propositions:

b': the distribution of electron hits on the screen is the arithmetic sum of those obtained from the electrons going through one slit when the other slit is open, plus those going through the other one with the first one open.

  n: the distribution of electrons going through one slit is not affected by the opening of the other slit.

And now indeed $b' \cap n \to b$. But while b' is difficult to doubt as long as particle numbers are conserved, there seems not much reason to accept n. Even in classical physics this proposition cannot hold unless the (macroscopic) particle interact only by colliding with the slit walls and not with the medium

---

[8] The argument is stated in this form by Heelan (1970) and in a similar one by Fine (1972); the latter paper explicitly mentions the physical stipulation we discuss but prefers the terrain of logical revision.

through which they move. Think of the extreme case of corks carried by a water current through the two slits, and then close one of them: will the corks going through the other move as they did before? Now it is possible to consider that the fluctuations of the electomagnetic background field provide just such a medium for quantum particles (Brody 1983); but whether such ideas are viable or not is a matter for research in physics, and it is undesirable that a purely logical argument should hide the need for this research. Yet just this results when we use the logical incompatibility of propositions as an explanatory concept.

The substitution of logical principle for physical problem in quantum logic (whether considered as a logic or merely as a calculus) has quite generally been an inhibiting influence on further research onto the foundations of quantum mechanics; understandably so, for doing logic instead of physics has the subtle ontological implication that things are the way they are, and further questioning is useless. The implication is, of course, wholly unfounded, but since it is never spelt out it has great psychological potency. That is why, in the author's view, quantum logic has not helped us to reach a deeper understanding of the many unresolved issues in quantum physics, and why an increasing number of writers on quantum logic never seem to ask what its conceptual scheme implies or where it leads to, a deplorably pragmatic attitude which nevertheless steadily pervades physics.

Such difficulties are not peculiar to the specific direction the development of quantum logic has taken; in the last analysis they are inherent in the exclusive preoccupation with logical form from which these developments spring. However, a physical theory is very much more than the skeleton composed of its formalizable elements; to reduce it to this creates distortions for which we pay a heavy price (Brody 1981); in the particular case of quantum logic this price has been to obscure the road: the conceptual problems of quantum theory are very real, but they can be solved only by research into the physics involved and not by logical reformulations.

Acknowledgments. I am grateful to Prof. John Stachel for sending me the manuscript of a very relevant paper, to L. de la Peña for several useful discussions, and to two anonymous referees for a number of helpful criticisms.

References

Achinstein, P., Barker, S.F. (eds.) (1969): *The Legacy of Logical Positivism* (Johns Hopkins Press, Baltimore, Maryland)
Amrein, W.O. (1974): in J.A. Lavita, J.P. Marchand (eds.) *Scattering Theory in Mathematical Physics* (Reidel, Dordrecht) p. 97
Birkhoff, G., Neumann, J. von (1936): Ann. Math. **36**, 823
Born, M. (1926): Z. Phys. **38**, 803
Brody, T.A. (1968): *Symbol-Manipulation Techniques for Physicists* (Gordon and Breach, New York)
Brody, T.A. (1981): Rev. Mex. Fís. **27**, 583
Brody, T.A. (1983): Rev. Mex. Fís. **29**, 461 (and references cited therein)

Bub, J. (1974): *The Interpretation of Quantum Mechanics* (D. Reidel, Dordrecht)

Cantoni, V. (1975): Commun. Math. Phys. **44**, 125

Carnap, R. (1928): Der logische Aufbau der Welt (Weltpreisverlag, Berlin)

Cooke, R.M., Hilgevoord, J. (1981): in E.G. Beltrametti, B.C. van Fraassen (eds.) *Current Issues in Quantum Logic* (Plenum, New York) p. 101

Cornfoth, M. (1946): *Science versus Idealism* (Lawrence and Wishart, London)

Deliyannis, P.C. (1976a): J. Math. Phys. **14**, 249

Deliyannis, P.C. (1976b): J. Math. Phys. **17**, 248

Fine, A. (1972) in R.G. Colodny (ed.) *Paradigms and Paradoxes* (Pittsburgh University Press, Pittsburg Ill.) p. 3

Finkelstein, D. (1969): in same book as Putnam (1969), p. 177

Greechie, R.J., Gudder, S.P. (1971): Helv. Phys. Acta **44**, 238

Gudder, S.P. (1967): J. Math. Phys. **8**, 1848

Gudder, S.P. (1970a): in Bharucha-Reid, A. (ed.) *Probabilistic Methods in Applied Mathematics* (Academic Press, New York)

Gudder, S.P. (1970b): J. Math. Phys. **11**, 431

Gudder, S.P. (1978): in A.R. Marlow (ed.) *Mathematical Foundations of Quantum Theory* (Academic Press, New York) p. 87

Gudder, S.P. (1979): *Stochastic Methods in Quantum Mechanics* (North-Holland, New York)

Gudder, S.P. (1980): Int. J. Theor. Phys. **19**, 163

Guz, W. (1978): Ann. Inst. Henri Poincaré **A 28**, 1

Hardegree, G.M. (1976) in P. Suppes (ed.) *Logic and Probability in Quantum Mechanics* (D. Reidel, Dordrecht) p. 55

Heelan, P. (1970): Synthese **21**, 2

Hempel, C.G. (1965): *Aspects of Scientific Explanation* (Macmillan, New York)

Hilbert, D. Ackermann, W. (1949): *Grundzüge der Theoretischen Logik* 3rd edn. (Springer, Berlin Heidelberg New York)

Jammer, M. (1974): *the Philosophy of Quantum Mechanics* Chapt. 8 (Wiley, New York)

Jauch, J.M. (1968): *Foundations of Quantum Mechanics* (Addison-Wesley, Reading, Massachusetts)

Kleene, S.C. (1952): *Introduction to Metamathematics* (D. Van Nostrand, Princeton, New Jersey), p. 82

Mackey, G.W. (1963): *Mathematical Foundations of Quantum Mechanics* (Benjamin, Reading, Massachusetts)

Maczyński, M.J. (1971): Rep. Math. Phys. **2**, 135

Maczyński, M.J. (1981) in E.G. Beltrametti, B.C. van Fraassen (eds.) *Current Issues in Quantum Logic* (Plenum, New York) p. 355

Mielnik, B. (1976): in M. Flato et al. (eds.) *Quantum Mechanics, Determinism, Causality and Particle* (D. Reidel, Dordrecht), p. 117

Mielnik, B. (1981) in E.G. Beltrametti, B.C. van Fraassen (eds.) *Current Issues in Quantum Logic* (Plenum, New York) p. 465

Mittelstaedt, P. (1960): Naturwissenschaften **47**, 385

Mittelstaedt, P. (1961): Fortschr. Phys. **9**, 106

Mittelstaedt, P. (1979): *Rendiconti Scuola Int. di Física "Enrico Fermi"* LXXII (North-Holland, New York) p. 264

Pearson, D.B. (1974): Helv. Phys. Acta **47**, 249;

Pearson, D.B. (1978): J. Funct. Anal. **28**, 182

Piron, C. (1976): *Foundations of Quantum Physics* (Benjamin, Reading, Mass.)

Popper, K.R. (1968): Nature (London) **219**, 682

Putnam, H. (1969) in V.R.S. Cohen, M.W. Wartofsky (eds.) *Boston Studies in the Philosophy of Science* (D. Reidel, Dordrecht) p. 216

Putnam, H. (1974): Synthese **29**, 55

Reed, M., Simon, B. (1979): *Methods of Modern Mathematical Physics* Vol. 3: *Scattering Theory* (Academic Press, New York)
Shimony, A. (1971): Found. Phys. **1**, 325
Suppes, P. (1966): Philos. Science **33**, 14

21. Resistance to Change in the Sciences: *
The Case of Quantum Mechanics

In the last few years there has been much discussion of the problem of resistance to change within the sciences, and we have now moved beyond the positions of Kuhn (1962), for example, who put forward the notion that scientists usually cling fiercely to their "paradigms" and resist novel ideas; only a new generation, then, could establish an alternative paradigm. A position closer to reality distinguishes between resistance stemming from the social environment, and resistance originating wholly from within the scientific community, and which is generally expressed through scientific argument.

In this work I will refer primarily to resistance of the second type, not because I would deny the existence of the first – which is evidently of fundamental importance – but simply because the second has received inadequate treatment, being often reduced to purely psychological factors. For analogous reasons, I will not treat the (in my view very different) problem of technological change.

There are various aspects to the problem I wish to examine, the most relevant here being the philosophical ones. If conceptual change can be rationally justified, then resistance to new theories is basically unscientific and as such cannot be defended. But if different conceptual frameworks are in fact incommensurable, as suggested by Hanson, Kuhn, and Feyerabend, then there are no rational arguments edging us towards one framework rather than the other, there can be no generally acceptable way to justify a change of viewpoint, and therefore those stubbornly opposed to such changes should be applauded (since we cannot say they aren't right). The discrepancy here is both epistemological and ontological, between a world which can be rationally understood and a world in which rationality is at best local.

We must ask ourselves which of these two possible descriptions of change in the sciences corresponds to reality, describes adequately what occurs in some concrete situation. I propose to examine here the case of quantum mechanics, both its introduction to physics and its current evolution; and the answer, as we shall see, suggests a rather different state of affairs, at least in this case. Firstly, the change which actually occured is not the substitution of one theory by another but the demarcation of the limits of the region of validity of the older theory, and the incorporation of its results (though not of all of its

---

* "Resistencia al Cambio Científico: el caso de la Mecánica Cuantica" (Translated by C. Brody)

concepts), as limiting cases, into the new theory. Secondly, the resistance to change may be sometimes strong, sometimes weak; and – more importantly – it may sometimes have excellent rational justifications, while sometimes not. The answer to our question, then, is not some unique conclusion applicable to all cases, but rather an indication of the need to study each case individually.

21.1 The Birth of Quantum Mechanics

The model of the atom conceived by Bohr and Rutherford between 1900 and 1913 made manifest what had by then already been felt for some time: classical theories were not enough to explain why atomic spectra are composed of discrete lines instead of a continuum. The quantum concepts of emission and transmission of electromagnetic waves, principally due to Planck (1900) and Einstein (1905), broke with classical frameworks but did not offer a coherent conceptual structure to replace them. The 1913 Bohrian theory (today known as "the old quantum theory") for the first time attacked what we call quantum problems (for example, the theory of atomic spectra) in a way which had great explicative power and provided correct results. However, it had a series of internal difficulties:

- It contained a conceptual contradiction: within "permitted" orbits, Maxwell's laws of electromagnetism were not applicable, while outside them they were. Taken literally, such discontinuities could violate the principle of conservation of energy, or (even worse) imply that radiation emitted from an internal transition could not go beyond an orbit with bigger radius.
- It also contained a methodological contradiction. The emitted photon took with it an amount of energy calculated as the difference between the energies of the two orbits involved – which was quite natural if it really was a particle; but associated with this energy there was a number which one had to interpret as the frequency ... of a wave. And such an interpretation was essential, since it was the frequency which was observed in the laboratory. As we can see, the peculiar contradiction and mixture of waves and particles preceded quantum theory proper.
- It lacked systematic procedures. To solve a problem, it was first cast in classical terms; then the quantization conditions, whose selection depended in a very bothersome way on the set of coordinates one had chosen to describe the problem with (which of course should have had no physical meaning), were applied; and lastly, one had to derive the desired results in a way which was different for each case and which had to be guessed at.
- Finally, while the theory had provided excellent predictions for a whole series of problems, it gradually revealed serious deficiencies. While being eminently satisfactory for the spectra of hydrogen and all hydrogen-like metals, for helium it explained only half the spectrum and for heavier metals it failed completely.

That Bohr's theory could not be a definitive answer was evident, even to its own author. But despite much effort between 1913 and 1926, no more adequate model of atomic processes was found. (It is noteworthy to observe here that the history of these models is still to be edited: there is little doubt that among the failed attempts much interesting material could be found, both for physicists searching for ideas which "nearly" worked, and which reinterpreted could become valuable and relevant, and for historians trying to understand the true process of scientific creation).

This state of affairs led Darwin (1919) to state that "I have long felt that the fundamental basis of physics is in a desperate state"; and some years later Lorentz (1925) declared that "La mécanique des quanta, la mécanique des discontinuités, doit encore être faite." Many other such quotes may be found; dissatisfaction and preoccupation were in fact general. To make matters worse, the malaise, rooted in the depths of physics, seemed to be no more than a reflection of the profound decomposition of the "outside world": a disastrous world war, an economic crisis of hitherto unimaginable proportions, political, philosophical, moral confusion, and a new generation which threw itself into drugs, sex and anarchy in disappointment with what their parents' generation had offered. Note that we are speaking of 1925, not the 80's ...

Heisenberg and Schrödinger's (Heisenberg 1925, Schrödinger 1926) presentations of their work, within a few months of each other, founding essentially new concepts while simultaneously creating the basis for procedures which actually worked, caused great relief. The great novelty in methods, arguments and type of concepts introduced took people many months to understand – that even had to learn new mathematics (matrix calculus; though Jacobi had developed it in mid-nineteenth century, almost no physicist was familiar with it at the time). But in less than a year there was almost unanimous agreement that these were the solutions to a series of profound problems. Note that while Schrödinger's procedures used differential equations of a well-known type which were easily understood by all physicists, as soon as both approaches were shown to be equivalent it was Heisenberg's concepts and his intention to use only observable quantities that gripped scientists' imaginations, innovative and difficult though these concepts were.

I believe the reason for this can be found precisely in the general atmosphere of the times described above: Heisenberg, with his implicit black box philosophy, with his idea (later to be elevated into principle by neopositivists) that theories should not give explanations but should only present formalisms which connect postulates to predictions, offered a way out of the reigning confusion of ideas. A resigned way out, taken almost with abandon; but a way out nevertheless, where many had ceased to expect one. Furthermore, it was a way out which worked, gave predictions of extraordinary quality, and which could therefore be presented invested with all the authority of the sciences.

One cannot speak of important resistance to scientific change in this case. Naturally, some physicists were slow to understand the new concepts, or even denied them validity; but they were an insignificant minority.

21.2 The Revision of Quantum Mechanics

Heisenberg (and the so-called Copenhagen school) garnered the notions mentioned above from current trends in academic philosophy and transferred them to physics: his extraordinary genius allowed him to explore new avenues in physics under the guise of what were properly metaphysical ideas. Philosophers of neopositivist or idealist tendencies saw in these successes the full confirmation of their theories, in spite of the fact that the concepts quickly led to a whole series of paradoxical (or at the very least extremely strange) conclusions. At this, critical voices were heard, some of them belonging to physicists whose personal contribution to the development of quantum mechanics gave them every right to speak. They should not be lumped in with that backward minority mentioned above (though several ill-informed authors have done so), since their criticisms were attempts to enhance the solidity of quantum mechanics, not to reject it. I refer to physicists like Einstein, Schrödinger, de Broglie or (recently) Dirac; and other, younger researchers, like Willis Lamb. None of them wished to throw quantum mechanics' immense successes overboard. Rather, they were searching for a way to eliminate conceptualisations drawn from a false and inadequate philosophy, conceptualisations which have today led to the stifling of progress in the theory.

There have been many attempts to reformulate quantum mechanics; the literature is swamped in work in which the "definitive solution" to the conceptual problems of modern physics has supposedly been found. Most of this work reveals more imagination than knowledge of physics, and it is difficult to draw anything of value from it; but there are three lines of research which do have real significance.

Chronologically, the first is Slater's statistical interpretation (Slater 1929; see also Ballentine 1970). For Slater, quantum mechanics is a theory about great "ensembles" of physical systems over which averages are calculated; the Copenhagen school (which is neither a school nor is based in Denmark, but is so-called due to its being associated – not always correctly – with Bohr's ideas) believes that though one speaks of probabilities, these are applicable to a single system. The discussion, older than half a century now, neither progresses nor is resolved, but it is notable that theoretical physicists on the whole tend towards Copenhagen, while experimentalists are "statistical" often almost unconsciously); the gulf between the two branches of physics is what impedes clarity. One of the most important consequences of the statistical approach is that, as it stands, quantum mechanics is incomplete, in the sense pointed out in Einstein, Podolsky and Rosen (1935); but, how can we complete it?

For many years it was believed that von Neumann (1932) had shown that quantum mechanics could not be completed. Only recently (Brody 1983) has it been shown that von Neumann's argument, though correct, does not lead to this conclusion – rather, his argument suggests that any reformulation of quantum mechanics must be made on a stochastic basis. This fully permits the development of alternative conceptualisations which would reformulate (rather then merely add to) quantum mechanics: formal stochastic theories, started by Fenyes (1952) and Nelson (1966) (see also de la Peña 1969), presuppose a stochastic markovian process which would explain quantum mechanical movement without justifying this choice or saying anything about the physical nature of this process. The easy mathematical tractability of Markov processes has given them a certain advantage over theories for which we still lack adequate mathematical tools, like stochastic electrodynamics, which is due to Braffort et. al. (1954) and Marshall (1963, 1965) (later developments are described by Brody (1983) and de la Peña (1983)). Stochastic electrodynamics explicitly takes into account the fact that any atom exists in a world of many other atoms, formed by moving electrical charges. This implies that an electron inside our atom moves not only under the influence of the nuclear field, but also under the influence of random, completely uncorrelated electromagnetic radiation, emitted by the movement of all other particles in the universe. The result is a stochastic theory which offers a conceptual clarity and elegance which is in stark contrast with the dualities, collapsing wave functions, indeterminacies and other such notions current in Copenhagen. Other successes can be seen to lie ahead, as soon as we can overcome the mathematical difficulties; meanwhile, the fact that this reformulation eliminates the usual conceptual confusion (the "Great Quantum Muddle", in Popper's terms), has earned it the attention of many researchers.

These have not been the great majority of physicists, educated with courses and texts which teach quantum mechanics as derived from a series of postulates (which have no justification), through procedures (which fell from the blue) to give results (the physical meaning of which one is not allowed to investigate). But they have been among that constantly growing group of physicists who are no longer satisfied with such teachings, of a more scholastic than scientific spirit.

Though there is much discussion, there is still no acceptance, nor even a consensus that this is the right path to explore, despite the many years that have elapsed since the first articles were published. This seems to be due to several causes:

- As I mentioned above, many unfortunately worthless attempts to reformulate quantum mechanics have been published. Besides published work, a veritable rain of manuscripts and self-published articles, constituting a monument to the ignorance fostered by our educational system, is distributed every day. Amid this plague of madness a minimum of resistance is absolutely necessary, so as to filter out work without promise. But such a "filtration" process is slow and not at all trivial.
- Then there is the fact that meaningful reformulations (and in particular – in this author's opinion – stochastic electrodynamics) are still not complete; for example, no experimentally accessible prediction which differs from "normal" quantum mechanics has yet been found. The theory's main success have been the conceptual clarifications which it allows. For physicists raised in the philosophical traditions inherited from the empiricists,

this is not much: as long as it "works", that is, predicts data which is verifiable in the lab, what do we care about the structure of the theory? What do we care about explanations, as long as we can calculate?

- Another factor which has come into play is that the urgency to find a new fundamental theory is no longer felt as strongly as it was in the 1920s. Quantum mechanics seems safe because we have gone beyond it: already we have quantum electrodynamics, other field theories, and even unified theories; pioneer problems are to be found in elementary particle physics. Certain difficulties with quantum mechanical predictions seem unimportant. After all, there are also problems with the foundations of thermodynamics, statistical mechanics, or even with those of a theory as solidly established as Newtonian mechanics.

- In fact, it could be said that a general atmosphere, opposed to new fundamental theories (specially of this type), is dominant. I must emphasise the word *fundamental*: that is, a theory which would allow us to better understand the structure of our world. Other types of theories (including vast generalisations, such as the inflationary models of the origin of the universe) are accepted without difficulty.

We should note that this last tendency already existed in the 1920s: it explains why Heisenberg's conceptualisation was preferred to Schrödinger's, even if the latter's was constantly used. Science is useful, without a doubt, but it is also dangerous, inasmuch as it explains, inasmuch as people accept that it explains, for then they could demand explanations of why there is so much misery, so much war, so much oppression, amidst so much scientific knowledge. It is thus necessary to reduce its explicative power, and if the opportunity is there, a philosophy in which science does not explain, but merely calculates, is perfect. Naturally, nobody argues explicitly thus; but many let themselves be slowly pushed in this direction, sometimes unconsciously, merely because of the great fear too-convincing explanations give rise to.

Nevertheless, it is also important to note the form which resistance takes in the present case: few scientists have stated contrary opinions; only a few have formulated criticisms (some of which have been well-founded ones); no social pressure against anti-Copenhagenists has been felt; their work has been published with relative ease; and though funding for their work has diminished, this has not been to a greater degree than the reductions to their colleagues in a time of profound crisis.

Resistance has mostly taken the form of silence. While defenders of the ideological status quo are given the opportunity to expound what are sometimes even fallacies and absurdities (e.g. Bell's famous inequality, with which space and time in newspapers and TV has been given to parapsychological nonsense), the fact that there are differing opinions is not even mentioned. Actually, such attitudes are only due to a minority; the majority does not even know that there are alternatives to traditional quantum mechanics.

21.3 Change and Resistance

In the cases we have referred to (which would both require much greater examination) we observe a process of profound change; yet reactions to each are very different. In one case, we see universal acceptance, almost without resistance; in the other case, there is resistance with certain ideological bases.

It is evident that the explanation partly resides in the fact that one of the changes, that of the 1920s, supported (or could be used to support) an ideology satisfying some of the profound needs of the moment; while the other change is almost diametrically opposed to this, directing itself towards more explicative, more scientific (in the non-publicity sense of the word) theories, it does not lose itself in speculations whose foundations and capacity to tell us something about the world are doubtful. This type of resistance has its origin in the general social atmosphere, even though it manifests itself within the scientific community.

But the explanation is also doubtless due to the other, internal, factors I have mentioned: the quantum mechanics which is to be reformulated seems to have been already overtaken and left behind by theoretical research (though one may have doubts about such progress), and new theories still lack the possibility of experimental verification. Before reaching definite conclusions on these two cases, we must wait some years to see how the situation with respect to stochastic electrodynamics and the other alternatives develops.

We can, however, firmly reach one conclusion: the process of change is not the one assumed by theories currently in vogue. Firstly, we observe that in the 1920s, change was anxiously awaited; secondly, the new theory did not replace the old one. Rather, it absorbed the old theory, so that current quantum mechanics explains the concepts of old quantum mechanics, allows its experimentally verifiable results to be deduced, and shows where its limitations lie. The last point is particularly important: knowing a theory's limitations, we can use it where we know it works, and only there. And it is precisely in this way (inexplicable to Kuhnians) that old quantum theory is still used. Analogously, stochastic electrodynamics, far from denying the validity of quantum mechanics, intends to shows why it is valid, thus allowing us to deduce where it is and where it is no longer valid. Even more than with the case of old quantum theory, the old theory would here survive: many quantum mechanical techniques and concepts are definitely here to stay.

For physicists this point is not new. Let us quote Bragg (1912): "The problem becomes, it seems to me, not to decide between two theories of X-rays [particle-like and wave-like], but to find, as I have said elsewhere, one theory which possesses the capacity of both." Not only is his extraordinary vision notable here, but so is the great clarity with which theoretical progress is seen.

This kind of process is most clearly exemplified with the case of classical mechanics. Far from being replaced by relativity or quantum mechanics (or a combination of both), it has seen a renaissance in recent years, through non-linear mechanics and through chaos: neither Kuhnian nor Popperian notions seem to be able to explain this phenomenon.

Evidently, I do not mean to say that there are no replacements in the history of science. On the contrary: they occur frequently, in sciences which are still immature, in the sense that the experimental data predicted by their theories is not yet safe or complete. This results in experimental criticism which may lead to the elimination of much material previously thought secure, and thus to the abandoning of the corresponding theories, well-established though they may have seemed. Here, then, there is replacement. This process has on occasion provoked resistance by some scientists, resistance which cannot always be classed as rational, and in which external factors, not examined here, commonly intervene.

We cannot deny that even in the mature sciences many not well-established theoretical proposals are simply eliminated. In the first phases of their construction, many theories or models are eliminated by their own authors, specially when they discover that their range, their scope, either is null or insignificant, or is in essence already contained within a well-established theory.

Of the two cases examined here, the first (the introduction of quantum mechanics in the 1920s) is the one with the greatest – external – appearance of change through replacement, since it allowed us to practically forget old quantum theory. Nevertheless, it is precisely this case where resistance is smallest. In the second case, resistance is healthy and useful inasmuch as the new theory has still not presented adequate proofs, according to the reigning concepts of what a scientific theory should be; but these reigning concepts have significant, and not always scientifically justified, ideological roots. "Resistance through silence", then, cannot be fully justified.

In conclusion, we can see that it is impossible to say that scientific change always finds resistance. There may be resistance when the change is one where a theory will be wholly substituted by another, but while it is important to recognize that such changes are frequent in science, we must note that they do not apply to well-established or significant theories. Resistance also appears against still unconfirmed or unconfirmable because confused theorisations, or (currently) against theorisations which are equivalent in their predictions even if their conceptualisations are different; this resistance is, in good measure, a necessary self-defense mechanism against the introduction of anti-scientific concepts. Such resistance – part of the indispensable mutual ciriticism between scientists – is based much more on "internal" reasons which refer to the theoretical structures, than on external factors. Reinforcing this type of resistance, whose role in scientific development is very positive, is important.

I have examined here only one form of resistance to change in the sciences, that which originates from within the scientific community. And I have only tried, through the study of two specific cases, to eliminate some manifestly inadequate ideas. We are still very far from a coherent theory of scientific change, one which would not only explain obstacles in the way of progress, but would also explain its drive and the factors which determine its direction.

References

Ballentine, L.E. (1970): Revs. Mod. Phys. **42**, 358
Braffort, P.B., Spighel, M., Tzara, C. (1954): C.R. Acad. Sci. Paris **239**, 157, 925
Braffort, P.B., Spighel, M., Tzara, C. (1954): C.R. Acad. Sci. Paris **239**, 1779
Bragg, W.H. (1912): Nature **90**, 360
Brody, T.A. (1983): Rev. Mex. Fis. **29**, 461
Brody, T.A. (1984): Found. Phys. **14**, 409
Darwin, C.W. (1919): (unpublished paper, cited in M. Jammer (1966): *Conceptual Development of Quantum Mechanics* (McGraw-Hill, New York), p. 171
Einstein, A., Podolsky, B., Rosen, N. (1935): Phys. Rev. **47**, 777
Fenýes, I. (1952): Zeits. f. Physik **132**, 81
Heisenberg, W. (1925), Zeits. f. Physik **33**, 879
   see also Born, M., Jordan, P.: ibid. **34**, 858; Born, M., Heisenberg, W., Jordan, P.: ibid. **35**, 557
Kuhn, T.S. (1962): *The Structure of Scientific Revolutions* (Chicago Univ. Press, Chicago)
Lorentz, H.A. (1925): in *Collected Papers* Vol. 7 (1934) (Martinus Nijhoff, The Hague), p. 285
Marshall, T.W. (1963): Proc. Roy. Soc. **A276**, 475
Marshall, T.W. (1965): Proc. Camb. Phil. Soc. **61** (1965), 537
Marshall, T.W. (1965): Nuovo Cim. **38**, 206
Nelson, E. (1966): Phys. Rev. **150**, 1079
Nelson, E. (1967): *Dynamical Theories of Brownian Motion* (Princeton Univ. Press, Princeton, N.J.)
Neumann, J. von (1932): *Mathematische Grundlagen der Quantenmechanik* (Springer, Berlin Heidelberg New York)
Peña, L. de la (1969): J. Math. Phys. **10**, 1620
Peña, L. de la (1983): in B. Gómez et al. (eds.): *Stochastic Processes Applied to Physics* (World Scientific, Singapore) p. 428
Schrödinger, E. (1926): *Vier Abhandlungen zur Wellenmechanik* (J.A. Barth Buch., Leipzig)
Slater, J.C. (1929): J. Franklin Inst. **207**, 449

Part IV

General

22. Epistemological Implications of Artificial Intelligence *

(i)

The term "artificial intelligence" covers a multitude of sins against last century's common sense, all of which propose exploiting electronic computers for effects which in some sense correspond to human intelligence. The field's coverage ranges from automatic language translation to automated mathematics, passing through chess playing, medical diagnosis, and satellite photograph analysis, for example. Initially little more than an academic diversion, artificial intelligence has recently found a series of important and considerably interesting applications. The construction and industrial exploitation of intelligent robots is not unlikely in the near future; the possibilities are good enough for companies little prone to pursuing chimeras, like General Motors in the United States or Hitachi in Japan, to invest considerable sums in them.

The emergence of a new technology which breaks so radically with millennial traditions is naturally accompanied not only by a concerted effort to create a theoretical basis for it, but also by much discussion and concern with issues which are properly part of philosophy. During the first years, the central question was simply the *possibility* of a non-human, purely artificial, intelligence. After some initial vacillations, the problem seemed adequately defined in Turing's (1950) formulation:

Let us place a person before a computer terminal (something like an electric typewriter connected to another machine), and let the person engage in dialogue with whatever is at the other end of the line; if after some reasonable amount of time the person cannot say whether she is communicating with a human being or a computer, then the program controlling the computer's activity deserves being called intelligent. Unfortunately we now have evidence that Turing's criterion is not sufficient: various programs which maintain dialogues and have a high percentage of interlocutors who consider them human have been written without incorporating any features of intelligence; what they simply do is select some elements of the phrases given to them at the terminals, manipulate them with some simple rules, and return them in the form of slightly more general comments or questions which request further information; the rest is done by man, with his fatal aptitude for accepting

---

* "Implicaciones Epistemológicas de la Inteligencia Artificial" in: Revista Mexicana de Física **35** (Supl.) (1989) p. 89 (Manuscript written in 1976, translated by C. Brody.)

some comon remarks and amusingly phrased questions as if they were real communications requiring intelligence.

In spite of this, discussion about the possibility of an artificial intelligence has quietened down – perhaps because concrete evidence of some activities which we would call intelligent being tractable by computers has been accumulating. It is worthwhile to cite some examples:

Programs exist which are capable of playing quite complex games, like checkers, go and chess; they can reach a quite respectable level, in the simpler games they can be champions, and even in the most complex of them, chess, though they are yet to reach international master standing, they can easily beat most human players (Greenblatt et. al. 1967; Samuel 1954, 1967).[1]

There are a whole series of programs dedicated to mathematical problems. One is used to prove theorems in Euclidean geometry; it deduces them from the axioms (Gelernter et. al. 1960). Another solves integrals in algebraic form (that is, it doesn't calculate the result numerically, but writes out the corresponding algebraic expression), and would be capable of passing all integral calculus exams up to university level (Slagle 1961; Moses 1967). Another evaluates the Feynman diagrams which describe the physics of elementary particle interactions (Hearn 1966). Yet another sums the complex trigonometric series found in positional astronomy, a non-trivial problem since the series are infinite (Iturriaga et. al. 1966a, 1966b). There are also programs which can prove logical theorems (see, for example, Cooper 1966).

In a different field, there exist programs which take a rough drawing of mechanical pieces or architectural constructions, transform it into an exact drawing which satisfies engineering and other requirements and finally produce the drawings, plans and projections needed by workshops or engineers. Still other programs are used in computer manufacturing; they optimize the design of individual electronic circuits, plan the construction of integrated circuits (that is, those integrally built on minute silicon or germanium crystals), and optimize the millions of interconnections necessary to make a computer out of the electronic pieces.

There are programs which can decipher texts through optical devices or transform them into data which can be manipulated by more conventional programs; or which can interpret aerial photographs and transform them into high precision maps; or which can recognize objects in a scene shot through a television camera.[2]

The list could go on almost indefinitely. It is more important, however, to point out some of the essential properties of these programs:

1) It is possible for them to produce highly original solutions to some problems. A single example will show this: a theorem-proving program was asked

---

[1] In chess, advances have been great. Being built at this time (1990), is Deep Thought II, whose designers predict will leave world champion Kasparov far behind. (Translator's note.)

[2] Owing to commercial and military secrecy, the published information about these two fields is as incomplete as it is extensive. I have therefore preferred not to give references. (Author's note.)

to deduce the geometric theorem which states that in an isosceles triangle the two angles at the base are equal. We can all remember the proof given by Euclid, in which it is necessary to build a line perpendicular to the base which goes through the apex of the triangle and then show that the two resulting triangles are equal. The program produced a much more elegant and simple proof:

$AB = AC$ (given that the triangle is isosceles)
$AC = AB$ (consequence of the above)
$A = A$ (identity principle)

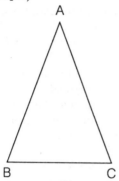

$\Delta BAC = \Delta CAB$ (that is, the triangle is congruent with its specular reflection)
therefore $B = C$.

That a relatively simple program should be capable of this level of inventiveness makes for a very promising future (Minsky 1971).

2) Many of these programs achieve their objectives faster or better than human beings: but none has flexibility. None could attack *all* of the problems I have mentioned – the human brain, however, can play chess, integrate, solve Feynman diagrams and design circuits, and people who can actually do all of these are not so rare. In the sense of a generalized capacity for problem solving, we still don't have intelligent machines.

3) This lack of flexibility, however, doesn't mean that we have not developed fairly general methods, to the point where it has been possible to create entire theories around them (see, for example Mesarović 1965) and create programs which can, in principle, solve any kind of problem (for example, Newell, et. al. 1960, and in slightly more sophisticated form Raphael 1964). The obstacle which prevents greater flexibility consists rather in that how to represent data coming from varying and unforeseen fields is not known. To put it another way, we know how the computer should handle its knowledge, but we don't know how to transmit and store this knowledge without excessive restrictions. Everything seems to indicate, however, that at least partial solutions to this problem can be expected in the next few years.

4) In almost all cases, the nature of the problem solved by these programs is such that heuristic methods are essential. Either the problem cannot be described with sufficient precision, or the number of paths to explore to find the

solution is astronomical (it is estimated that for chess it is of the order of 10^{120} ..., or both, so that it is not possible to conceive an algorithm which goes in a relatively direct manner to the goal. Among heuristic methods the one of most interest for us is that of learning: in its simplest form, the program has a series of criteria for evaluating each possible step, and on the basis of accumulated experience the program adjusts the priorities with which these criteria are used, emphasizing those which lead to success; more elaborate versions of the same idea are capable of creating new criteria, later submitted to the evaluation process, and of taking into account the particular circumstances of each success or failure. The program is therefore learning from its own experience. It is indispensable for this that it be informed of the results it obtains – that there be feedback from the environment with which the program interacts, in other words.

5) In the process of solving problems these programs (or at least those with the capacity to learn) take actions in their environment and receive information about the results obtained. Through this process, they form an important store of knowledge about the environment; part of the information can be found in tables of data, but a more essential part modifies, and is integrated into, the program itself.

6) A considerable part of these programs corresponds, in general, to logical deduction: but it is only a part. For example, the generation of new criteria to be employed in the learning process, the random selection of paths to explore, certain incomplete solution evaluation techniques – there is a whole series of methods which do not conform to the traditional rules of logic, which nevertheless play a decisive role in making the achievements of artificial intelligence programs possible.

(ii)

There is no longer any doubt that in these programs we have demonstrations, albeit partial ones, of intelligence; there is not even any doubt that sometimes the intelligence exceeds that of the originator – at least in the restricted sense that some of the game-playing programs can beat their programmers.

Another point which deserves attention is that of creativity. This concept is almost impossible to circumscribe clearly, maybe even more so than "intelligence". But if we consider creative an entity which produces more than one idea for each idea fed into it (assuming that the number of ideas can be counted), then many already existing programs are creative – and many human beings are not. The type of originality shown by the geometrical-theorem-proving program described above, for example, firmly supports this conclusion.

A philosophical problem of a somewhat different nature stems from the fact that intrinsic limits to the logical-deductive process are known; these limits are consequences of Gödel's well-known theorem (1931; see also Kneebone 1963) about the impossibility of proving the internal consistency of sufficiently complex axiomatic structures. One of these consequences is that we cannot write a program which could discover all the errors in other programs. The proof is both simple and instructive. If the program to be examined never terminates its execution, then it never produces any results, much less correct ones; termination, then, is a necessary condition. Now, let us assume there exists a program P which can determine whether any program will terminate: that is, it will answer T if the program P' which it is examining terminates, and N if it doesn't. We can easily add some extra instructions to P so that instead of returning T it continues running indefinitely (goes into an infinite "loop", as programs say), while still returning the answer N when appropriate. Let us call this new version Q. We have assumed Q will work on any program: this of course includes Q itself. But now we are in the contradictory situation of having a program Q (acting as the examining program) which terminates if and only if Q (the same program, but now acting as the examinee) does *not* terminate. This contradiction is enough to show that Q (together with its original version P) cannot exist.

This result and a series of other, more complex ones have given rise to a broad theory (see, for example, Davis 1958); what is most relevant here is that they constitute limits to what a computer can in principle do, so therefore intelligence is impossible in machines. Or, more or less, so the argument goes (a good example is Taube 1961). A first answer has already been given by Turing (1950): the same limitation is presumbably also valid for human beings. A second answer can be derived from the points mentioned above: both humans and machines use heuristic methods which go far beyond logic (ranging from the simple "try-it-and-see" to extremely elaborate techniques), while they pay the price of not *guaranteeing* the solution or even a good approximation to it; of course such methods must be apt and efficient, but one does not demand rigor or consistency from them – so they escape the limitations implicit in Gödel's theorem. Whether these limitations are truly important in practice, and whether the methods employed in heuristics not only go beyond logic but also contradict it are problems I will discuss on another occasion; for the moment, it is enough to point out that in my opinion, we can see in these methods which overtake logic yet are nevertheless rational, possible starting points for future extensions and transformations of logic.

(iii)

Among the properties pointed out above, one in particular gives rise to more profound conclusions. Point (5) summarizes the experience that programs with the capacity to learn gather information about their environment by interfering with it and observing the resulting changes. This goes directly against one of the principles of academic epistemology: information from our environment is received by our senses and passively absorbed. This principle is so deep-rooted in empiricist and positivist philosophies that its explicit statement is not even felt to be necessary; but it forms the backdrop to descriptions of how "sensory perceptions are impressed on our brain" (Berkeley 1710), to descriptions of how "an impression first falls on our senses and makes us percieve" (Hume 1739), to the sensory data that Russell and Moore attempt to identify, or to "the given" with which Carnap is satisfied "protocol phrases refer to the given and describe directly given experiences or phenomena" (Carnap 1932). The origin of the concept of passiveness is much older, stemming from the "simple ideas" with which Locke tried to eliminate the notion of innate ideas so as to start from his famous *tabula rasa;* and before him, it can be observed in Hobbes and even in Bacon.

Although almost undiscussed, this notion of our passivity with respect to a parade of sensory perceptions, whose analysis is therefore the only route to knowledge of the world, is fundamental. Once this has been accepted, we are led to the dilemmas of traditional philosophy, and most of all to the great question of whether the world we are trying to construct even exists, behind the barriers which the senses have erected between it and us: the terrible ghost of solipsism, which so many philosophers have tried to exorcise in vain, and which others, braver or perhaps more naive, have accepted, now threatens us.

On the other hand, if what the experience with cognitive mechanisms suggests is also valid for human perception, the world opens freely to our gaining knowledge of it. If our interference with the "external" world, our *actions,* are an essential element in the perceptual process, if our information about the world comes not so much from static, accessible observations but from inferences we draw based on deliberately induced transformations, then the results of a perceptual process contain within themselves their own confirmation. For example, if we know that a bottle no longer contains any wine, we truly *know* because we have opened it and tried to pour wine into our glass; and only experience accumulated over many years will let us reach the same conclusion merely by a quick glance at the bottle. The example is trivial and obvious; but it strongly suggests that any perceptual process must be an active one. It is thus worthwhile to examine in more detail the achievements of artificial intelligence with computer programs, where the structure of the process is accessible.

The electronics needed to connect a TV camera through appropriate circuits to a computer in a manner which simulates the connection from eye to brain presents no problems. The information received by the computer is a sequence of light intensities associated with the coordinates corresponding to each one of them in the visual field. A first method for extracting information from the observed scene could be the one Michie (1971) calls the monadic one: compare clusters of neighboring points, after some more or less preestablished transformations, with a table of patterns. In automatic language translation – a problem whose intellectual structure is very similar – this method corresponds to looking up each one of the words in the text in a dictionary and simply translating on a word-for-word basis. Little effort was needed to convince linguists of the absurdity of this procedure; but in visual image interpretation work along these lines continued for a number of years.

A second level is a structural one: instead of measuring distances and areas in the image, one measures angles and treats the problem with projective geometry. This is much more natural, since there is evidence (Johannson 1973) that via this route at least the human eye compensates for constantly occurring head movements, eye movements and object movements; but it implies we need to use at least *two* images (simultaneous if with binocular vision or successive if with aerial or satellite photography) to determine the projection point, where the observing apparatus is located. In linguistics this method has its analogue in the use of generative grammars introduced by Chomsky (1957). These second level techniques allowed a great leap forward in the quality of the results (Guzmán 1968; Clowes 1971), and many military projects in the United States used them for aerial photograph analysis. Enormous sums were also invested in automatic translation; but after a series of initial successes, progress came to a halt, for reasons analyzed, for example, by Bar-Hillel (1964) for the linguistic problem. For the perceptual case, they could be stated as follows: the method allows the analysis of very simple images; but as their complexity grows, the number of different possibilities which must be examined grows so rapidly that it soon exceeds the capacity of the biggest machines; and no possibilities for reducing the flood to manageable proportions can be seen.

In the United States this was considered for a time to be an argument in favour of building even more gigantic machines. A fact which should interest the historian of science is that the first indications of a way out of the rut came from the artificial intelligence group at Edinburgh, where economic restrictions had limited computers to at best medium sizes. The fundamental idea, which has taken root, is that instead of trying to get the maximum possible out of an individual image, one must instead simply obtain indications of how to modify the camera-world situation so as to obtain a second image which would better the interpretation, and then continue this alternation until an adequate amount of information has been gathered. Michie (1971), quite rightly in my opinion, calls this the *epistemic* level. It is worthwhile pointing out that at this level we obtain a description of the observed world which has an additional dimension: time, and in it, causal relationships. At the monadic level we compile dictionaries with no notion of the structures we have found. At the structural level we search for individual and static structures, without worrying about their connections and temporal evolution. At the epistemic level we proceed after the fashion of scientific research: a first glance suggests some possibilities of structures and links between them; the machine provisionally postulates some ideas of how these structures would change if this or that action were taken, and on this basis selects an action; a new image, from the modified world, is obtained, and it is now relatively easy to determine how far the predictions were correct; in case of unacceptably large errors, the process is repeated – as many times as necessary to obtain in the computer's memory a description of its world which is capable of *simulating* the world and its behaviour, and therefore capable of predicting what will happen if such and such an action is taken.

The epistemic method thus tries, as soon as possible, to obtain a certain amount of understanding of the world in which the computer and its camera can operate; understanding at least in the sense implicit in the possibility of reasonably correct predictions. Once this principle was established, it became obvious that the perceptual process could be bettered through giving the system a greater amount of possibilities of acting in its world. To changes in position and direction of the camera, changes in the lamp's position were added, which allowed extracting conclusions about the alteration of shadows, then an arm which could move individual objects was added, and this was provided with senses to judge weight and hardness of objects. Michie (1974) summarizes the experience as follows: "structural linguistic methods are incapable of taking scene analysis beyond a certain depth. Even at this depth, their exhaustive use is excessively cumbersome. Ambiguities must be dynamically resolved during the analysis, when they occur, through reference to an epistemic model. In fact, this semantic resolution of ambiguities must be done as soon as possible at each stage, instead of waiting until the end of the process." Other studies which reach similar conclusions are due to Minsky and Papert (1972) and Lighthill (1973).

The speed at which the epistemic level allows the cognitive process to proceed is high. Once the general structures of the world model have been well established, it is possible to isolate within the model those characteristics needed to make the necessary discriminations for each case. It is thus possible to develop, through practice, a series of short cuts, and to choose the shortest one; only if this fails do we backtrack and try some slightly longer route until we reach the goal. But we have paid a price in that the model and its structures must already be formed, and this process may take time; it is a learning process which even in the very simplified world we place the computer in must be organized with care to avoid turning the machine into a "psychiatric" case. Learning from the world is of course a never-ending process, since this world never ceases to change. If this aspect is not yet programmed into computers in general, its fundamental importance for human beings is evident; we can nevertheless include it in our considerations, since the obstacles are of a practical nature: in the very restricted applications of intelligent machines which are for the moment foreseen it is not needed; the principles and even the majority of the details of its realization are well known, since they are used in the initial learning which already forms part of so many existing, working programs.

Now, could we say that human perception employs similar methods? Unfortunately the evidence is quite incomplete; partly because experimentation with humans has its problems and in particular makes establishing physiological correlations very difficult, and also partly because psychologists as much as other scientists have been under the spell of empiricist epistemology and have simply failed to search in the appropriate directions. Nevertheless, there is already a collection of data which reveals part of the perceptual structures of animals and of their operation; J.Z. Young's work with the octopus, Ewert and others' with frogs, and most of all Hubel and Wiesel's (1959) with cats have amply demonstrated the extraordinary flexibility and adaptability of these mechanisms, even to the point of anatomical restructuring of the nervous system's connections in very young animals. Recently, Creutzfeldt and collaborators (1975) have extended the range of this flexibility quite considerably. Some of the more relevant results in human visual perception are due to Julesz in Bell Labs (see, for example, Julesz 1973), to Harmon (1971) and to

Johannson (1973); an excellent discussion of the results mentioned above can be found in Gregory (1966). Much work remains to be done, and the overall picture is still incomplete, but at least up to now nothing contradicts the epistemic model as valid for human perception while various features agree well with it. The situation is too complex for detailed discussion here. Possibly one of the best arguments available today is precisely the fact that in the field of artificial intelligence, after starting from models of a completely different nature, people have been forced to change their paths and to develop the method we have called epistemic.

(iv)

Without a doubt, the most fundamental implication obtained from the nature of the perceptual process just described concerns the relationship between the perceiving entity and the perceived environment: it is a complex process, in which error (that is, discrepancies between answers expected according to the model and those in fact provoked by the entity's action) play a decisive role in achieving the gradual adaptation of the model, and this cannot be conceived if the environment thus perceived does not possess a reality on which the entity can have an influence but which is independent of it. In other words, the environment *exists,* in the same sense and at the same level as the perceiving entity. What's more, the model which is gradually built and adjusted is a relatively faithful representation of the environment, not, of course, in the sense that if we see a mountain then we must have another mountain in our brain, perhaps smaller, but in the sense that the model adequately reflects the properties and evolution of the environment and allows us to obtain precise and specific predictions of how different elements of the environment would behave under different circumstances. What is reflected in the model, then, are the dynamic relations of the perceived environment, according to the language of physicists; that is, the relations (normally given in terms of causal interactions) which determine how the environment changes over time.

That this model provides an adequate representation of "external" reality is something that obviously cannot be demonstrated logically through deduction from a set of axiomatic bases; firstly because its construction and continual adjustment finds its justification *a posteriori* in its application – as with scientific theories; and secondly because the model is never perfect, always having approximations and errors, whose very existence is the motor that stimulates the ulterior adaptation process and is therefore vital, in the most literal sense of the word. "Errare humanum est"; if we wished to exaggerate, we could say that it is indispensable to err before being correct. But it would't be much of an exaggeration!

Therefore, the materialist ontology that we obtain from this epistemological concept is still not proven in incontrovertible fashion from universally acceptable first principles; such pretensions would be absurd; but I do consider that careful examiniation of what artificial intelligence is achieving helps to make this ontology plausible.

A second point of philosophical importance is the absence in the perceptual process we have described of a basic element that is definitely accepted without being subject to revision; there are no "atoms" in Russell's sense of the word, or "basic particulars" as Strawson postulates. Worse yet, each time we examine the reasons for accepting part of the analysis offered by the model, we see that it is based on the rest of the model – and these other parts are in turn based on what we are examining. Each part of the analysis is pushed just as far as necessary to direct our practical activities, and in other moments or contexts, this analysis may go farther or be pursued less. What in one instant is a starting point in another is a result.

The satisfaction of seeing atomistic epistemology vanish, arbitrary in its postulates and exposed to unsolvable contradictions, should not hide the fact that in consequence we are faced with a much less simple task: the justification of any perceptual model may no longer be found in its impeccable deduction from incontovertible bases, but only in the success it may have in guiding our activities. This formulation of the problem implies that we must examine what each model has allowed us a posteriori; it also implies that if success is not merely fortuitous but enduring, then we have reached a structural similarity between our perceptual model and the world which surrounds us: consequently, neither aprioristic concepts nor pragmatic simplifications will be acceptable. Both would cut out an essential part of the perceptual process, be it on one side or another.

There is another web of problems, which has obsessed a good part of philosophical discussion for the past half century, that fades away. Even if it is the case that we can observe certain phases in the process we call perception, and that it really extends in a continuous fashion into the cognitive, these phases are not ordered either logically or chronologically. In the programs employed by artificial intelligence laboratories we can see different levels; it is possible to speak about form isolation, structure recognition, the discovery of laws of change, the development of causal laws and connections, and the construction of predictions; and each phase involves different appropriate forms of interference and interaction with the environment whose interpretation is sought. But in terms of execution, switches occur from phase to phase which are not predictable, dictated only by the moment's convenience. This fact explains why it is that all attempts to isolate an ordered process, from the reception of the perceptual data to the crowning of the edifice through a deduction of the essential world structure, have failed. This is not to say that analysis of the process is not possible; it is, but not in the evidently inappropriate terms of concepts which depend hierarchically on others which are lower down the structure. We must accept the interdependence of all the elements, in both directions, and understanding the workings of this process is a difficult task in which we have progressed little; but it is nonetheless entirely feasible. What is most important is not to establish arbitrary demarcations: perception up to this point, from here on logic. The brain has greater common sense, in that it allows these two full collaboration.

In the computational program this interdependence is reflected by an initially surprising phenomenon. The structure of the program contains what, with some oversimplification, can be called a subroutine for each of the elements revealed by perceptual analysis; but these subroutines, in turn, require the use of a whole series of subroutines which represent the diverse categories of properties and the ensuing forms of dynamic evolution; and these subroutines, which we might call descriptive, contain features which can only be described through "elementary" subroutines. We are faced with a stituation where there are a whole series of subroutines linked among themselves because they employ subroutines from a different level, whilst they are also subroutines of the subroutines which employ ... This appears confusing at first, but it is not. If we consider two objects in the visual field of the apparatus, then their description, containing not only their momentary aspects but also their future interactions, must refer to the other object. Technically, we are dealing here with *co-routines,* each of which calls the others into execution; involved computational structures are part of recursive programming (Barron 1968; see also Brody 1968). What is most notable about them is that one part of a program can use any other part of the program – even itself, or the whole program – in such a fashion that here the part is greater than the whole. Of course, the paradox is no more than apparent, because it is realized only in the program's execution, whilst statically the set of instructions which makes up the subroutine is a smaller set than the whole program; but it is nonetheless useful to keep this pseudo-paradox close at hand, as it helps to avoid many of the confusions which emerge.

One of them involves the problem of consciousness. If the kind of perceptual mechanism of computers can be applied to human beings, then the model which he constructs for himself must include a representation of himself. We usually believe that the knowledge one has about oneself is as good or better than that possessed by other people; in the model, then, the representation of oneself must be quite complex, and it must include a representation of the representation of the world that one has. This is the start of an infinite regression which is unacceptable in a finite mechanism. But recursive techniques reveal that there is only a confusion here; regression is realized only dynamically, and its infinity is only potential, because in practice programs, (and presumably brain mechanisms) reach a variable and unpredictable level of recursion which is nonetheless always finite, limited simply by purely practical necessities. But, on the other hand, we might perhaps speculate that in the mere *potentiality* of an indefinitely profound recursion we have found traces of the kind of structure which might allow us to understand the elusive phenomena of human consciousness and self-awareness.

(v)

Naturally, the epistemological implications of the development of techniques in artificial intelligence which we have pointed out here are not new. Not even the model of the cognitive process as the structuration of hypotheses which are successively fine-tuned is new; it was sketched in by Pierce in

1903, surprisingly. Of course, Pierce did not provide an explicit and detailed description, such as can be provided today; moreover, he completely ignored the need we have underlined to verify these hypotheses through active interference with the world around us, and thus he fell easy prey to the extravagant notion that hypotheses are self-confirmed by their "pull" – the most peculiar essence of pragmatism.

What is entirely novel is the possibility that in the not too distant future we will be able to explore epistemological problems experimentally. We might wonder if the philosophical institutes of the year 2000 will have to install laboratories, or if, as we have seen many times, we are witnessing the birth of a new experimental science, soon to separate from its mother philosophy.

References

Bar-Hillel, Y. (1964): *Language and Information* (Addison-Wesley, Reading, Mass.)

Barron, D.W. (1968): *Recursive Techniques in Programming* (Macdonald, London)

Berkeley, G. (1710): *Treatise Concerning the Principles of Human Knowledge* Part I (Everyman, London), p. 1910

Brody, T.A. (1968): *Symbol Manipulation Techniques* (Gordon & Breach, New York)

Carnap, R. (1932): *Die Physikalische Sprache als Universalsprache der Wissenschaft* (Erkenntnis) p. 112

Chomsky, N. (1957): *Syntactic Structures* (Mouton, Hague)

Clowes, M.B. (1971): Artificial Intelligence **2**, 79

Cooper, D.C. (1966): in L. Fox (ed.) *Advances in Programming and Non-numerical Computation* (Pergamon, Oxford) p. 155

Creutzfeldt, O.D., Heggelund, P. (1975): Science **188**, 1025

Davis, M. (1958): *Computability and Unsolvability* (McGraw-Hill, New York)

Gelernter, H., Hansen, J-R., Loveland, D.W. (1960): Proc. Western Joint Computer Conf. **17**, 143

Gödel, K. (1931): Monatsc. f. Math. u. Phys. **38**, 173

Greenblatt, R.D., Eastlake, D.E., Crocker, S.D, (1967): Proc. Fall Joint Computer Conf. (Thomson Book Co., Washington D.C.) p. 801

Gregory, R.L. (1966): *Eye and Brain* (Widenfeld & Nicolson, London)

Guzmán, A. (1968): Proc. Fall Joint Comp. Conf. (Thomson Book Co., Washington D.C.) p. 291

Harmon, L.D. (1971): Abh. 4. Kongress d. Deutschen Ges. f. Kybernetik (Springer Berlin Heidelberg New York) p. 277

Hearn, A.J. (1966): Comm. Assoc. Computing Mach. **9**, 573

Hubel, D.H., Wiesel T.N. (1959): J. Physiology **148**, 574

Hume, D. (1739): *A Treatise of Human Nature* (Penguin Harmondsworth), p. 1939

Iturriaga, R., Standish, T.A., Krutar, R.A. (1966a): Proc. Spring Joint Computer Conf. (Spartan Books 28), p. 241

Iturriaga, R., Standisch, T.A., Krutar, R.A., Early, J. (1966b): *The Implementation of Formula Algol* (Carnegie Institute of Technology Memorandum, Pittsburgh, Ill.)

Johannson, G. (1973): *Perception and Psychophysics* **14**, 201

Julez, B. et al. (1973): Perception **2**, 391

Kneebone, G.T. (1963): *Mathematical Logic and the Foundations of Mathematics* (Van Nostrand, London)

Lighthill, M.J. (1973): *Artificial Intelligence. A general Survey* (Science Research Council Report, London)

Locke, J. (1690): *Essay on Human Understanding* Part II

Mesarović, M.D. (1965) in Sass, Wilkinson (eds.): *Computer Augmentation of Human Reasoning* (Spartan Books, Washington D.C.) p. 37

Michie, D. (1971): Experimental Programming Report No. 22 (Department of Machine Intelligence and Perception, Univ. of Edinburgh)

Michie, D. (1974): *On Machine Intelligence* (Edinburgh Univ. Press) p. 123

Michie, D., Ambler, A.P., Barrow, H.G., Burstall, R.M., Popplestone, R.J., Turner, K.J. (1973): Proc. Conf. in Industrial Robot Technology (Univ. of Nottingham), p. 185

Minsky, M. (1971): private communication

Minsky, M., Papert, S. (1972): *Artificial Intelligence: Progress Report* (MIT Artificial Intelligence Memo **251** (MIT, Cambridge, Mass.)

Moses, J. (1967): Ph. D. Thesis (MIT, Cambridge, Mass.)

Newell, A., Shaw, J.C., Simon, H.A. (1960a): Proc. Int. Conf. Info. Processing (UNESCO, Paris) p. 256

Newell, A., Shaw, J.C., Simon, H.A. (1960b): in Yovits, Cameron (eds.) *Selforganizing Systems* (Pergamon, Oxford), p. 153

Pierce, C.S. (1903): "The Reality of Thirdness and Some Consequences of Four Incapacities" in *Collected Papers of Charles Sanders Pierce* Vol. V (Harvard Univ. Press, Cambridge, Mass.) p. 63

Raphael, B. (1964): Report TR-2, Project MAC, (MIT, Cambridge, Mass.)

Samuel, A.L. (1959): IBM J. Res. Div. **3**, 210

Samuel, A.L. (1967): IBM J. Res. Div. **11**, 601

Slagle, J. (1967): Ph. D. Thesis (MIT, Cambridge, Mass.)

Taube, M. (1961): *Computers and Common Sense* (Columbia Univ. Press, New York)

Turing, A.M. (1950): Mind **59**, 433 (reprinted in Feigenbaum, Feldman (eds.) (1963) *Computers and Thought* (McGraw-Hill, New York) p. 11

23. Artificial Intelligence: Possibilities and Realities, Hopes and Dangers

(i)

In the last few decades, computers have begun to make a serious impact on our world. They started out as immense, delicate and extremely costly inventions, available to at most a few specialist mathematicians; now all banks, all government offices, and most middle-sized or large commercial offices have them. And the small computer, of a size that will fit onto the top of any desk and that has twice the capacity and ten times the speed of the biggest ones of 1950, – the so-called "personal computer", is being sold by the thousands and hundreds of thousands to anyone who wants the most marvellous toy of them all at a tenth of the price of an ordinary car.

Yet the real effects of these extraordinary machines will not appear until the way they are being used has changed in fundamental ways. At present they are mostly used for large jobs, things like calculating the weekly wages of a factory's workmen and employees, keeping the accounts of an insurance company, and taking care of the complex problems in the design of a new aeroplane. Scientists use them extensively to carry out in minutes computations that by hand would take centuries. But in none of these applications is the real power of the computer being exploited to the full; mostly we employ them to do vast (and sometimes nebulous or even useless) numerical calculations. Computers can do more, much more, than this. We are in fact beginning to discover how to make them intelligent. And this is creating a large number of new problems. Some of them are merely technical, and I will mostly avoid their discussion; some of them are more philosophical, and these are fascinating and have many and far-reaching implications; and some are concerned with the social impact of computers, which seem able to penetrate into every corner of our lives, for good or for ill.

But before entering into these discussions, we must talk a little about what computers really are and what they can do, for the great propaganda machines, from science fiction of TV to salesmanship of all kinds, have managed to surround them with every possible illusion; and not unexpectedly, people's reactions go from the extreme of thinking that they can cure all our social ills to the other one that they are the cause of all our troubles. The result is, inevitably, that most people feel that they are so far removed from any possible understanding of what this computer business is all about that they had better leave it to the experts. The experts, however, are those who work for the big organizations, the manufacturers, the banks, the military; who else could have paid for the enormous amount of engineering that went into developing computers? And so, consciously or not, the experts commonly represent the interests of these big organizations. Those of us who do not "belong" are left out in the cold, to suffer willy-nilly the consequences.

Some realisations of these questions has in recent years stimulated many groups into trying to make people aware of what computers are and can do, through all sorts of courses on "computer appreciation"; the word "numerate" has been invented, in analogy to "literate", in order to designate those who feel at home when talking about bits and bytes and binary basis. Unfortunately much of this effort is misplaced. True, people can and even should learn how computers work; but that does not enable them to tackle the serious problems that uncontrolled computer usage can create. Just as it is not necessary to know how a car engine works in order to understand and solve the endless problems that the excessive car traffic has brought into our towns.

(ii)

What, then, do we need to know about computers?

The first thing is that they do *not* deal in numbers. It is true, numbers are what one first thinks of; it was to handle numbers that they were first of all built. But it is not numbers that move through their complicated circuitry. Indeed, what is a number? Is 4 a number? No, it is only a symbol that represents a number, the same number as δ' represents, or simply four strokes, as $////$. Another representation is just a sequence of four electrical pulses; and it is this sort of representation that computers use. They handle pulse sequences, or patterns of pulses stored in their memories; at the output, in the printer connected to the machine, this representation is turned to the more familiar one, 4 in this case. But it is only we who associate the concept of the number 4 with this symbol. And when the computer "handles" numbers, it really does nothing but transform pulse patterns into each other, according to rules we have given it.

This is a philosophical point, but of enormous practical importance. Because it is now clear that we could give the computer other rules, make it print out other symbols, and then associate other meanings with them. And indeed, we already do this. If the very first computers built could only handle numerical rules and symbols, this was certainly not a limitation implied by the basic structure; it was only a failure of imagination on the part of the builders. Indeed, already the very first to think of computers, Charles Babbage and Ada Lovelace (who was the daughter of Lord Byron, then already dead at Missolonghi[1] had foreseen these possibilities in considerable detail. And today almost all computers can freely handle any kind of bit pattern (a "bit" is just a single pulse – or its absence). Among other symbols they can print alphabetic symbols, in a variety which is limited only by the amount of money we are willing to spend on it.

---

[1] In Greek in the original

The second point concerns what kind of "handling" our computers do. If we think of them simply as large and fast numerical calculators, the handling would be very restricted. "Add the number in memory cell 1107 to that in cell 4352" – that kind of thing. This would speed up long uniform series of calculations. But even the human calculator has to stop and think every now and then, and decide whether to go one way or another in his work. If the machine had no such power of choice, it would lose all its speed advantage when it has to stop and wait until the user tells it what to do now. Fortunately it is easy to build in decision handlers: "If the result is zero, then do so-and-so, otherwise just keep going". The criteria used in the decision are always some internal condition – whether a number is zero or not, whether some result has already been obtained or not, and so on. This is why the choice operation is simple to carry out. Moreover, it generally is a choice between two possiblities only. This seems altogether a very innocent, almost trivial, affair. But if we remember that the machine is also very fast, then it is clear that it could carry out a very long and complex sequence of such two-way decisions, and so arrive at a well-grounded decision between many different paths of action.

I said "well-grounded", in the sense that the machine could state in every case just how and for what reason it took a particular way. Indeed, by making it keep an internal record of what it did, we can get it to tell us how the choice came about. It might nevertheless be not at all well-grounded in the sense that we would have wanted it to make that choice under those particular conditions. It might, for instance, have written a cheque for Mr. A.X. $Μαρκόπουλος$ in Athens instead of Mr. A.X. $Μαρκόπουλος$ in Patras, simply because it found the right name in the Athens list, and no check in other lists was foreseen. Programmer's error. Of course. But the point is that the machine did exactly what it was told to do, not what the programmer wanted it do. But perhaps the man's name really was different, say $Μακρ^πουλος$, and the typing error was not found out in time. In that case the machine would have done no worse that the human being, but the error may be more difficult to correct; and the man who did not get his money in time is equally annoyed.

Two different points appear here. The first is that the computer makes mistakes. It is not the perfect, all-knowing super-brain that science fiction paints. It makes mistakes for three different reasons: because it breaks down, because there are mistakes in the program, or because it was given erroneous data. The first is a very rare occurrence; not that they do not break down, they do quite often, but they are engineered so that they can continue from just before the breakdown, without losing any data. A machine error that is not detected thus happens extremely rarely. But program errors are frequent, and errors in the input data are a constant headache. Of course one takes precautions, one uses the computer itself to do as much checking as possible – but there are limits to what one can do, both because it takes time and because it costs money.

In the second place, we have this business of the program. This is really nothing but the sequence of instructions that the computer gets. The program must be stored inside the machine, for if at every step the operator had to tell it what to do, all the speed advantage would be lost. And this advantage is one of the decisive points in favour of the computer, of course. But to write the program, one has to think out very carefully beforehand what the machine should do, one has to foresee every possibility that might arise and deal with it correctly, one has, finally, to test the program in every possible way. Once one has grasped the basic principles, it is easy to write a small program, with maybe a hundred or so instructions; yet an individual instruction does only very simple things, and so a program that deals with an intricate problem may be very long indeed: programs with tens and even hundreds of thousands of instructions are no longer exceptional today. Writing such a program is something that no one person can undertake, it must be a group effort, and organizing this work in such a way that everything works and all mistakes are weeded out – that is indeed a difficult business. The computer itself is used for this purpose; bit patterns are now no longer number representations, they are instructions, initially in a way that the human programmer finds easy to understand, and then – after the machine (under a special kind of program known as a compiler) has translated them – instructions of the form the machine understands. This implies that the instuctions have the same format as numbers in the machine; a point that many people who have not grasped the fact that the machine does not handle numbers but only symbolic representations find difficult to understand.

If instructions look like numbers, can the machine do calculations on them? It can. It can therefore modify its own programs. This, of course, is just what the compiler does, for a program that will be run afterwards, when it has finished. But the machine can also modify the program that is running at the moment. And now we come into deep waters. For a program that does not modify itself, we can (at least in principle) foresee just what it will do; it may be very difficult in actual fact, and large programs often have such a varied and sophisticated behaviour that the machine user, sitting at his terminal, gets the feeling that he is battling with a sentient and quite human being "in there", so that his applying a human psychological terminology to it is very natural: the program "chooses" to do this or that, "remembers" something and "forgets" something else, and so on. Yet if the program can modify itself (of course in ways that do not destroy it), then the picture changes. Now these modifications can depend on the data the program is given to work on; and since it is not difficult to provide the computer with detailed control over the choice of input, we can now no longer, not even in principle, foresee what the combination of computer plus self-modifying program will do.

One simple point, however, before we go into these questions. It is that the computer's memory can receive for storage a finite but often quite large number of data, in the form of bit patterns. Each storage place is known by a different number, which is called its address, exactly as houses have street addresses by which they can be found. But the address in a computer memory is itself a number; we can therefore store it somewhere else and can manipulate it. In fact, an address can point to a place where we have stored two addresses side by side; one points to where some specific information is stored, the other points again to another such address pair. The first address in a pair might also point to the beginning of a different such sequence of address pairs, and somewhere in this sequence there might be another such pointer to a subsequence. And so on. There is no limit to the complexity of this sort of structure we can build up – except, of course, the memory capacity available. And these structures may be manipulated by a program; they can grow and they can change – and any part can also be wiped out again. Such structures are one very commonly used way (but not the only one) for representing in the machine's memory notions much more complex than simple numbers, notions, in fact, that themselves are built up from more elementary notions. And it is possible to "forget" them again. One particular kind of structure of this type is a program; and this program, which can manipulate such structures, can now clearly manipulate itself, too.

(iii)

From elements of this sort, one can now build up more complex ones that come much closer to intelligent behaviour. Of the many technicalities involved, I will mention only one: recursion. Normal, so-called iterative, programs contain many sections that are repeated over and over again; indeed, if this were not so, solving a problem by computer could well take much longer than doing it by hand; but usually the results of one repetition are used simply to start off the next one and then are forgotten (this next repetition simply writes the new results into the same places where the old ones were stored, wiping them out). In recursive programming, the old results, even the intermediate ones, are kept; one can therefore at any point break off and do something else; one can even, with other data, restart the same program section. If the previous execution of the program section was not complete, then one keeps a record also of where it was broken off, so that when the interruption is finished, the computer can go back and complete the earlier work.

The chief reason why recursion is so important is that it allows unexpected breaks in the program execution to occur; unexpected, that is, in the sense that the writer of the program had not foreseen them.

With this and other programming methods, one can now write programs that can "learn": at the lowest level, these are programs that receive from the outside information about the results of their actions, whether they were a success, just where they went wrong, and so on, and that then use this information to evaluate the rules they have been using. They eliminate rules that make them fail, they give greater weight to better rules, and more elaborate programs of this kind also evaluate how the rules combined: "Do A first, and then B" gives very good results, "Do B first, then A" is disastrous, "Do B first, then C, and finally A" is still better. At the next level, the evaluation and readjustment of rules goes on while the program is running. Such techniques have given us draughts programs that are world master, chess programs that beat perhaps 99% of all players (without yet reaching the Grandmaster class), and so on.

One can also write programs that have at their disposal a very large stock of rules and other kinds of information, and which have the ability to reason about all this. For if we can represent both numbers and letters, addresses and instructions by means of bit patterns, we can also represent "true" and "false" and combine them with the appropriate rules taken from logic. In fact, this is particularly easy, because the two bit values, 0 and 1, can just be used for the two logical values (in some machines the 1 is "true", in others 0 is "true": it does not matter which convention one chooses – so long as one sticks to it). This kind of program has become known as an "expert system", because the detailed information it works with is gathered from one or several experts in a given field. Some of them, such as *Dendral* for the evaluation of spectrographic data for organic chemistry, have had a remarkable success; others, such as Mycin, used in medical diagnosis, has not received so much applause, in this case because the way the program "thinks" is unfamiliar to the medical man not used to artificial intelligence and its peculiarities. But how does he discover the way the program thinks? He can ask it, and it will tell him, in as much detail as he wants, just how it drew its conclusions. Various such expert systems are becoming commercially available, though not many are in fact as complete as Mycin is. Others, mostly still in process of development, improve on Mycin in that they can learn, – not merely acquire more factual knowledge (as Mycin does, too) but find new ways of handling this knowledge. Nowadays this mostly means that the user shows them how to; but the possibility that the program itself discovers entire rule patterns that often occur in successful runs, and proceeds to set them up in their turn as "super"-rules, this possibility may well appear implemented in the next generation of expert systems.

The higher levels of learning, where by a process that is half analogy, half intuition, with a dash of sheer guessing, we create new concepts and new theoretical structures, is as yet out of the reach of artificial-intelligence systems. There are two kinds of obstacles before this level can be reached. One sort is essentially technical; this ability requires that an immense amount of previous knowledge be made available, what an erudite scientist accumulates in a lifetime of dedicated work in his field. We have so far no methods by which such huge knowledge stores could be created in a machine or, if available, be made accessible in a reasonable time. The real difficulties lie in the intricate interconnections of a large body of knowledge. Take ten facts: we could pair them in 45 different ways, and some of them might be relevant to what we are doing. But a million facts could be paired in something like 10^{10^6} ways – a number with one million zeroes at the end! And an average human being probably knows a lot more than just a million facts ... Thus we must need find a way to pick out only the relevant connections, and no such way is yet known. The other sort of obstacle is more conceptual in nature. Let us suppose that somehow we have selected a likely-looking set of data connections; how do we now formulate a new conception that explains them? Is it merely a generalisation that can be deduced from the facts and their links, or is it the creation of new ideas that possess genuine originality? This is what is known, philosophically speaking, as the problem of induction. Some attempts have been made to program the first notion, but with only mediocre success. Though the philosophical community is still debating the matter, most scientists would probably agree that there is no such thing as induction, and the necessary ideas must be newly invented. And this, likewise, is something we do not yet know how to do, even though many basic principles have already been elucidated. At least in the case of simple game playing, computer programs have been developed that possess this second-level ability to learn which consists in not merely evaluating the rules of play but evaluating the rules of evaluation for the rules of play.

But playing games or generating expert systems does not tackle one very important problem, the problem of making use of sensory information, in particular of visual one. It is not difficult to connect a TV camera to a computer, through a circuit that converts the output signal into a bit pattern representing the light intensity at every point in the image; but it is very much more difficult to interpret these bit patterns as the images of the different objects one could see. As a first stage, one might pick out areas of more or less uniform illumination, and mark as possible object boundaries the lines where the light intensity changes abruptly. And then one tries to combine a few lines, to see if they make sense as outlining an object. Here we come up against a problem very much like the one of finding interconnections between facts in a memory system: to recognize a simple object – a cube, say is easy; but when we have several different bodies, we must discriminate between these bodies as well as the shadow areas. With ten bodies in the image, there might be 80 points where two or more lines meet, and each such point "belongs" to up to three bodies or a shadow, so that there are $4^{80} \cong 10^{48}$ possible combinations to explore. At the rate of a nanosecond each (a speed which we are very far from being able to achieve) this would take some 10^{30} years: many many times the age of the universe. The solution to this difficulty – or the beginnings of the solution, at least – came with the recognition that instead of trying to get every possible piece of information out of every image, one should merely get enough to suggest what to do before the next image is looked out. Thus if the computer, after getting this list of 80 points and the lines between them, moves the light illuminating the scene a little to one side, it might find that 30 of the corners have moved and therefore delimit shadows. For the remaining 50 points there are now only three possibilities at most, and so we have $3^{50} \cong 10^{24}$ possibilities to explore, which could be done in "only" about three million years. We now make a guess: the uppermost 7 corners remaining belong to a cube. If the computer now moves the TV camera, we get a third image that, together with the second one, gives a stereo view and fixes these corners in the three-dimensional space. If the thing really is a cube, the computer can now get its mechanical arm to take it and put it down somewhere. Should this work, we have not only found out that we had a cube at the top of the picture, we even know its weight (from the force the mechanical arm had to use). Continuing in this way, the whole complex picture can be analyzed in a matter of seconds.

Two things are to be learnt from this. Firstly, mere observation presents us with an impossibly complex set of data that is wholly uninterpretable; but if we interfere in a well thought-out way with what is being looked at, each such interference leads to confirming or disconfirming a guess at some component of the picture, until gradually the whole immense confusion that first appeared is sorted out. And secondly, this "thinking out useful interferences" needs a model inside the computer (or whatever is doing the observing) which is piece by piece built up to represent the objects being looked at. And when the process is complete, the model not only contains the needed information about the shape of the various objects, but also a lot of further information – as for instance their weights, how to handle them, perhaps how soft or hard they are. It is not merely a static picture, it has become dynamic, in the sense that it allows us to foretell what will happen if we do this or that. And this is something absolutely vital: it is the core of the capability for flexible planning which must surely figure at the centre of any conception of intelligence that we might have.

Incidentally, the attempt to do these things on the computer (and we have been able to analyse scenes made up of a few simple objects, like cubes, cones, pyramids, spheres in much the way I have sketched) shows us how much the human eye – with a brain behind it, of course – is superior. One gains a very healthy respect for the apparently commonplace activity of looking at a cluttered desk, picking up idly one or two objects, and recognising hundreds of different things, papers, books, pens and pencils, a small tray full of clips and so on: and all of this is going on in a brain that at the same time is controlling the multiple internal doings of the body, keeping it balanced upright while one shifts from one leg to the other, and even maintaining an abstract discussion with other people. We are enormously far away from being able to build a mechanism able to do such things. Yet we apparently have made enough progress to let us suppose that the possibility is real.

Given the ability to interfere appropriately with its surroundings, then the computer can build up a model of the "world" it inhabits. And the model is dynamic; it can be run on the computer as a subprogram in the general program that controls the computer's behaviour. It will then foretell how things are going to change in this little "world" if this or that is done. But who does these things? The computer itself, through its various mechanical arms and other appendices under the program's control. This means, however, that if the model is to be useful in planning future actions by the computer, it must include a representation of the computer itself. And the most characteristic property of the computer, for the purpose of such a dynamic model, is of course that it contains this model; therefore the model must include a representation of itself; and this representation of itself must in its turn contain a representation of itself. And so on, ad infinitum ...

Fortunately the technique of building lists of address pairs, one of the pair pointing to the next pair of the list, the other to whatever information one wants to hold, is quite adequate to the problem of realising such representations nested inside representations. *One* model of the world is built up, and it contains *one* model of the computer's behaviour; but these models really are more like programs than anything else, and when needed they can break off, store away the information required to continue again, and then restart themselves but using the information stored away, so that now they behave like models of the models; and then they can do it all over again, becoming now models of the models of the models. This can go as deep as necessary; once the lowest needed level is reached, the program finishes at that level and returns to continue where it broke off at the preceding level; it can then either start a new lower level, or return to the next higher level; and so on. This technique is known as recursive programming and has a considerable body of theory behind it.

One interesting consequence – philosophically speaking – of this possibility is that it spells out an explicit mechanism which answers one of the most puzzling questions in psychology. Human beings, in this like the higher animals, are conscious; that is to say, they are aware of the world that surrounds them. But unlike the animals, they are also aware of their own awareness; they know that they know, and they also know that they know this. As well as being conscious, they are self-conscious. Do these psychological mechanisms serve any purpose? Opinions in the past have widely differed but few have maintained the usefulness of consciousness, the need for it. Yet here, insofar as one can say that a computer program is conscious because it can build up a model of its surroundings, we see that this consciousness is an essential part of the mechanism of intelligence, of the capacity to plan. Moreover, as long as this capacity remains at the level of immediate reactions, only a first-level modelling is required, but if one needs to plan ahead, to foresee eventualities and problems that could crop up out of one's own actions, then one must be aware of what one will do oneself; and a central part of what one will do is precisely to build up a model of the world, including oneself. One must, in other words, be self-conscious. This is an essential part of the mechanism, not only of human but – and here the suggestion arising is indeed remarkable – of any kind of intelligence.

A second and perhaps even more significant philosphical conclusion is implied by the way the mechanism I have sketched for seeing works. It is not simply seeing, not pure observation. Almost all epistemological thought in recent years, certainly all of that has followed the long empiricist tradition that began with Hume, has always been based on the view that seeing (or indeed any kind of perception through whatever sense organ) is mere observation, mere taking in of sense impressions. What the computer simulations suggest is that this is not a reasonable account. Pure observation soon gets bogged down in the attempt to decide between the endlessly and rapidly increasing number of different possible interpretations of what is being seen. One must break through the barrier by using a model, however crude, to take some planned action – planned in the sense of foreseeing the probable result – and then comparing one's observations with these results as foreseen. Thus only a few elements in the rich field of observation are picked out as relevant, and here it is the correlation between what was expected and what actually happened from which we derive any knowledge, any genuine interpretative elements. By continually modifying the model, one rapidly arrives at a quite satisfactory one; and for the next occasion one has a much better starting point. But the essence of the process is the interference with the part of our world that we want to know about. Acquiring knowledge is an active process, not the passive, armchair sort of business that most philosophers appear to think it. Once this is recognised, many of the central questions of school philosophy simply evaporate: Berkeley is proved wrong and the world is shown to exist, quite independently of our thoughts, simply because we cannot find, as we constantly do, unforeseen consequences of our planned actions unless this world is not just a figment of our imagination; similarly, Hume's problem about the reality of the causal process is resolved, because we do not merely observe that B always after A and conclude that A causes B, we must ourselves make use of causal connections in order to bring about the actions we have planned and so cause the effects we have foreseen.

Simulating intelligent behaviour on a computer has remarkable consequence for philosophical thought ...

Yet we have not even dealt with one of the big questions that are always asked, the question whether it is even possible. For most of the elements I have sketched above, we do possess programs that carry them out, even if generally only at a rather primitive level. We have not yet achieved any program, however, in which these elements are brought together and interact so as in fact to exhibit intelligence. Thus the question about the possibility of actually achieving genuine intelligence through suitably programming a computer remains open.

Most people actively engaged in working with computers consider the possibility to be real, largely because of what has already been done; but it could be argued that through a sort of professional optimism they greatly underestimate the difficulties still to be faced. On the other hand, many other writers have had serious doubts. Very roughly, these doubts may be grouped into two categories:

There are those who react emotionally to the idea of intelligent computers. They commonly feel that if this were possible, human dignity would be downgraded intolerably. Or they are afraid that if the computers really became intelligent, they would combine against humanity and either destroy us or else reduce us to mere slavery. Such arguments owe a little bit too much to science fiction, one is tempted to answer, and this is of course a perfectly justified answer. Yet there remains an element of uncertainty even after the overactive imagination of science fiction authors has been duly discounted; we come back to the question below. An altogether different question is raised by those who are afraid that relying excessively on the use of intelligent machines will reduce our own capacity to deal with problems by allowing us to sink slowly into a too comfortable kind of half life, dedictated only to pleasure and refinement; but here it is tacitly supposed that intelligent machines are in fact possible, which is just what we are at the moment asking. Another variety of emotionally tinged response comes from those who every time that a newly written program does something not so far achieved simply say "Oh, of course, but that is not what we mean by 'intelligent behaviour'." *Some* responses of this type are justified, because the concept of intelligence is indeed confused and ill defined; but to define it implicitly as that which we do not yet know how to do on a computer seems to load the dice in a not very acceptable way.

The other category of doubters argues that artificial intelligence is impossible as a matter of principle. Some objectors of this variety simply define intelligence as what the human brain is capable of, so that if any other sort of material object exhibits the same sort of behaviour, it cannot be intelligence. This argument, together with certain religious ones, hardly needs refuting; there is, however, one serious argument which is worth examining because it is based on a very widely held but almost surely erroneous conception. This is the idea that a computer program necessarily conforms to the underlying logic which it realises, and cannot exceed its limits; but Gödel's well-known theorem limits severely what any logical structure is capable of, while human intelligence certainly exceeds these limits. We need not here go into the details of Gödel's very remarkable theorem, which at the time of its discovery brought about great consternation among those interested in the foundations of mathematics and logic and has given rise to much unresolved controversy concerning its implications. Suffice it to say that it shows quite indubitably that in any logical structure (i.e. a set of axioms and the infinite set of theorems that may be deduced from them) there are some theorems which cannot either be proved or disproved, and among these theorems there are the two that demonstrate the consistency and the completeness of the initial axioms. What is wrong here is its application to the case of artificial intelligence. In the first place, the term *logic* as used by computer specialists does not mean the same as logic in the logician's sense; it refers only to the basic notions about the problems tackled by the program and the way the solutions are determined, and no questions concerning either completeness or consistency are involved. Indeed, since only methods for transforming bit patterns are involved, such questions seem rather irrelevant, even though one can in certain cases actually prove that the program does what is expected of it. In the second place, Gödel's theorem applies to any one particular logical structure, in which the axioms and the theorems deduced from them treat of a definite set of concepts (finite or infinite) which the propositions making up the theorems and their proofs are about. This set is technically known by the somewhat pompous name of "universe of discourse". Now the intelligent computer constantly does something which makes Gödel's theorem irrelevant: it takes in new information, obtained by guessing at the interpretation of what is new, acting on this first guess, comparing the results with what this guess implies as expectations, and using the discrepancies to improve the guess; repeating this cycle a sufficient number of times yields an excellent approximation to the needed new information, as we have seen. But here the computer extends its universe of discourse by incorporating new concepts, and moreover does not expect absolutely exact results. Thus, just as for human intelligence, we are not restricted by the limits given in the Gödel's theorem. Artificial intelligence, for good or for ill, is possible.

(iv)

For good or for ill.

So far we have only talked about artificial intelligence in abstract, as it were, what it is, how it works, if it is even possible. But the work on artificial intelligence is going on here, not on Mars; expert systems are already being sold commercially, and robots with just sufficient primitive intelligence to be able to see could well appear before the end of the century. Thus the questions how this will affect us, affect our human society and what we should do about it is becoming very urgent.

But if the beginnings of computer intelligence are to be expected in the near future, anything like a genuine intelligence in a machine, like for instance a capacity to hold a conversation with a human being in which new ideas appear (not just the typical exchange about the weather, which is so stereotyped that we could program it even now if it were thought useful) is still very far ahead, a hundred years or perhaps two hundred. Let us then look at the question by considering first the long-range questions, and then those we have to face in the near future.

As we mentioned, there are many thoughtful people who are very much afraid of the long-range perspective that "real" artificial intelligence would offer us. There are also the endless science-fiction enthusiasts who hope for robot paradises. Which side is right? Or will there be a mixture of the two? Or perhaps something altogether different? I am inclined to think that the last is the more likely.

Let us, first of all, ask if computers, intelligent computers, will be inimical to man. This kind of fear is perhaps only natural in a society in which technological change has always meant – even if later great benefits derive from it – unforeseen and undesireable troubles that fall largely on the working population: unemployment, whole groups of specialized workers thrown on the scrap heap, at the best lengthy and difficult periods of readaptation. And the cheapest kind of science fiction plays on this fear in an uncontrolled and irresponsible way, through its pictures of enslaved humanity or robots gone mad or on the contrary so much more advanced than we are that they keep us as pets.

In reality things will in all probability be very different. Such imaginings forget that the learning process must play a central part in the development of artificial intelligence, just as it does in that of human intelligence. To put it crudely, an artificial intelligence, just like a natural one, must be educated before it actually is intelligent. And this education must, of course, be carried out by human beings: remember that we are not talking about formal education merely, about schooling, but about learning to see and hear and talk, to move around and handle things to form notions and communicate them; in brief, to live with other intelligent beings in a society. Whatever level of intelligence machines will have achieved at any one point, they can only realize it by continuous interaction with human beings in ways that will no doubt differ radically from what at the present time we call education but that will nevertheless be exactly that; and it will therefore also have the same result: the intelligent computers will form part of human society. If this society were to be our present-day society, they would of course share all its contradictions and splits and divisions; fortunately by the time such things as the intelligent computer come to exist, all these class divisions and other opposing groupings will have gone. This does not mean, of course, that all diversity will have disappeared: on the contrary, and the computers will only add to it.

Such participation of the intelligent computers in human society also means that any aberration in what the machines do will be observed and corrected in time. The fears of those who are beguiled by science fiction thus have very little basis.

There is, moreover, a very good chance that this participation will go much further. This is extremely speculative terrain, for obvious reasons, but since we are considering a remote future, let us in fact speculate. Now computers, intelligent or not yet, are in fact only the most recent in the long series of tools that mankind has developed in its long history in order to extend its powers. The computer possesses two features which make it useful to us. One is the capacity to develop a proper initiative on specific tasks; at present this is still extremely limited but in the future it will grow to an extent we are quite unable to foresee; the other is its capacity to communicate, which we have not yet taken into account.

Human beings communicate with each other by talking and listening. The amount of information in one man's voice is of the order of some 10 000 bits per second (or 10 kilobaud, as the technical man would say); but of this only perhaps a hundred or so bits per second transmit the meaning, the remainder is either irrelevant information, such as the particular dialect he uses, or that he has a small speech defect, or else it is redundant; this extraordinarily high redundancy means that we can interpret what the man is saying even in very difficult circumstances – say in the middle of a cocktail party with a record player turned up high. A computer, on the other hand, can communicate at rates that even now reach one million bits per second – and the redundancy needed to ensure reasonably correct reception can be kept very much lower. This rate is admittedly still very much smaller than the rate at which signals travel inside the brain; but if we found a way to tune into some of the brain's signals (not merely, as the encephalograph does now, to observe the gross features of these signals), then we might break through the barrier of communication between man and machine. If this were possible, then the machine could actually serve to extend our own intelligence; even a much increased memory would greatly augment our capacity for thought, but what a genuine addition to our "mental equipment" could achieve is at present quite beyond our imagination.

Yet we need not stop even here. For it we could increase our intellectual abilities by connecting a machine to our brains, we could clearly build machines that would accept a connection to *two* different brains. But why stop at only two? Even now, when a group of people work together closely enough and intensely enough to achieve a kind of automatic and almost unconscious identification, they soon arrive at a point where they understand each other's ideas with only half a word spoken; and the result is an extraordinary ability for creative thought processes which goes greatly beyond what any one of them could have achieved by himself. What could we not do with a far more effective, indeed intelligent, communications link among a group of close collaborators ...

This is pure speculation, of course. The reality is almost bound to be entirely different. Yet it serves to underline how very wide of the mark are the frightening (and frightened) perspectives the science-fiction merchants offer us.

Unfortunately things change very drastically when we come from the distant future to what we have to face in the next ten or twenty years. We live in a world in which computers – quite without intelligence, so far – have already caused considerable change. They have been involved, in the first place, in the creation of several new technologies: neither nuclear reactors nor present-day aeroplanes could have been developed and designed without them. Secondly, they have altered almost every other existing industry in numberless ways. And thirdly, all our administrative work has been profoundly affected by them. Lastly, they form themselves the object of a very considerable industry. In all these areas they have brought about endless improvements in the way things can be done; they have made possible new ways of tackling problems, they have provided new insight through their extraordinary data-collecting capacity, they have improved – and made safe – many industrial processes, they have eliminated or at least speeded up an endless variety of stultifying routine jobs. They have done all that. But the cost has not been negligible. They have left millions of people without a job[2], they have totally altered the working conditions of countless other millions, and since their benefits appear only when they are properly used – which is very often not the case – they have also caused a lot of unnecessary misery. They have also created several quite new problems. One of them is the very widely held idea that what the computer says must be right. Yet thought its malfunctions can usually be detected with no difficulty (and are much rarer than a human being's), they produce the desired results only when they are properly programmed and also

---

[2] Some years ago, there was much discussion about this question. There were those who said that the increase in computer use had caused much unemployment, and quoted figures of total unemployment to prove it; others showed that this was not the case, by examining individual firms where computers had been introduced without creating unemployment, sometimes even generating new jobs. The obvious solution (that the successful introduction of computers in a firm gave it an edge on the market over its competitors, so that any unemployment caused by the computer would appear in those firms which had not yet installed any) does not seem to have been discussed; but this is not the place to examine the reasons for this silence. What is more astonishing is that today, when the unemployment has reached catastrophic levels in almost every country, the role played by computers in bringing it about should go almost unmentioned.

when they are given the right information. Since this point, however obvious it may be, is commonly ignored by whatever management has the responsibility, we are often faced with computerised systems in which there is no possibility for checking for possible errors or – even worse – for correcting them speedily and adequately when they are brought to the attention of the administration. A further problem is created when the not always faultless information about someone cannot be seen and verified by him, even though it may affect him in quite unexpected ways. He may have been fined as a youngster for driving through a red light – and years later he has difficulty in getting a job because the records held in one or another computer wrongly show him to have a criminal record; this and similar cases have happened far too frequently for comfort. Nor can we feel any happier when we remember that in recent years most national police organizations have acquired large computing systems in which they hold information concerning thousands and thousands of people – not only proved criminals, but also of those merely "suspected". Suspected of what? Were it merely a matter of dealing with the drug traffic, few would complain; but it is only too evident that many of these records concern people who are *politically suspect* – a very convenient phrase which tends to mean "does not agree with the government". Or even worse, "does not agree with what we (i.e. the police) think the government should be".

Other problems arise in areas where at first sight the use of the computer is an unmixed blessing. Consider their educational use. Here they do not create significant unemployment problems, since in very many countries there is an acute shortage of properly trained teachers, particularly at the secondary and university levels. Making use of the facility with which a computer can be programmed to hold a dialogue simultaneously with hundreds of students, offering them the opportunity to work at any hours that suit them, often even at home, and for as much time and at a rhythm that can be different for every one of them. Moreover, and even more important, the computer can adjust the level at which it teaches to the abilities and previous knowledge of each student individually – something that is quite beyond what the human teacher can do when he is faced with a large class. There is thus everything to be said for using computers to help out the human teacher. But there is another side to the medal. It is due to the extreme unevenness in the technical and industrial developmenmt of the nations of this world. We thus find that it is the highly industrialized countries that develop this sort of system and propose to sell it in the poorer countries. Typically IBM – I need hardly explain that they are the largest computer manufacturers in the world – has combined with McGraw-Hill, an important publisher of school and university texts, for the purpose of developing the so-called computer-aided instruction systems: they will provide the machines, while McGraw-Hill undertakes the programming and publishes all the incidental material. And they have publicly stated that once they dominate the US market, they plan to attack the Latin-American one. Using, of course, whatever methods and material they have developed for Kansas City, roughly translated into Spanish or Portuguese. Even supposing that their products really suit the educational needs within the US, how well will they fit the entirely different background in small Mexican or Bolivian town? We need not even think about the possiblities of deliberately influencing these people's cultures; quite unconsciously and with the best will in the world, the work of the US pedagogues in this area will adversely affect the children of countries with very different traditions.

It is vital to realize that mere protest, even for the best reasons in the world and with the clearest arguments to justify it, will not help here. The only decisive argument possible here is to be the first. Fortunately this is possible here. It is entirely possible for quite a small group of knowledgeable teachers to write a program for computer-aided instruction that will at least equal in quality any commercial product, that will properly take into account the peculiar needs of their country (or even of a much smaller unit, of a particular school, perhaps) and that can be offered at a better price. And it can have one tremendous advantage which no commercial service could ever offer without effectively destroying its market: it could be open, so that the users themselves, learning from how well it performs, change and improve it.

So far we have discussed only the "unintelligent" uses of computers. Will the bare beginnings of computer intelligence that are possible today make much difference? Yes. In the first place, they are bound to step up the process of computer penetration into all aspects of our life; at first only by making them more effective but within a very few years by making possible considerable extension. And in the second place, there are evident new uses of computers that they will make possible, and these could have a very deep impact on our society.

Computer-aided instruction belongs to the first category. The presently existing methods, though effective and properly used quite adequate, are difficult to apply to subject matters where subtle appreciations are needed and even substantial differences of opinion may be accepted provided that the pupil has genuinely thought about the matter and can give a well organised opinion, with facts to back it up: say, the teaching of literature. No computer program exists at present that could really supply this flexibility; but some first steps are already possible, and could well lead in a very few years to quite usable systems. The reader will be able to imagine for himself the enormous potentialities of such development.

Here belongs also the possibility, still very far from any implementation but being thought about by some specialists in the field, that the ordinary administrative programs might be supplemented by an "intelligent" package of programs which in case of doubt or error engages in a dialogue with the irate user. This is treacherous ground to tread, no doubt. The extremely low levels of machine intelligence which is the most we can hope for in the next two or three decades will no doubt be adequate for writing programs that will defend to the end the correctness of whatever the standard administrative program produced and that are intended only to make the unfortunate sufferer from the mistake feel that instead of being given a runaround he has been attentively listened to and treated courteously; and it might be only when it is too late that he wakes up to the fact that his problem has not been solved after all.

In the second category of the entirely new applications come the uses of expert systems. For these are intended to replace or supplement, not the routine or low-level workers, but the people at the very top, the most highly qualified specialists. Thus Dendral does an important job of helping the research chemist in the interpretation of spectroscopic data, and Mycin helps the clinical physician to arrive at a correct diagnosis in case of bacterial infections. Other such expert systems exist to take over at least a part of the work of the petroleum geologist, of the aeroplane designer, and so on. Here we must face a doubly dangerous situation; not only do those systems present the problem of unemployment in a wholly new range of jobs, those at the top of the scale rather than at the bottom, but there is a more subtle and in the long range perhaps more worrisome side: at the specialist level, personal opinion, based on long experience, and the resulting discrepancies between experts and their extensive discussion are what ensures their remaining up-to-date and constantly striving to better their knowledge and performance; but expert systems do not discuss with each other – indeed, they do not even "know" of each other's existence – and though they can be (but not always are) engineered to learn from their accumulating experience, this process is limited to the sifting in more or less mechanical ways of their increasing stock of data. There is therefore a real danger that if such computer programs replace to any extent the specialists in a profession the exercise of that profession will gradually lose the fine edge of constant revision and improvement of knowledge and slowly slide into a steadily less appropriate routine. And when finally people become aware that the quality of the expert systems' expertise has fallen to dangerously low levels, there will be no more human experts to revise the machine expert and bring it up to scratch. Now it is possible in principle so to engineer the expert systems that they work *with* the human expert and stimulate him rather than work against him; but so far as I am aware, no research on this problem is being done. Why? Perhaps the answer is simply that we live in a competitive society, where such research has to be paid for; and who would pay for it?

Finally, there is the obvious problem of monstrously increasing unemployment. To the extent that we can build low-level intelligent machines, they can take over the less qualified jobs, both in industry and elsewhere. This has already started, but we may expect the problem to grow as such robots become better at their jobs and cheaper. The great advantage about such a machine is that if it does the job at all then it will do it all the time, twenty-four hours a day and three hundred and sixty-five days a year (with an extra day thrown in for every leap year). It will not go on strike for a better salary, nor does it ever get tired and want a coffee break, and it never answers back when you shout at it. And if by chance anything goes wrong with it, you simply call in the service engineer. When it is worn out, you can still use the pieces to repair another one. It is the ideal workman – according to the employers ... Thus once they become cheap enough, we may have to face massive unemployment on a scale we cannot even imagine today; and the consequences of the social upheaval this will bring about cannot be foreseen.

We have to recognise that artificial intelligence could within relatively few years create major problems for us. Nor can we afford to do what almost all progressive political organisations have tended to do, namely deal with problems on a day-to-day basis and let the longer-range future take care of itself; a policy must be worked out and appropriate action taken now.

Fortunately there are certain steps that almost suggest themselves. For one, as in the case mentioned above of computer-aided education, it is not so much the actual machines (the "hardware", as the technical jargon has it) but the programming, the "software", that is in question. And here even the poorest country can afford to mobilise the relatively few experienced programmers needed to set afoot a national enterprise of writing suitable programs that can help to avoid some of the worst abuses. But clearly this is not enough; many of the difficulties cannot be dealt with merely by technical solutions, they have too many social aspects. So a second step must be to alert the public, to help them understand the problems and to get them to exert pressure on our governments to deal with the issues. This will not, of course, come about smoothly and easily; there will be many cases of hardship suffered by individuals that in one way or another are linked to the increasing use of intelligent computers. And so a third step seems reasonable: open advisory centres where help and explanation can be given by experts. Such centres would serve a double function; they would help the public and thus also make the parent organisation better known for taking seriously its social responsibility. At the same time such a centre would be an invaluable source of information about the effects of the growing computer involvement; from the incoming reports an inside knowledge would be gained that is indispensable for developing concrete and politically feasible proposals for common action, and – once such action has been initiated – for checking on how well it works and so allowing us to adjust it to unforeseen difficulties.

And no doubt there are endless other possibilities that only require that the whole subject be opened for discussion. Thus, while the problems we will have to face in not too many years will be extremely serious, there is much that can be done. The only thing that we cannot afford is inaction.

24. Philosophy and Physicists *

(i)

Physicists need certain philosophical notions in order to work. The experimental physicist sets out with the firm conviction that the objects he studies through manipulations in the laboratory really exist – sometimes even coming to suspect that they are malevolently recalcitrant. The theoretical physicist has ideas which are generally less clear about the relationship between data produced by the experiment and his own work. Even the teacher of physics needs certain philosophical concepts if he is to justify the way he presents physics to his students.

Unfortunately, few physicists support their work with consistent philosophical principles, produced specifically as an aid to their labours. Most limit themselves to notions which they have absorbed from their work or family environment, read or heard by chance, which might be current as folklore in such a circle, but fail to conform a coherent philosphical conception. Even worse, the conceptions which guide a physicist's ideas can be unconscious: researchers are often not aware of the roots and implications of what they do.

This situation makes up the general framework within which certain specific problems with particular philosophical aspects emerge. Here are some more or less representative examples.

1. When we measure some physical quantity, the observed values can generally be found concentrated within a relatively small interval around a certain average – let us say twice the standard deviation which can be considered our estimate of experimental error; but often, a few values appear which are far from the average. In this case, should we eliminate such values or take them into account? Statistical theory provides us with a clear answer; experimental physicists often decide on the basis of intuition. "Cookbook" recipes have also been provided, such as "ignore all measurements further than 4σ from the average."

How can we justify these diverse means to solve the problem? The question is one of method, but no physical theory (or mathematical theory, which precludes experimental consideration) provides a feasible basis for the decision: only some philosophical principle, a basis for scientific methodology, can guide us. But as yet, we do not know what this principle might be.

---

* "La Filosofía y Los Físicos", Rev. Mex. Fís., **27**, 393 (1981) (Translated by C. Brody)

The example appears to be trivial (although it is not) and only marginally philosophical, although it is a good representation of a whole series of difficulties which arise from the use of statistical methods in physics and the correct interpretation of their results. Let us examine a second example:

2. For centuries, physics has had both an experimental and a theoretical aspect. We know that the relationship between the two must be close if investigation is not to lose ground; if theory is insufficient, experimentation becomes a craft; without an adequate experimental basis, theory becomes sterile. Precisely what the nature of this relationship is and how to improve it are problems which neither physicists nor philosophers have been able to solve. Perhaps due to this lack of clarity, the recent introduction of a third element into the situation creates serious problems for the future development of physics. This third element is the growing dependence on computers. Not only are there special, and sometimes highly complex, programs which are used to accumulate, clean up, transform and interpret experimental data before human beings even see it, thus removing the possibility that an unexpected irregularity might suggest new goals for investigation; but also, we have an increasing number of programs which make up a sort of neutral territory between theory and experiment, serving as a theoretical model for the experimental physicist and as an experimental field for the theorist. On the one hand, such programs reduce the direct contact between theory and experimental material; on the other, they pose problems for the sane criticism of investigation (what can the referee of a scientific journal do when faced with results obtained through the extensive use of computers?). The effects of what has been termed computational physics bear upon the branch of philosophy known as epistemology. What this discipline might have to tell us is something which would be of immediate and profound interest to the physicist who is concerned about the future of his science; unfortunately, epistemologists seem to be as yet unaware of the problem.

These two problems evince the lack of a philosophical basis for their solution. In the following case, the difficulty pivots, rather, around the excess of philsophical considerations surrounding the problem.

3. The most commonly accepted basis for quantum mechanics has led to a discussion which has now lasted for more than half a century. Known as the Copenhagen interpretation (perhaps because it is neither an interpretation, nor was it developed in Copenhagen), it has generated apparently insoluble paradoxes and contradictions. From the first important result of this kind, which was Einstein, Podolsky and Rosen's 1935 demonstration that quantum mechanics is incomplete, up to the latest paradox, discovered recently by Sudarshan and his collaborators, there has been no contribution to the debate which has not referred to its philosophical background. Academic philosophers have participated extensively, although with few results, here; in fact, detailed analysis reveals that a very specific philosophical tendency – neopositivism – has deformed the problem, leading to the formulation of highly confused conceptions, and hindering the healthy development of investigations.

The lack of substantive advances in studies which are crucial to quantum mechanics, in both its physical and its philosophical sides, is evident and underscores the conclusion which has already been indicated: we have set out from inappropriate philosophical considerations. Only recently have we been finally able to achieve some advances, on the basis of very different philosophical notions. The situation does not seem so bad in another problem:

4. University careers usually include a much discussed and despised topic: the history of physics. For a long time, the teaching of history was completely divorced from philosophy, and was thus practically reduced to the level of anecdote; naturally, students, and sometimes even their teachers, viewed it as secondary and useless. Today, the situation has changed. Due to an increased consciousness on the part of student bodies, we have come to realize that, in Kant's well-known phrase (as reinterpreted by Lakatos) "the history of science without philosophy is blind; the philosophy of science without history is vacuous." This same impulse has led to a reconsideration of other subjects taught in schools of physics; the effect we seek when we include historical subjects in study plans – which is the mellowing of dogmatisms which are inherent in our pedagogical methods – must be increased. What must be taught is a science and not a doctrine, that is to say, an activity of search and investigation, full of problems and difficulties, discussion and discrepancy, success and contradiction. In order to achieve this, we must discuss the history of our science and its philosophical implications along with the ideas themselves, in order to give them life.

The scope of this essay will not allow for a detailed discussion of this example which includes references to how different historiographies can lead to different philosophies of science, nor to the different styles of teaching which result. It is more important to point out that this problem cannot be easily divorced from the following one:

5. Physics, like every science, is a social activity, developed within a complex series of institutions, and paid for by society to provide certain services, in this process transforming society itself: this is something which is neither foreseen nor desired by those who pay for it. But the details of this interrelationship and its historical evolution have remained in good measure unexplored, with the consequence that the two main forces in charge of developing and implementing specific policies have erred. One proposes achieving positive effects upon the economy by stimulating science; the other acknowledges the investigator's social responsibility, and seeks to direct his efforts towards goals which benefit society, especially its most underprivileged sectors. Both currents suffer frequent disappointments when their goals are not reached; and in many cases their disappointment originates in their lack of knowledge about epistemology, in that they do not know or understand the workings of the process whereby scientific knowledge is gained and taken advantage of through technological change. Most of all, those who believe in the stimulation of science as a direct route to the betterment of the economy often harbor simplistic ideas, even believing that all that we have to do is to install a few well-known physicists in the laboratories that call for them for the economic

crisis to recede; and when this does not happen, they are often quick to imply that science is totally irrelevant to economic change.

These five examples provide some evidence, we hope, of the full and complex gamut of philosophical problems which are relevant to the physicist; from technical difficulties in the laboratory to aspects of politics of science. Of course, there are many other problems. If the selection we have provided here underlines the diversity of the problem on the side of physics, we could provide an analogous series which demonstrated that all aspects of philosophy are also involved, from ontology to epistemology through ethics and even aesthetics.

But there is something else which this selection reveals: in general, the problem is not solved. Philosophical issues either are ignored or have been taken for granted along directions which are of no use to physics.

(ii)

To what extent is academic philosophy in a position to shed some light upon this wide variety of problems?

This is a question which quite simply could not have emerged three centuries ago. Physics was at that time barely distinguishable from philosophy as an independent discipline; it then bore the name "natural philosophy" (which is the case even now in certain European universities), and even if it had achieved a clear separation with theology, it was still possible for Berkeley, as a philosopher, to confront Newton with serious criticisms, which received a satisfactory answer somewhat later through the work of Cauchy. The separation between physics and philosophy came about as a consequence of the Industrial Revolution: specialists were then needed who were capable of creating new techniques and solving the problems inherent in the expansion of old ones, without, however, becoming involved in metaphysical doubts. Physicists were required who could provide the young and dynamic bourgeoisie with the power to control nature, without questioning their political dominion. What this implied was an increase in physical knowledge regarding nature without a corresponding study of its implications, both philosophical and political. The increasing professionalization of physics, and of all the other sciences, which was in good measure a result of the enormous accumulation of new knowledge, forced the pace of separation. By the mid-19th century, this process had become irreversible:

> "Philosophers regarded physicists as diffident, and the latter regarded philosophers as senseless. Scientists began to emphasize the lack of philosophical influence in their work, and shortly, many of them, even outstanding figures, came to condemn philosophy not only as useless, but even as a fatal influence."

> Helmholtz 1865

Helmholtz himself deplored and fought this tendency, along with many other important physicists, but in vain. To this day, communication between philosophers and physicists is strained, even in encounters organized for this very purpose.

The consequence for physics has been that all its problems of a philosophical nature remain unsolved and even unapproached; at the same time, its evident ontological or epistemological roots have been examined only on the level of personal reaction, unexamined tradition, and uncriticized confusion. This lack of clarity on philosophcial issues has reached the point where it demoralizes physicists, as can be understood from the despairing words of one of the best ones:

> "I lost (as a consequence of the contradiction between classical electromagnetic theory and the new quantum mechanics) the certitude that my scientific work was bringing me closer to objective truth, and I no longer know, why I continue to live. I am only sorry not to have died five years ago, when everything appeared clear."
>
> H.A. Lorentz in 1924, as quoted to A.F. Ioffe 1963

Nothing could provide more poignant evidence of the profound disturbance which a physics without an appropriate and articulated philosophy can cause. It is sad that a man of Lorentz's quality – blinded, presumably, by an excessively simplistic epistemology – was unable to perceive that his work not only paved the way for quantum mechanics, but also that one day it would participate in a new synthesis which transcends quantum mechanics; this is something we now begin to see. Of course, it is not the investigator's psychological problems which concern us here, but even physicists are human
. . .

What has become of philosophy without physics? Let us briefly examine the fields covered by philosophy today. There are three central branches:

– Ontology, which studies the problems inherent in the constitution of our universe. Basically, it offers two distinct answers: Idealism, (according to which the material world is an illusion, or at the very least secondary, being generated by the spirit – divine or not –, abstract ideas or ideal forms, or, according to some, by our concepts) and Materialism (according to which the material world is both logically and chronologically prior to the human mind, which is the source of all ideas and of itself in the working processes of the human brain – a highly material organ). Each of these conceptions can be subdivided according to the attitude taken regarding what happens in time: Mechanicists (or Metaphysicians, in different terminology) reduce everything to the movement of different kinds of objects, or even to a single kind, which are unalterable except possibly through the external influence of other object's movements; in the Dialectical conception, objects are not immutable, which means that the subdivision of our world into objects is temporary and relative, much like a photograph which registers a temporal slice of what can better be seen as the interweaving of many processes with differing time-scales. These differences involve diverse attitudes concerning the problems of determinism, causality, probability and also a series of more detailed metaphysical conceptions, which attribute "fundamental reality" either to atoms or other discrete objects, or to something continuous, such as energy.

– Epistemology or gnoseology, which studies the nature of our knowledge, the sources which originate it, the means that could exist to acquire it, the justification which we might have for accepting it, and a whole series of associated problems. The range of different formulations is even more elaborate than in the case of ontology. Depending on the formulation, knowledge can be real, providing us with sufficient representation of the world, or it can be illusory, it can be the cause of events, or it may allow us little more than fortuitous associations, it can be absolute and certain, or no more than probable and chance-governed, it can rest on eternal principles which we grasp through intuition ("categories", in Kantian language) or it can be purely empirical ... This effervescence is due to the dichotomy between Idealism and Materialism, which is further complicated by many Positivist and Neopositivist postulations, which either consider this a fundamental problem, one which is insoluble by definition, or one which is devoid of sense. Each conception, in turn, is the origin of a maremagnum of attempts to refine or refute them.

– Logic, which was originally defined as the science of "the laws of thinking", is today more often thought of as the "principles and methods of valid thinking". Its first version, which was Aristotelian logic, had the great advantage of being unique; but today, we are faced with a disconcerting variety of different and incompatible forms of logic, which are usually a formal derivation upon an axiomatic scheme: several multivalued logics, modal logic, the logic of relevance, quantum logics, and intuitionist logic, amongst those which have been formalized; there are also certain logics which deny the sense or validity of formalization. If the purpose of logic really is providing us with the requisite tools for valid thinking, then it is hard to conceive that there could be more than one – unless we can develop a 'metalogic' in order to determine which particular logic applies to each specific case; this metalogic, of course, would have to be unique. The proliferation of incompatible logical systems in fact indicates that the simple identification of what is rational with what is logical, which has always been tacitly presupposed in the past, must be abandoned; but we do not yet know what should replace it.

The more specialized branches of philosophy (ethics, aesthetics, the philosophies of history and of politics, as well as that of science) are constructed upon these fundamental principles. But they are not independent of one another. Science – and particularly physics – presents a number of ethical problems, some of great importance and of a political character, whilst even aesthetics is relevant to certain aspects of the methodology of theoretical physics.

Given the dispersion of ideas regarding the three fundamental branches of philosophy, we should not be surprised by the vast proliferation of conceptions of what constitutes a philosophy of science in general, and one of physics in particular. Now, had these branches of philosophy been more or less closely linked to the study of those sciences about which they purport to philosophize, the gradual elimination of those notions which do not "function" adequately – which do not stimulate and guide investigation in a fruitful direction, rather than limit or distract it – and the corresponding development of those notions which do "function" in this sense, would have led to a more intelligible form of philosophy.

But such was not the case. The consequence has been that much of what is done in academic circles interested in philosophy is badly directed. Or else, perfectly real problems are tackled with unreal methods and concepts; a good example is the endless discussion regarding the criteria which must be employed in order to decide between alternative theories on the basis of experimental evidence – the so-called confirmation problem. This discussion is made irrelevant by the tacit presupposition that theories necessarily displace one another, i.e. the idea that the theory which "loses" is definitely discarded. This occurred sometimes, during the historical infancy of the sciences; but in physics, for example, what happens is that the empirical scope of a theory is determined without its invalidation. Thus, Newtonian mechanics, apparently "refuted" both by relativity and quantum mechanics, is today alive and productive. Or else, pseudoproblems are created; the most notorious case is that of the so-called inductive logic, whose aim is to calculate the probability (which is relative according to some, and absolute according to others) that different scientific theories might be valid; this form of "logic" has been put forward as a working tool for confirmation theory, in order to designate as "truthful" that theory which displayed the greatest probability of being valid, discarding all the rest. This logic has no basis in reality; it can be shown that the concept of probability cannot be applied to a theory's validity. It is clear that a mature science does not usually discard theories; rather, it knows how to combine them and – as we might expect – there is no concrete example of any scientific investigation which resorted to such an inductive logic. This, however, has not detained the spread of inductivism, to the extent that it is now beginning to appear in the introductions to certain scientific textbooks.

On the other hand, the real problems generated by scientific investigation are rarely touched upon within the circles of academic philosophy. Some have already been mentioned; it seems worthwhile, however, to point out two more: although much debate has turned around the question of up to what point an experiment designed to test a particular theory is impregnated by that theory, no mention has been made of a more complex and meaningful problem, which is that the design and execution of this experiment depends on other theories, which are often in open contradiction with the one under scrutiny. Another issue worthy of philosophical attention is the re-emergence of classical mechanics as one of the most fruitful and active areas in physics today. That these investigations display points of interest for the philosopher is evinced, for example, in the title for a summer school which will be devoted to them throughout the coming months: "Chaotic Behaviour of Deterministic Systems" ...

We may conclude, then, that the development of physics and its applications generates a whole range of philosophical problems whose solution is urgently required, but academic philosophy is not really in a position to provide them. It seems, therefore, inevitable that physicists must initiate certain critical investigations of a philosophical character. The situation here is no different from that which reigns in other areas of physics: we have never been slow to venture into mathematics, engineering or chemistry when it is required; why, then, not into philosophy?

(iii)

The list of problems of a philosophical nature pointed out above would serve as an ideal starting point for just such an array of physical and philosophical investigations. What is less clear is the reply to two questions: who should carry out these investigations? and, under what organizational framework?

It seems fairly obvious that investigators dedicated to such problems must make up interdisciplinary teams of physicists and philosophers, with the occasional specialist from other areas. As has been pointed out, a large part of the problem pivots around the profound division which has formed between the two disciplines; nonetheless, specialization has borne much fruit which cannot be eliminated. The solution, therefore, lies in collaboration. Physicists are necessary, for it is in our field that the problems appear. Philosophers are needed in order to provide the accumulated knowledge of three or four centuries of work in their field (part of which is of questionable value, although even knowledge of mistakes made in the past can be of immense value so long as we recognize and analyze them as such). In physics, too, the confusions and misconceptions of the past have frequently served to signpost a better road. Physicists must also ensure a firm contact between the investigation and physical reality, whilst philosophers will contribute methods, criteria and doubts which could, at more than one point, shed light on the problems at hand.

How to organize such interdisciplinary investigations? It would be absurd to put forward here a general plan for an acitivity whose modalities have yet to become fully apparent, and where local particularities will play a decisive role. Even for our own country, we can do little more than suggest projects aimed primarily at awakening an interest in these issues: "workshops" organized around wide-ranging discussions on some paper or research. We must firstly set to work, without institutions and titles. Once we have something worth organizing, the form we must give to the organizing body will appear clearer.

References

Helmholtz, H. von (1865): Populärwissenschaftliche Vorträge, Bd. I (Braunschweig) p. 7
Ioffe, A.F. (1963): *Begegnungen mit Physikern* (Moscow) p. 77

25. The Axiomatic Approach in Physics [*]

Abstract. In this paper it is shown, firstly, that physical theories contain not completely formalizable elements, the most significant being meaning assignments to formal components, approximations, unspecified elements fixed by suitable determination in specific models, and the scope (or region of validity) of the theories; secondly, that the importance of these elements depends on the type of theory as specified by a number of characteristics; and thirdly, that theories can – and must – contain various kinds of inconsistencies that are amenable to rational manipulation but preclude axiomatization within the framework of formal logic. Finally a number of aims for the axiomatic approach are outlined which are compatible with these non-formalizable elements and moreover do not exist for axiomatic methods in mathematics.

The axiomatization of theoretical physics was included by Hilbert (1900) in his famous list of twenty-three outstanding mathematical problems. Though since then various attempts at axiomatizing physical theories have been made, notably for analytical mechanics and for quantum mechanics, only one has had much significance for the development of physics: Carathéodory's (1909) work on (classical) thermodynamics. At the present time one can find both passionate defenders[1] of the axiomatic method in physics and determined attackers[2]. Yet the mainstream of physics has ignored the problem.

In this no doubt the innate conservatism of physicists has been a factor; so also has been the abuse of axiomatic techniques in some neighbouring fields such as systems theory, where they have served to throw a glamorous mantle over the otherwise too evident theoretical poverty. But there are also more solid reasons, rooted in the essential differences between the two fields, why axiomatics has contributed so much less to physics than the mathematics,

---

* First published in: Rev. Mex. Fís. **27**, 583 (1981)

[1] "The essential or primitive concepts of a theory cannot be discerned with clarity and certainty unless the theory is axiomatized ... And as long as there is no clarity concerning the building stones (primitive concepts and axioms) of a theory, discussions on fundamental problems are likely to be confused, because immature". (Bunge 1968; see also Suppes 1954)

[2] "In particular, any "axiomatization" of the theory can at best help to avoid trivial contradictions or redundancies in its formal apparatus, but is incapable of throwing any light on the adequacy of the theory as a mode of description of experience. This last problem belongs to what the scholastics pointedly called "real logic", in contradistinction to "formal logic" ... Kept within proper bounds, axiomatization of the formalism would not make any difference ... When, however, exaggerated claims are made about its powers, disastrous results follow." (Rosenfeld 1968)

and it is the aim of this paper to have a look at these reasons and their more relevant implications.

After outlining, in Sect. (i), the nature of the axiomatic approach in mathematics and why it is of significant importance there, we examine in Sect. (ii) how far the different non-mathematical elements in a physical theory are formalizable; Sect. (iii) looks at various relevant characteristics of physical theory and draws conclusions concerning which types of theory lend themselves to axiomatization; Sect. (iv) studies the specific problem of the inconsistencies in physical theory and their role; finally, in Sect. (v), some new aims for the axiomatic approach are outlined which are relevant to physics but not to mathematics.

(i)

The axiomatic method has, ideally, two phases.

In the first phase, the aim is to exhibit the structure of a theoretical edifice in systematic form as what we shall call an "axiom-based deductive structure", or ADS. From its axiom base, consisting of primitive concepts, axioms and rules of derivation, the theorems comprising the body of the ADS are derived by successive formal deduction, without making use of elements outside the ADS[3].

Of all the many questions associated with the axiomatic method in mathematics, only one point need be briefly examined here, namely the formal rigour it is intended to help establish. Now its desirability is not disputed even by the intuitionist school for whom the logic employed in a mathematical method is something to be discovered *a posteriori*. What is perhaps less generally appreciated and yet would seem to be inescapable is the quite practical reason that underlies this drive to maximal rigour. For a mathematical theory (or, for that matter, a logical system) is constructed in independence of any particular application; but if it is to work correctly in any situation where its central concepts and their interrelations adequately represent the essence of that situation, then it must have a level of internal consistency and closeness of argument which is adequate to cover the needs of any foreseeable application. The growing variety and sophistication of the applications of mathematical techniques in physics, engineering, and a rapidly increasing multitude of other fields has meant that the users of mathematical theories have become steadily more exigent over the last few centuries: it is without a doubt this need that has stimulated the continuing effort to achieve greater rigour and firmer foundations for the mathematical edifice. But what is of central importance for our present purposes is that no one of the possible applications of a given mathematical structure or theory may be permitted to fix the meaning (i.e., to establish a definite correlation between objects external to the structure and objects belonging to it) of its elements; or rather, none of the possible meanings can be used in establishing the validity of the

---

[3] A rigorous description of axiomatic techniques will be found in many textbooks, e.g. Hilbert & Bernays (1934/39) and Kneebone (1963).

structure. Thus the elements in the structure are reduced to the status of referenceless symbols and their interrelations must be stipulated with seeming arbitrariness: in other words, in pure mathematics we can have only formal structures, and the validity of deductions can only be based on those properties of the formal elements that have been explicitly stated. Thus the axiomatic method becomes indispensable[4].

Though in the course of the last hundred years various conflicting views on what, precisely, should constitute the basis of an axiomatic reformulation of mathematics have been propounded, this much at least is common ground. And even the intuitionists have implicitly given recognition to this state of affairs in that they have themselves attempted to formalize their methods (e.g. Heyting 1930).

The second phase of the axiomatic method consists in the exploitation of the ADS constructed during the first phase. In mathematics, the reformulation of a theory as an ADS offers a double advantage: on the one hand it allows the verification that is needed of the internal logical consistency of the theory – or at least, as far as the implications of Gödel's theorem and present-day limitations of technique allow; and on the other hand, the separation of presuppositions or postulates and deductions or theorems greatly eases the task of checking to what extent a proposed application is effectively possible. This second point is of importance because a large proportion of the applications of many mathematical theories is to other mathematical theories. Here exact equivalence of the axiom basis is both possible and desirable, and where it is achieved we shall speak of realization of the theory[5]. Where a mathematical structure is applied in physics or another natural science, however, there can be no question of an exact adaptation; one could at best speak of approximate realizations. For here the relationships among the relatively few concepts involved in the mathematical construction cannot do more than reproduce the central features that characterize the multifarious richness of any particular segment of nature. It is only through a more or less extensive effort of abstraction that we can arrive at a morphologically simplified version which the mathematical description will fit; and in this process of abstraction certain features are lost, others exaggerated, still other distorted, so that the fit is very far from exact. If the theoretician has done his work well, the fit will be adequate for the purpose in hand – no more.

And this purpose is, of course, quite different. The realization of one mathematical structure in another has as its aim the reduction of the consistency problem for the one to that of the other; the aim of a physical theory, however, is an understanding of the underlying part of nature, or at least sufficient understanding to enable us to use and control it. It is this difference of purpose – and the resulting differences in procedures – that makes it desirable to con-

---

[4] This must not blind us to its limitations in mathematical research; these are lucidly discussed by Hao Wang (1963).

[5] The metamathematical literature uses the term "model". I prefer here the word "realization", to avoid confusion with "model" as used below in connection with physical theories.

sider the questions of axiomatization in physics from the standpoint of that science rather than from the mathematical one.

(ii)

In what follows, we shall consider the problem in physics – with a few glances at related sciences – essentially because physics of all natural sciences has the most developed formalizable core[6]. We shall moreover, restrict our considerations to theoretical physics, which is the part containing whatever formalizable elements there are; nevertheless it must not be forgotten that physics could not exist without its experimental half, whose influence on theoretical development is decisive, however complex and ill-understood it may be.

A first and all too often forgotten point is that physics is a "bootstrap" activity – one, that is to say, in which the revision of any part depends on the others, which themselves are remodeled on a basis that includes the part being revised. It is thus the complete negation of the Cartesian method of universal doubt, for it concentrates its effort on a single point at a time. As a consequence, every part in the structure is dependent on every other part, though the precise form of the dependence (which is by no means purely logical) changes dynamically. An ADS thus misrepresents the situation by singling out certain concepts and axioms as basic. Nevertheless, if we concentrate our attention on a particular theory and neglect its connection with the rest of physics, its structure at a given moment of time can be to some extent be represented by an ADS; and it is this relative and limited validity that we shall examine in what follows.

Since in physics the mathematical structure of a theory is always applied to one specific kind of object and this application is fundamental in the construction of the theory, it is clear that the theoretical edifice must contain a lot more than just the elements formalized in an ADS of the mathematical type. The kind of object to which the theory applies will here be called a physical system (or simply system) and will be taken to be a segment of the universe and therefore to have real existence. Clearly a theory applicable to only one particular system can have genuine interest only if that system is in itself sufficiently complex and important; this will be so if the system is unique – as is the case for cosmology – or if we have no direct access to others of the same kind – the situation in geophysics[7]. But equally clearly this is a fairly exceptional state of affairs; usually the number of possible systems is large and hence their variety is considerable also. The physical theory is expected to be applicable to all of them, and this condition, too, has important consequences that we examine below.

---

[6] That the author is, professionally, a physicist may also have something to do with it ...

[7] Note, however, that the situation is not welcome to the geophysicist: the very small amount of information satellite exploration has so far yielded concerning the other planets has had a major impact on geophysics, essentially because it removes the restriction of having only one object to study.

Besides the elements which correspond in kind to those in a mathematical axiom base, a physical theory will then contain a number of other components, among which the following are the most significant ones:

1) Assignments of physical meanings; that is to say, the establishment of correspondences between quantities (or sometimes non-quantitative elements) in the theory and properties of or relations among the entities composing the type of physical system described by the theory.
2) Assumptions as to what entities (elements or properties) in the system are to be neglected or treated only approximately in the theory.
3) The scope or region of validity of the theory; that is to say, the region of phenomena over which the theory provides satisfactory explanation and sufficiently accurate prediction.
4) Certain open positions in the theoretical structure where specific information (which can be both quantitative and qualitative) can be inserted. The presence of these "holes" guarantees the required generality of the theory. Filling in all the missing information completes the building of a specific model, intended to describe one physical system (or a restricted range of systems): the information thus charcterises the system, and includes data concerning its constitution as well as the initial conditions.

That these additional components are not sufficiently formalizable and yet indispensable to the formulation of a theory may require some discussion. Thus it has been considered (e.g. Bunge, 1973) that the semantics required to establish the connection between mathematical structure and physical reality should form part of the formal axiom base. Now it is perfectly true that we can include among the axioms whatever stipulations are needed to assign clear meanings to the basic entities (i.e., the undefined concepts and quantities appearing in the axioms) of the theory. But if the semantic axioms are to play their proper role in the theory, we also need methods for deducing the meanings of derived quantities – methods, moreover, which can be stated as rules that reduce these deductions to formal schemes; for stipulating axioms is of no use unless such formal rules integrate the content of the theory into an ADS. Now there are certain aspects of semantic constructions which can be formalized in such a way; an important instance is the calculus of dimensions; and one can be misled into thinking that this holds for all the varied components of meaning. But this is not so, as we can see by means of some examples. In equilibrium thermodynamics the distinction between extensive and intensive properties[8] is extremely useful, and may moreover be formalized without difficulty, so that one can determine from the rules whether a derived quantity is extensive or intensive. But the dichotomy is valid only so long as surface effects may be neglected; when very small systems are considered – and also in many other branches of physics – there arise quantities which are intermediate or depend in more complex ways on the size of the system; and

---

[8] An extensive quantity is proportional to the size of the system, i.e. to the total amount of matter in it, while an intensive one is independent of the size.

while the exact mathematical dependence may be obtained once a quantity is defined, it is by no means clear what this contributes to the *meaning* of the quantity. Extensivity restricts strongly the possible combinations of the quantities that have the property – as long as it must characterize all quantities appearing in the theory – and thus gives one part of the meaning; but in a wider theory these restrictions disappear. The calculus of dimensions – which, as we mentioned, is formalizable – contributes a far from negligible ingredient to the meaning of all those concepts whose quantitative expression has dimensionality. But physics employs a good many purely qualitative ideas; there are also many dimensionless constants and quantities of differing and unrelated significance; and that the dimensions are inadequate for specifying the nature and meaning of a physical concept is made evident by such cases as angular momentum and action – which have the same dimensions.

It will be clear from these considerations that we have tacitly extended the implications of "meaning" from the simple stipulation of a correspondence between a theoretical quantity and an experimentally characterizable component of systems, i.e. an extensional definition, to embrace the intensional aspects; but few will deny that these belong to "what we mean by meaning". The intensional aspects are, however, essential, all the more so because without them an unresolvable paradox arises: we cannot assign an extensional meaning to the basic entities of the ADS without stepping outside the bounds of the ADS itself, in a manner incompatible with the self-contained nature we require of it. This paradox reflects, of course, the misrepresentaion that we mentioned above of a dynamic physics by static ADS's.

In constructing a theory, only a finite and relatively small number of elements and characteristics of the physical systems to which the theory is to apply is taken into account; all others are either completely ignored or else incorporated in the theory in an approximate, condensed and hence distorted form. Explicit statements to this effect will essentially be of two kinds: firstly, those that stipulate the complete elimination from the theory of certain elements, and these need not figure in the axiom base for obvious reasons; and secondly, the statements that specify how to approximate or resume other elements which are not to be treated in full explicitness. Such statements are, of course, required in the theory; but since they are made use of in ways that are not very easy to formulate in a general and abstract way and are rather meant to guide the physicist's judgment, their inclusion in the axiom base does not seem justified.

Of course both kinds of statements will become necessary when we attempt to set out how the theory relates to other theories; but whether or not we consider such relations to belong to the theoretical structure, there is no place for them in the axiom base.

The scope of a theory likewise cannot be dealt with by axiomatic methods, particularly so because it is (except for certain special cases) determined experimentally, and therefore unknown at the time the theory is first formulated. It might therefore be argued, with some justice, that the scope is not properly part of the theory, and this is certainly true so long as we restrict ourselves to the purely formal aspects of the theory. But – in this unlike a mathematical theory – a physical theory is a theory *of something*; and to know precisely what it is the theory of we need to know how far its validity extends, we need to know its scope. More practically, without information about its scope a theory is useless; we do not know when and where to apply it; thus a well established theory is always described together with its scope. There is, further, a much more subtle point: as we shall see below, the ADS of a physical theory is not normally taken to include all logically possible deductions from the axiom basis, but only those that fall within its scope; for the others are in some sense aberrant, are irrelevant to any use we may make of the theory for understanding the world we live in and controlling it, and should therefore be excluded (though they would not be in a mathematical ADS). In other terms, the scope belongs to a theory much as a man's skin belongs to his body.

Now, knowledge of the scope is evidently required for establishing the (extensional) meaning of the concepts a theory organizes into a whole. If then we propose to include meaning in the theory's ADS, the same should be done with the scope; however, in view of its empirical nature, this seems peculiarly absurd: a scope can hardly be seen as axiomatizable. Nor do we have, so far, any general rules for combining scopes in order to derive the scope of a composite theory; indeed, it may be doubted whether such rules are even possible, though an upper limit for a composite theory's scope can of course be found. We conclude that scope cannot form part of an ADS, and with it the meanings it helps to delimit must be excluded; they are integral elements of a physical theory, but not of its ADS.

Finally, physical theories are, in a sense, incomplete; in order to derive definite models from them, we must complement them with the information that specifies the details of the particular physical system to which they are to be applied. These details must cover everything from the qualitative and quantitative description of the system itself to the initial values of the various quantities involved – energies, positions and velocities, and so on. The "holes" left for this purpose within the theoretical structure are, once again, not susceptible to axiomatic treatment; though we can axiomatize about numerical values not yet specified, we cannot do so for unspecified structural information; nor do we know how to do so with purely qualitative information that we do not have, and even the very way in which the theory uses some of the numbers may depend on their values. It is thus evident that the part played by the "holes" for system specification and initial conditions is very much dependent on the uses of the theory and on the approximations made in it and therefore cannot sensibly be incorporated in an ADS.

(iii)

As well as containing non-formalizable elements in an essential way, the theories of physics differ among each other in ways which are relevant to our problem. There are several unsolved questions connected with this point which are worth stating, if only to stimulate attempts at answering them.

Theories differ in their degree of generality, in their phenomenological character or profundity, in the degree of quantitativeness, in their independence with regard to more fundamental theories, and in their structural exactness.

A theory is more general (or more fundamental) if it covers a wider range of phenomena than another theory – not merely in the sense of having a wider scope, but rather in the sense of applying to a larger variety of *types* of phenomena. The less it depends on the specification of the sort of physical system it describes, the more fundamental it will be; its generality is given by the range of differing models that one can build with it. Thus quantum mechanics is more general than, for instance, the theory of the solid state, and the latter is more general than the theory of metallic conduction. It should be noted that there is in no sense more merit or even greater profundity attached to a more general theory: the Bardeen-Cooper-Schrieffer theory of superconductivity is a highly specific one, covering as it does only the behaviour of electrical conduction electrons in certain metals at very low temperatures, – yet it is profound indeed in that it opens up a first chapter in what may well be central to the physics of the 21st century, the behaviour of systems of "not very many particles". In fact physics needs both general and specific theories, because only the combination of both types will yield useful predictions (whether usefulness is here of an applied kind or to another research problem is quite irrelevant); thus value judgment are very much beside the point.

One misconception that may be mentioned here because it is frequent even among physicists is that the various fundamental theories of physics must be independent of each other in the sense of being closed off and self-sufficient. In fact, of course, the unity of physics is not merely one of subject matter and method but also one of mutual connections among *all* its theories. Thus the most general theory of all, classical mechanics, is also in a sense the most dependent on the others, for the forces that enter into Newton's equations (or the potentials in a Lagrangian formulation) are not explained or even described by it but originate in phenomena discussed by other theories. And one of the most elegantly rounded theories, Maxwell's electromagnetic theory, cannot account by itself for the stability of extended elementary charges; this has in the past led many theoreticians into accepting the view that the elementary charges must be point-like, though it is quite well known that this leads to contradictions and paradoxes.

A profound theory may be contrasted with a phenomenological one. This distinction is significant in physical research, but not easily circumscribed. The extreme of phenomenology is the abridged description of experimental results: a theoretical formulation which does not go farther has no predictive power beyond what is already known from the laboratory. Such an extreme is almost unknown to natural science, though it has its place in social science where no clear account relating the various factors may be available. Historiography, for instance, is of this sort. One step further up the theoretical ladder leads us to theories which generalize beyond experimental results and so predict new observations, but only for very similar situations. Laws such as Boyle's law connecting the volume and pressure of a gas at constant temperature were

– at the time of their discovery – of this sort, though now they appear as consequences of a more elaborate theory. Situations which do not go much further than this exist, of course, in physics; in elementary-particle physics, a large number of such experimental generalizations are known – and the fact that we can often achieve an elegant and economical description of such laws by group-theoretical structures must not blind us to the fact that classifying the known particles according to SU(3), for instance, yields no understanding of *why* this works; and the predicted discovery of the omega particle is evidence that it works remarkably well.

This *why* is an essential question, for the ability to answer it, what is commonly called its explanatory power, is precisely what makes a theory profound. The distinction between profound and phenomenological theories is not only common knowledge among physicists, it is indeed a useful and widely employed concept when formulating research aims. But among philosophers of science the concept is often held to be meaningless. The problem appears to lie in a certain confusion about the answer expected to that question, *why*. Let us consider an example. A violin string, suitably excited, will vibrate at certain selected frequencies, while motion at other frequencies is rapidly damped out. One kind of explanation that may be offered runs like this: the second-order differential equation describing the string's motion has stationary solutions only when suitable boundary conditions are satisfied, and this happens only for certain specific frequencies. True, but not illuminating. Another sort of explanation considers the possible ways a string fixed at either end can oscillate; each kind of oscillation has its wavelength, determined by the frequency and the mechanical properties of the string, and clearly there must be an integral number of half-wavelengths in the length of the string, or else the string will either snap or transfer all the motion's energy to the end-blocks. Hence only some frequencies correspond to vibrations that can last. Again true, but now we gain some insight into the mechanism that stabilizes certain frequencies and not others; we see what forces are at work and we can extend the model to account for frictional effects and so on.

Thus an explanation may be based on the mathematical structure of the theory alone, or it may derive from the physical significance of the concepts involved. In the latter case it will not only furnish a causal structure and hence a dynamical account for the phenomena covered by the theory, it will also provide a framework that links the meanings of the concepts in the theory, exhibiting them as far from arbitrary: in a physical theory of any profundity the meanings of its concepts, which we saw above to be an essential ingrediant, cannot be assigned at will but only in such a way as to yield confirmable causal nexuses; in a phenomenological theory, this is not the case. Such distinctions depend, of course, on the explicit premiss of a world that is both real and independent of what we may happen to think; certainly it is among those who accept this premiss that we find the recognition that explanatory power is relevant – e.g., in Bhaskar (1978). Only on this premiss can we accept that the theoretical physicist's constructions are able effectively to mirror the behaviour and relations of the entities he studies: that we are, in other words, able to build functioning, dynamical models of selected aspects of our world. To deny such distinctions, as logical positivism obliges us to do through all too well known arguments, not only deprives the physicist of a useful tool but can create serious confusions. Thus the common view (repeatedly stated for instance by Bohr, as Scheibe (1973) brings out clearly) that classical physics is based on the point particle, with zero extension, and the field, with infinite extension, usually leads to the conclusion that such easily visualizable (!) models are inappropriate to quantum mechanics, and therefore only explanations of the mathematical type should be sought; and this is meaningful only if the profound/phenomenological distinction is abandoned.

The necessity for a theory in physics to be basically quantitative is by now well established, and none of the major theories are chiefly qualitative. Yet no theory is *purely* quantitative, and its qualitative features are essential to an understanding of its meaning and also to its applications. These qualitative aspects of a theory tend often to be forgotten; yet commonly they determine the field of usefulness or scope of the theory – as for instance in the case of the distinction between continuum and corpuscular theories.

The independence of a theory is connected with how fundamental it is, but is by no means identical with this property. A fundamental theory of wide scope may form the basis for many more specific theories, but need not therefore be independent. Thus a great deal of the present-day theory of the solid state, itself the generator of many detailed theories, is directly dependent on quantum mechanics on the one hand, and on statistical thermodynamics on the other; the latter depends in its turn on quantum or classical mechanics and on statistical theory. At another level, the theory of general relativity is perhaps less fundamental than quantum mechanics (in the sense that its scope is more restricted and that it has generated far fewer dependent theories; we repeat that this does not mean that it is less profound); but it is more independent in that it creates the basis for its own formulation of mechanics, while quantum theory requires such a basis from outside – either Newtonian mechanics or special relativity.

Lastly, a theory may be said to be structurally exact if nowhere in its deductions the need for approximations arises. We must distinguish here between the sort of approximations that are used in deriving specific models because we do not have suitable mathematical tools, and the approximations made in order to be able to neglect what we judge to be inessential factors. Only the latter are relevant here, since they enter in an irremovable way into the framework of the theory and therefore must be considered when we attempt to create an ADS for it.

Structurally inexact theories are often theories in process of development: their central features may already be clear, but many details are lacking and with them a fully developed mathematical apparatus. In other situations the limitation is essentially experimental, as when we have quantities whose values are important but which we do not know how to measure.

Of course these various distinctions among theories are not independent of each other. Thus a general theory is mostly also profound, quantitative, and structurally exact; but as we have noted, there are exceptions, and the contrary is not usually valid.

The importance in research of the formal or formalizable part of a theory depends very much on where the theory falls along the scales of these different characteristics. To the extent that a theory is general and structurally exact, its formal part is central, and its profundity will then lend importance to the attempts at creating an ADS for it. But it is by no means clear (at the present moment, at least) how far a very specific and rather phenomenological theory can be axiomatized in a satisfactoy way – i.e. without trivializing it; and in fact there is reason to doubt whether the exercise would be at all useful. A similar caution is needed with still undeveloped theories, because they tend to be structurally inexact in crucial places. For when a theory contains approximations in an essential way, all attempts at formal description will distort it beyond recognition; in fact, using an approximation is equivalent to an open invitation to use one's intuitive judgment about the validity of the procedure.

It is this situation which both justifies many of the fears that have been expressed by opponents of the axiomatic method in physics because its proponents have seen it, quite absurdly, as universally applicable – and creates the basis for selectively axiomatizing those theories for which it is meaningful. And it cannot be denied that such theories as classical mechanics or thermodynamics offer a very suitable field. But in quantum mechanics the attempts at axiomatization, though in many respects extraordinarily useful, have intensified the problems of interpretation rather than helped to resolve them.

We must conclude that the axiomatic approach is by no means universally desirable in physics; while it offers definite advantages (which we discuss below), there are clearly also some dangers that threaten.

(iv)

But before entering into these advantages of the axiomatic approach, we must mention an important matter which has by no means received the consideration it deserves. This is the presence of inconsistent and sometimes openly contradictory elements within the framework of physics. Not all the varieties of inconsistency in physics are directly relevant to our problem of the axiomatic method; but since they are all fairly intimately linked, it seems worthwhile to list the important ones.

A first kind of inconsistency arises because of the need to connect theory and experiment: for the construction and operation of experimental set-ups, and the interpretation of the results obtained from them, require a set of concepts that usually go well beyond what the theory under test can offer. As a result, we employ an astonishing mixture of theoretical notions with very different and often incompatible basic assumptions in order to do experimental work and link it to theory. Thus the experimental verification of relativity theory makes use of instrumentation designed on the basis of classical mechanics, of optics, of quantum theory, of electromagnetic theory, and of other branches of physics as well.

The experimental physicist is quite at home in this situation; he knows that for this purposes any theory that yields a sufficient approximation is good enough, and he has developed into a fine art the technique of combining incompatible theoretical constructs. This is not the place to examine the various epistemological and other presuppositions that make this "fine art" possible; suffice it say that one can indeed work in a consistent fashion by combining inconsistent elements, provided certain intuitively understood constraints are observed.

A second variety of inconsistency, more directly internal to theoretical structures, is closely connected to this one: it arises whenever an approximation is made within the framework of a theory, for the basis on which such approximations are accepted is either another and usually incompatible theory or a set of experimental data, likewise obtained on theoretically unrelated foundations. The various "semiclassical" calculations so beloved of the quantum chemist are of the first sort; they exemplify the combination of incompatible theoretical contributions at its best, for they are both ingenious and remarkably successful. A second case is that of the experimentally justified estimation of relative magnitudes which allows us to neglect a small but theoretically bothersome term; again this is of frequent occurrence in physical theory.

There is a sense in which both these kinds of inconsistency are irrelevant or at least of reduced importance: they do not strike at the central core of a theory's structure – or at least they do not appear to do so. Yet only the second kind can be attributed to our human limitations; the first is clearly essential in the nature of things; and the systematic way in which both crop up places some doubt on their secondary character. A different kind of inconsistency, linked in quite central ways to the theoretical structure, arises because – as we saw above – the physical meanings incorporated in a theory are not in general the realizations (in the model-theoretic sense specified above, note[9]) of corresponding formal elements. Hence the formal part of the theory can imply consequences which go beyond what the nonformal part may justify; such consequences are dubbed "unphysical" and simply thrown out. This is the case when equations, written to describe a physical phenomenon, have solutions we do not want but are unable to get rid of in a mathematically satisfactory way. For instance, Maxwell's equations for the electromagnetic field have an advanced and a retarded solution, and we ignore one of them on the basis of a heuristic causality argument, not otherwise germane to the theory. Another case of a similar nature is that of the phase of a wave function in quantum mechanics, the absolute value of which is quite without physical meaning. Such cases appear to arise only rarely, because the majority of them is avoided by an important practical limitation we apply to an ADS: we do not allow it to include explicitly *all* possible consequences of its axiom base, but only those

---

[9] The metamathematical literature uses the term "model". I prefer here the word "realization", to avoid confusion with "model" as used below in connection with physical theories.

that do not obviously fall outside the theory's scope. This limitation is of profound significance. In the physicist's practice it is what allows him to combine incompatible theoretical constructs into one argument, as we exemplify below. From the epistemological point of view, it is quite as interesting; it is because of this limitation that the empirically expected consequences of a new conception do not extend throughout the whole edifice of knowledge, so that researchers can work each on his own problem without having to consider all the possible repercussions of any particular new idea. Moreover, one could use it acutally to *define* a theory's scope as the range of phenomena over which deductions from the starting postulates may meaningfully be made. From a logical standpoint the limitation is likewise of considerable relevance; apart from the fact that it points toward a clear discrimination between the logical and the rational, it opens up a promising new field of study: logical systems in finite universes, where logical operations may connect propositions from different such universes. But to the author's knowledge, this has not yet been explored.

But the mismatch between the formal elements and the physical significance may have much more serious repercussions. Again, we find a case in classical electrodynamics, which is an excellent example of a finished and elegant physical theory. If we attempt to calculate the energy a pure electric charge has because of its interaction with the electromagnetic field it creates around itself, we get into deep trouble: if we take the particle to be point-like, then this energy (usually called the self-energy) becomes infinite, with no apparent source to provide it; if on the other hand we take a particle of a small but definite size, then we need forces to avoid its being broken up by the field, and where would these forces come from? This difficulty (which we cannot remove by any of the tricks for getting rid of unwanted "unphysical" solutions) is not improved when we go over to the much more sophisticated theory of quantum electrodynamics: here these infinities turn up in just as disturbing a fashion. (We shall not enter into the thicket of renormalization theory here.)

The mismatch may also appear between different parts of the formal structure. Such a situation may even be deliberately created. Perhaps the most famous case was Niels Bohr's 1913 theory of atomic structure, in which he simply postulated that there are certain orbits possible for the motion of an electron within the atom where no radiation is emitted; this is in flat contradiction with what classical theory, based on Maxwell's equations, predicts, namely that every charge when accelerated radiates away some of its energy. Bohr, of course, was quite well aware of this; there is in fact so much sound theory and experimental verification behind it that his introduction of a contradictory postulate was an act of remarkable physical intuition and – the term does not seem misplaced – great moral courage. For the astonishing thing is that the hybrid theory worked surprisingly well. It could naturally be bettered as soon as quantum mechanics grew of age, and above all reformulated in a more consistent way. But of course quantum mechanics has in its turn introduced some very extraordinary contradictions into physics, without altogether resolving this one (see Claverie and Diner 1976).

The last type of contradiction – or at least inconsistency – which constantly appears in physics is that between theories. Since this may surprise those who are not specialists in physics (and perhaps some who are), let us examine examples.

The first – and conceptually the simplest – is offered by the plethora of theories that make up the attempts to build models of the atomic nucleus. In principle, we could solve the Schrödinger equation describing the motion of the particles within the nucleus; in practice this proves impossible, partly because we do not know the force acting between the nuclear particles with sufficient detail, partly because the equation is too complex to yield to presently known mathematical techniques – even with the aid of computers. So we construct models on the basis of simplifying assumptions. Perhaps the best known of these is the shell model, where each particle is taken to move in a common field of force created by its seeing, so to say, the average effect of the other particles; because of this averaging, the motion of each particle appears independent of that of the others. The other extreme in nuclear model-building is the liquid-drop model: here the nucleus is treated as if it were a continuous fluid, and we forget about the existence of individual particles. Both these models have been very successful, each in its own sphere. A number of nuclear properties which neither could explain easily have been treated by means of yet a third model, the so-called collective model, a betwixt-and-between construction which takes many features of the shell model but allows the motion of the nuclear particles to deform the shape of the common field of force. There are still other models, each with its own usefulness: the optical model, the alpha-particle model, the cluster model, the statistical model[10].

Though these are full-fledged theories, the nuclear physicist calls them models, to signal his awareness of the unsatisfactory state of affairs the need for such a multiplicity of theories represents. They are not merely different, they are incompatible to such an extent that the basic assumptions of no two of them agree; thus some even ignore the essentially quantal nature of the nuclear particles, though most draw central features precisely from quantum mechanics. And in spite of the intense and ingenious efforts that have gone into all this model building, every nuclear theorist would welcome the appearance of a genuine nuclear theory that could sweep them all into the dustbin of history. Yet so long as no such theory is visible on the horizon, we must go on using these diverse models. And here lies the awkwardness of the situation for the philosopher of science: for in many cases the theoretical explanation offered for an observed behaviour of nuclei is built on the judicious combination of several such models, in spite of their conflicting bases. In fact, some of the models themselves might be described as just such mixtures of theoretical oil and water. Yet they work, and often very successfully indeed.

---

[10] Details concerning these models will be found in the textbooks of nuclear physics e.g. Bohr and Mottelson (1969/75) or Brown (1971)

A second example is furnished by the relation between thermodynamics and statistical mechanics. Here the inconsistency is much more subtle, and may indeed for most practical purposes be ignored; yet it amply repays analysis. The situation is as follows. Thermodynamics is a theoretical structure whose remarkable internal clarity – put in evidence by the axiomatic reformulation first achieved by Carathéodory (1909) (see also Falk and Jung 1959)– cannot hide what we might call its lack of intuitiveness. A significant aspect of this is that it does not appear to have a direct link to other physical theories, while at the same time it is so fundamental that for instance the direction of the flow of time for all of classical physics, at least, is derived from it. The link to the rest of physics is established by underpinning it with statistical mechanics. This enterprise is complete but for one small loophole: the proof that the ensembles of statistical mechanics actually have averages of the required kind depends on the so-called ergodic hypotheses; but for all physically significant types of system this hypotheses remains no more than a postulate with *a posteriori* justification. A related difficulty arises for the concept of equilibrium: in thermodynamics the notion seems quite clear, but in statistical mechanics its exact meaning is very hard to pin down.

Now because of the practical success of statistical mechanics, and because all discrepancies between the predictions of thermodynamics and statistical mechanics are well understood, many and perhaps most physicists are content to accept without further ado the validity of the ergodic hypotheses. Yet there remains this small but very deep conceptual gap in our understanding; and such problems as the origin of the macroscopic irreversibility for many-body systems in which each body follows a fully reversible microscopic mechanics are evidently related to it[11].

These two examples concerned situations where more than one physical theory existed within the same field; to complete the picture, we will briefly mention a third example of contradiction between theories in different fields, though this case is not really relevant to our theme. It concerns the relations between quantum mechanics and relativity theory – the two great generalizations that twentieth-century physics has to offer. The need for combining them is obvious: there are too many situations of physical (not to mention astrophysical) interest in which subatomic particles, subject to quantum behaviour, move at speeds or through distances such that ordinary Newtonian mechanics is no longer a good approximation and relativity theory mut be invoked. But the difficulties in the way of this endeavour have so far won out. So long as we remain within Einstein's special theory the technical problems have largely been solved; at least in its applications to electromagnetic radiation and its interaction with matter – quantum electrodynamics, that is to say – the combination has proved spectacularly successful and has provided us with

---

[11] Ergodic theory (see e.g. Farquhar (1964) is the most widely studied but not the only way to create a physical justification for the use of ensembles in statistical mechanics; others – which generate different but no less intricate problems – are those of Tolman (1938), Jaynes (1957) and Penrose (1970). For the question or irreversibility see e.g. Mehra and Sudarshan (1972)

some of the most accurate predictions of any theory in physics; yet it is still true, as was written eighteen years ago, that "the fusion of these requirements (of special relativity and quantum mechanics) into a non-contradictory theory (in four dimensions) is well known to be a problem whose solution has not been achieved in a non-trivial way even in a model" (Jost 1960). And when we go over to the theory of general relativity, the picture is much bleaker: in spite of an enormous amount of effort concentrated on the problem, no satisfactory way of quantizing it in any of its forms has been found. For there is here a basic conceptual conflict: relativity theory is essentially a non-linear theory of continuously variable quantities. Quantum mechanics, on the other hand, yields discrete spectra and is basically linear; this is pointed up by the central role in it of the superposition principle, which states that the sum of two solutions of the wave equation is again a physically meaningful solution[12]. To conclude this long discussion of conflicting elements within physical theory, it must be noted that the situation is in no sense only temporary. Admittedly, any one of the inconsistencies or contradictions we have mentioned will only persist for a certain time, until the moment, in fact, when theoretical development makes a decisive step forward; so that indeed the elimination of these difficulties may be said to be one of the aims of the physicists' work. Yet every advance, every new idea, every reformulation of theory brings in its wake a number of new difficulties and conflicts which it substitutes for those it solves: for this contradictoriness is at the very least essential to our understanding of nature, built up as it is of a series of part views, each appropriate to its own purposes but not suited for any other one. Hence different theoretical structures must be incompatible to achieve their incompatible aims. It can be argued that this fact merely reflects, at the cognitive level, even deeper reasons; but we shall not pursue this matter here. Certainly these inconsistencies are so much a part of the physicist's way of life that he sees them as normal and may even be unaware of them[13]. And in the present context they are of course relevant, because they make otiose the attempt to axiomatize any theory that exhibits them.

(v)

What, then, will be the tasks for the axiomatic approach in physics? From what has been said above, it should be apparent that the question must be asked explicitly, since the aims usually proposed for mathematical axiomatization cannot be relevant to physics. Thus the completeness of an axiom base is not of practical interest because, as we have noted, a physical theory is of limited scope; hence only a limited range of consequences drawn from the axiom base is meaningful, and the physicist knows that he is on speculative

---

[12] A more detailed discussion of inconsistencies among theories, though from a different point of view, will be found in Tisza (1963).

[13] To a physicist the assertion that, say, classical mechanics formally contradicts quantum theory, even if true, will appear trivial or even ridiculous. He will prefer to see classical mechanics as a limiting form of quantum behaviour – though the problems of going to this limit are far from trivial.

grounds when he steps outside these limits. If the adding of further axioms to the base causes inconsistencies outside the scope of the theory, this is merely irrelevant; their elimination may be justified under Occam's razor within the scope but cannot constitute a logical requirement. In fact, of the mathematically interesting purposes of axiomatization only the somewhat pedestrian need to check the deductive soundness of the theoretical structure survives. Important though this may be in practice, we need hardly discuss it further here; not only does everyone agree on this matter, but it is essentially a matter of scientific technique rather than fundamental principle.

Are there any other purposes, specific to physics (or, perhaps, the natural sciences), for which the axiomatic method could prove useful? Most writers on axiomatization appear to take it as read that such purposes are self-evident and need not be stated; what follows then are suggestions for the future rather than conclusions from already existing studies.

Firstly, exhibiting the formal structure of a theory – and in particular making explicit its fundamental postulates – may be of great help in understanding it. What is important here, of course, is formulating adequately the theory rather than the axiomatic approach as such, which is only one way of achieving this, though a convenient one. The significance of the Carathéodory approach to thermodynamics lay largely in the fact that it stated unequivocally the concepts fundamental to the theory. This was of great usefulness in determining where it would apply and where not, and led to many fruitful extensions. In a similar way the two distinct but related axiomatizations that von Neumann (1932) proposed for quantum mechanics retain their significance in all the heated discussion about the interpretation of this theory.

Note here that the recognition of alternative axiomatizations opens the way to a many-sided and flexible understanding. Thus in the case of classical mechanics, we may build an ADS along the lines laid down by Newton; we start with space, time and particles of fixed mass, and take his three laws to make up the vital part of the axiom base. In this way we obtain a very direct access to solving the simpler problems, at the price of facing non-trivial difficulties when going over to, say, continuum mechanics. Alternatively, we may adopt a Lagrangian formalism. This is less intuitive in that the necessary axioms are no longer easily linked to everyday experience; the simpler mechanical problems already require some mathematical sophistication for their solution, in consequence. But we achieve a convenient way for solving the more difficult problems, we are led to new insights concerning the conservation laws by way of Noether's theorem, and the precise meaning of the transition to relativity or quantum mechanics now becomes clear. Both approaches thus have their justification. We also see that there can be alternative axiomatizations in a new sense, wider than the mathematical one which requires demonstrable logical equivalence: the "physical" equivalence of two ADS's is also possible, meaning thereby that within the scope of the theory all deducible consequences are indistinguishable.

Such a situation arises, for instance, in statistical mechanics (see Jaynes (1957), Farquhar (1964) and Penrose (1970), where several essentially different axiomatizations have been given. The present argument suggests that each of them highlights certain features of the theory: so far from being regarded as competitors, among which a "best one" should be chosen, the alternatives complement each other, and their relationships ought to be studied from this point of view. In quantum mechanics we also have a number of competing axiomatizations (see, for instance, Gudder 1977); but the situation is not quite the same, for they have all been constructed for fairly similar ends, and thus the study of their interrelation (though important enough in its own right) will not shed much light on the vexed questions of the interpretation of quantum mechanics.

Related to the increased understanding of a theory that axiomatizing it may bring about is the usefulness of exhibiting its formal structure for determining its scope, or region of applicability. As I have attempted to show in a forthcoming paper, talking about the probability of a theory's being true is not helpful; the problems raised by this approach are removed or turned into something useful by the scope concept; and the scope is of course essential in discussing any practical application of a theory. Let us look at this a little more closely.

The question whether building an ADS may be of assistance in finding the scope of the theory cannot arise, naturally, in axiomatic mathematics and is peculiar to the experimental sciences. This and certain other aspects of the question where axiomatizing a physical theory could prove important have not so far been taken very seriously; it may be suspected that this is because the great differences between physical and mathematical axiomatics have not been fully appreciated.

How, then, does an ADS help in determining the scope? What in practice *can* be directly found is the scope of the individual laws deduced from the theory (we are using "law" in the sense of a relation between experimentally measureable quantities which derives from the theory). Experimental work directly decides where the law "works" and where it does not. Now different laws in the same theory need not have the same scope; and we can clearly define an inner scope for the theory which is the intersection of the scopes of all its laws, and an outer scope which is the union – both these terms in their set-theoretical meaning. The inner scope is the range over which all of the theory is valid, the outer scope is the region where at least some of the theory works. But we cannot say that we have examined all possible laws to be derived from the theory, for their number is presumably infinite. Yet if we exhibit the place of those laws whose scope we know is the ADS, it becomes at once evident from this hierarchical structure where a new law might alter the theory's two scope limits and where we may go on deducing as many new laws as could be interesting without affecting the situation. Thus the ADS aids us in determining how far we can state that a theory's scope (either inner or outer) is already well established.

But we can go further. The ADS allows us to see which of the axioms are involved in the deduction of each of the laws whose scopes are experimentally known. We can therefore also find the scopes of the axioms individually (here only the inner scope, the intersection of the scopes of those laws that require the axiom for their deduction, appears to be meaningful). In general the axiom scopes will not coincide, and the "topology" of the situation can be quite complex. Let us examine only one case: where the inner scope of the theory as a whole coincides with the scope of one of the axioms (or perhaps a small set of the axioms). Here it is clear how a better theory, one of wider scope, could be found: by substituting for the limiting axiom another one of ampler scope. In this case, then, the axiomatization turns out to be useful because it suggests a direction for further development; just how this improved theory could be built is, of course, not to be answered by axiomatic or any other techniques, for this requires an effort of the creative imagination. But at least we know along what line to search.

This case will not always be the one we have: but to achieve it we may make use of the freedom we have in setting up the axiom base for the ADS and search for alternative ones which are either logically or at least "physically" equivalent (meaning thereby that within the theory's scope we cannot distinguish their consequences). If in one of these alternatives we can pick out a scope-limiting axiom of the sort just described, then we can obtain a hint towards improving the theory.

But there are certain questions which require further investigation. Does the existence of a scope-limiting axiom have some epistemological significance? Is it desirable that the scopes of the axioms in an ADS should approximately coincide? This last cannot of course be a primary criterion in judging a theory, since it is satisfied for a theory all of whose axioms have null scope; but if its relevance were understood, it could be helpful as a secondary criterion.

In conclusion, then, we see that the axiomatic approach does have a definite role to play in physics, but it differs markedly from the one it has in mathematics, and we must clearly understand the nature of the theory under study before deciding whether to axiomatize or not. Where it is appropriate, axiomatization can achieve significant results, perhaps more so than has been realized so far; but to apply it indiscriminately is to invite disaster. The situation might be summed up by saying that in mathematics we can axiomatize in order to understand; in physics we must understand in order to axiomatize.

References

Bhaskar, R. (1978): *A Realist Theory of Science* (Harvester Press, Sussex)

Bohr, A., Mottelson, B.R. (1969/75): *Nuclear Structure* 2 Vol. (Benjamin, New York)

Brown, G.E. (1971): *Unified Theory of Nuclear Models and Forces* 3rd edn. (North-Holland, Amsterdam)

Bunge, M. (1968) in I. Lakatos & A. Musgrave (eds.) *Problems in the Philosophy of Physics* (North-Holland, Amsterdam) p. 120

Bunge, M. (1973): *Philosophy of Science* (Reidel, Dordrecht)

Claverie, P, Diner, S. (1976): in O. Chalvet et al. (eds.) *Localization and Delocalization in Quantum Chemistry* Vol. II (Reidel, Dordrecht) pp. 395, 449, 461

Carathéodory, C. (1909): Math. Annalen **67**, 355

Falk, G., Jung, H. (1959) in S. Flügge (ed.) *Handbuch der Physik*, Vol. III/2 (Springer, Berlin Heidelberg New York), p. 119

Farquhar, I. (1964): *Ergodic Theory in Statistical Mechanics* (Wiley, New York)

Gudder, S.P. (1977) in W.C. Price & S.S. Chissick (eds.) *The Uncertainty Principle and Foundations of Quantum Mechanics* (Wiley, London), p. 247

Hao Wang (1963): *A Survey of Mathematical Logic* Chapt. III (North-Holland, Amsterdam)

Heyting, A. (1930): Sitzber. preuss. Akad. Wiss., phys. math. Kl. (Göttingen) p. 42, 57, 158

Hilbert, D. (1900): Nachr. K. Ges. Wiss., math-phys. Kl., 253

Hilbert, D., Bernays, P. (1934/39): *Grundlagen der Mathematik* (J. Springer, Berlin)

Jaynes, E.T. (1957): Phys. Rev. **106**, 171

Jost, R. (1960) in M. Fierz, V.F. Weisskopf (eds.) *Theoretical Physics in the Twentieth Century* (Interscience) p. 107

Kneebone, G.T. (1963): *Mathematical Logic and the Foundations of Mathematics* (Van Nostrand, London)

Mehra, J., Sudarshan, E.C.G. (1972): Nuovo Cimento **11B**, 215

Neumann, J. von (1932): *Mathematische Grundlagen der Quantenmechanik* (Springer, Berlin Heidelberg New York)

Penrose, O. (1970): *Foundations of Statistical Mechanics* (Wiley, New York)

Rosenfeld, L. (1968): Nucl. Phys. **A108** (1954), 241

Scheibe, E. (1973): *The Logical Analysis of Quantum Mechanics* (Pergamon Press, Oxford) p. 14

Suppes, P. (1954): Philos. Sci. **21**, 242

Tisza, L. (1963): Rev. Mod. Phys. **35**, 151

Tolman, R.C. (1938): *The Principles of Statistical Mechanics* (Oxford University Press, Oxford)

List of Publications of T.A. Brody

1. "Charles Babbage and computing", Modern Quarterly **27** (1945), no. 3, p. 117
2. *Mesure de la période radioactive du samarium par la méthode de la plaque photographique* (thesis), University of Lausanne 1950
3. "Condiciones de discriminación entre fondo y trazas en emulsiones nucleares", Rev. Mex. Fís **3** (1954) 217
4. "Alcance aparente de trazas en emulsiones nucleares", Rev. Mex. Fís **4** (1955) 224
 – Reprinted in: Anales del Instituto de Física **1** (1955) 1
5. "Estudio de las reacciones $_6C^{12}(d,p)_6C^{13}$ y $_8O^{16}(d,p)_8O^{17}$", Rev. Mex. Fís. **4** (1955) 207 (with F. Alba Andrade, M. Mazari, M. Vázquez B., V. Serment and A. Fernández)
6. "Primer informe sobre estudios de la lluvia radioactiva", Rev. Mex. Fís. **5** (1956) 153 (with F. Alba Andrade, V. Beltrán, H. Lezama, A. Moreno M., A. Tejera and M. Vázquez B.)
7. "Formación y extensión de los conceptos científicos", Seminario Problemas Científicos y Filosóficos **27** (UNAM, México) 1956
 – Second reprinting: UNAM 1988
 – Reprinted in: Mathesis **5** (2, May 1989) 193–213
 – Russian translation: Voprosi filosofii **2** (1957) 83
 – Reprinted in: I.V. Kuznetsov and M.E. Omelianovski (eds.) *Filosofskie voprosi sovremennoi fiziki* (Gosizpolit, Moscow) (1958), p. 147
8. "Segundo informe sobre la precipitación radioactiva", Rev. Mex. Fís. **6** (1957) 97 (with F. Alba Andrade, A. Tejera, M. Vázquez B. and H. Lezama)
9. "La nomenclatura física en castellano", Bol Soc. Mex. Fís. **2** (1957) 16
10. "Coloquio sobre el problema ético del científico", Cadernos del *Seminario de Problemas Científicos y Filosóficos* no. 10 (UNAM, México) 1958
11. "Mitología del hombre de ciencia", Revista de la Universidad de México, vol. 13, April 1959, p. 1
12. "Tercer informe sobre estudios de la precipitación radioactiva", Rev. Mex. Fís. **7** (1959) 8 (with F. Alba Andrade, A. Palacios, G. Rickards, A. Tejera and E.G.B. de Velarde).
13. "Cuarto informe sobre estudios de la precipitación radioactiva", Rev. Mex. Fís. **8** (1959) 6 (with F. Alba Andrade, A. Palacios, G. Rickards, E.G.B. de Velarde and A.M. Martínez)
14. "Métodos de cálculo de la precipitación radioactiva", Rev. Mex. Fís. **8** (1959) 43 (with G. Rickards and E.G.B de Velarde)
15. "Métodos de determinación del estroncio 90", Rev. Mex. Fís. **8** (1959) 27 (with A. Palacios and A.M. Martínez)
16. "Contador 4π con circuito de anticoincidencia doble", Rev. Mex. Fís. **8** (1959) 117 (with F. Alba Andrade and I. Castro V.)
17. "Brackets de transformación para funciones de oscilador armónico", Rev. Mex. Fís. **8** (1959) 139

18. "Matrix elements in nuclear-shell theory", Nuclear Phys. **17** (1960) 16 (with G. Jacob and M. Moshinsky)

19. *Tables of Transformation Brackets* (with M. Moshinsky) (UNAM 1960) pp. lxxiii, 175
    - Second edition: Gordon & Breach, New York 1964

20. "Simetrías y reglas de suma para los paréntesis de transformación", Rev. Mex. Fís. **9** (1960) 181 (with M. Moshinsky)

21. "Quinto informe sobre estudios de la lluvia radioactiva", Rev. Mex. Fís. **11** (1962) 1 (with J. Calvillo, S. Bulbulián and A.M. Martínez)

22. "Diseño de contadores 4π con anticoincidencia mutua II", Rev. Mex. Fís. **12** (1963) 139 (with J. Calvillo)

23. "El concepto de información", *Simposio sobre información y comunicación*, del XIII Congreso Internacional de Filosofía, Centro de Estudios Filosóficos, (UNAM, México) 1963

24. "Determinación del estroncio 90 en leches y aguas de lluvia", Rev. Mex. Fís. **13** (1964) 1 (with A.M. Martínez and S. Bulbulián)

25. "Marie Sklodowska-Curie", in *Homenaje a la Universidad de Cracovia en su Sexto Centenario*, Coordinaciones de Humanidades y de la Investigación Científica (UNAM 1964) p. 156
    - Partially reprinted in: *El Día*, May 21, 1989, p. 8

26. "Un circuito de anticoincidencia simplificado", Rev. Mex. Fís. **13** (1964) 75 (with A. Nava J.)

27. "Recursion relations for the Wigner coefficients of SU(3)", J. Math. Phys. **6** (1965) 1540 (with M. Moshinsky and I. Renero)

28. "Some matrix elements and normalization coefficients in $SU(n)$", Rev. Mex. Fís. **15** (1966) 145 (with M. Moshinsky and I. Renero)

29. "A closed formula for the $3nj$ coefficients of R3", Rev. Mex. Fís. **17** (1968) 301 (with J. Flores V. and P.A. Mello)

30. *LISPITO, a LISP processor for the IBM 1620* Centro nacional de Cálculo IPN (México 1965)

31. "Sobre la reproducción y distribución central de prepublicaciones", in: *Memorias del Primer Congreso Latinoamericano de Física* (Soc. Mex. de Física, Mexico 1969), p. 461

32. "The structure of LISPITO", in: *The Second LISP Book* (ed. Information International Inc., Boston 1969) p. 67

33. "Sum rules for the Moshinsky brackets", Rev. Mex. Fís. **19** (1970) 303 (with V.C. Aguilera and J. Flores)

34. *Symbol-Manipulation Techniques for Physics* (Gordon & Breach, New York 1970), pp. iii, 97
    - Reprinted in: M. Moshinsky, T. Brody, G. Jacob (eds). (1970): *Many-body problems and other selected topics in theoretical physics* (Lectures delivered at the 1965 Session of the Latin American School of Physics, México D.F.), (Gordon& Breach, New York), p. 715

35. "Display-to-plotter conversion", DECUS **11** (1970) 122 (with W.D. Hay)

36. *La Teoría de Grupos en la Física*, Lectures delivered at CURCAF, UNAN, Managua, 1–27 Feb. 1971 [mimeo]

37. "An ergodic property of orthogonal ensembles of random matrices", Phys. Lett. **37A** (1971) 429 (with P.A. Mello)

38. "Monte Carlo studies of a class of real symmetric matrices", Rev. Mex. Fís. **20** (1971) 217 (with O. Bohigas, J. Flores V. and P.A. Mello)

39. *Tratamiento de los Errores Experimentales*, Lectures delivered at UAP, junio 1972 [mimeo]
    - Reprinted: Facultad de Ciencias, UNAM 1974 [mimeo]

40. "A different proof of the Maxwell-Boltzmann distribution", Amer. J. Phys. **40** (1972) 1239 (with P.A. Mello)

41. "On hidden-variable theories and Bell's inequality", Lett. Nuovo Cim. **5** (1972) 177 (with A.M. Cetto and L. de la Peña)
42. "Recent developments in random-matrix theory", Rev. mex. Fís. **21** (1972) 141 (with J. Flores V. and P.A. Mello)
43. "A statistical measure for the repulsion of energy levels", Lett. Nuovo Cim. **7** (1973) 482
44. "Doorway states and nuclear-spectrum statistics", Lett. Nuovo Cim **7** (1973) 707 (with P.A. Mello, J. Flores V. and O. Bohigas).
45. "The Brashinsket and its computation", Rev. Mex. Fís. **22** (1973) 185
46. "Probability: a new look at old ideas", Rev. Mex. Fís. **24** (1975) 25
47. "La relación entre filosofía y física: algunos problemas", in *La filosofía y la ciencia en nuestros días* (Primer Coloquio Nacional de Filosofía, Morelia, 4–9 ag. 1975), (Grijalvo, México 1976), vol. 2, p. 9
48. "Spectrum fluctuations and the statistical shell model", in E. Sheldon (ed.) (1976): *Proc. Int. Conf. Interactions of Neutrons with Nuclei* (Lowell, Mass.) (invited paper) ERDA Publication CONF760715-P1/P2, vol. 1, p. 495
†49. "Implicaciones epistemológicas de la inteligencia artificial", Reporte interno, IFUNAM 76–402, 1976. Included in the Supplement dedicated to T.A. Brody, Rev. Mex. Fís. **35** (1989) 89 (Chap. 22)
50. "El método científico en las ciencias naturales" (Simposio *El Método Científico* Colegio de Bachilleres, julio 1976), Rev. Colegio. Bachilleres **1** (1976) 27
51. "La mecánica cuántica y sus interpretaciones", Rev. Mex. Fís. **25** (1976) 393 (with R. Cid, J.L. Jiménez, D. Levi, J.R. Martínez, P. Pereyra, R. Rechtmann and M. Rosales)
52. "Further comments on Bell's inequality", Epistemological Letters, Sept. 1976 (with A.M. Cetto and L. de la Peña)
53. "Level fluctuations: a general property of spectra", Nuclear Phys. **A259** (1976) 87 (with E. Cota, J. Flores V. and P.A. Mello)
54. "Statistical Theory of Spectra in Nuclear Physics" (Course delivered at the Latin Amer. School of Physics, México, 1977), Notas de Física **1** (3) (IFUNAM) (1978) 1 (with J. Flores, P.A. Mello, J.B. French and S.S.M. Wong)
55. "Sobre la inconmensurabilidad de teorías", Simposio de la Asociación Mex. de Epistemologia, March 1978. Included in the Supplement dedicated to T.A. Brody, Rev. Mex. Fís. **35** (1989) 103 (Chap. 6)
56. "Political Conditions for an Independent Scientific Development", *Proc. 29th Pugwash Conference on Science and World Affairs*, Mexico City, 18–23 July 1979, "Development and Security", México 1979, 164–168
   – Political Conditions for an Independent Scientific Development, Unpublished complete version of the above paper
57. "Hacia una formulación causal de la mecánica cuántica", Rev. Mex. Fís. **26** (1979) 59 (with L. de la Peña and A.M. Cetto)
58. "Materialismo y avance científico" in M.H. Otero (comp.), *Materialismo y Ciencias Naturales* (Tercer Coloquio Nacional de Filosofía, Puebla, Dic. 1979), Inst. Inv. Filosóficas (UNAM 1984)
   – Reprinted in *Elementos* Univ. Aut. de Puebla, año 3, vol. 2, no. 11, junio de 1987, p. 29
59. "El para qué de la historia de la ciencia", Naturaleza **8** (1977) 41
60. "La revolucion actual en la física y sus problemas filosóficos" in *La Filosofía y las Revoluciones Científicas* (Segundo Coloquio Nacional de Filosofía, Monterrey, Oct. 1977) (Grijalbo, México 1979) vol. 3, p. 49
61. "Conciencia, autoconciencia y computadoras" in A. Fernández Guardiola (ed.) *La Conciencia: el Problema Mente-Cerebro* (Trillas, México 1979), p. 133
62. "The ensemble conception of probability", Abhandl. VI. Int. Kongr. f. Logik, Methodol. u. Philos. d. Wiss., (Hannover 1979), vol. 7, p. 222

63. "Real and imagined non-localities in quantum mechanics", Nuovo Cim. **B54** (1979) 455 (with L. de la Peña)
64. ¡Ayuda!, *Boletín de Difusión*, Fac. Ciencias (UNAM, enero 1980)
65. "Zero-point term in cavity radiation", Amer. J. Phys. **48** (1980) 840 (with J.L. Jiménez and L. de la Peña)
66. "The double slit revisited", Reporte interno IFUNAM 80–01, 1980. Included in the Supplement dedicated to T.A. Brody, Rev. Mex. Fís. **35** (1989) 1
*67. Problems and promises of the ensemble interpretation of quantum mechanics, (Symposium on the Philosophical aspects of quantum theory Dubrovnik, April 1980) (Chap. 13). Included in the Supplement dedicated to T.A. Brody, Rev. Mex. Fís. **35** (1989) 19
†68. "La filosofía y los físicos" (Invited paper, Congreso de la SMF, Guadalajara Nov. 1980) Rev. Mex. Fís. **27** (1981) 393 (Chap. 24)
69. Comment on d'Espagnat's *Quantum Theory and Reality*, Epistemological Letters, April 1981, p. 7 (with P.E. Hodgson)
*70. "The axiomatic approach in physics", Rev. Mex. Fís. **27** (1981) 583 (Chap. 25)
71. "Random-matrix physics: spectrum and strength of fluctuations", Rev. Mod. Phys. **53** (1981) 583 (with J. Flores, J.B. French, P.A. Mello, A. Pandey and S.S.M. Wong)
72. "Foundations for quantum mechanics: results and problems", Rev. Mex. Fís. **29** (1983) 461 (review paper)
73. "La educación de nuestros hijos: ¿mercancía o derecho?", in: Rosario Green (comp): *Los mitos de Milton Friedman*, (CEESTEM-Nueva Imagen, México, 1983) p. 119
*74. "On quantum logic", Found. Phys. **14** (1984) 409 (Chap. 20)
75. "La historia de la ciencia en la enseñanza", Quipú **1** (1984) 195
*76. "A minimal ontology for scientific research". Included in the Supplement dedicated to T.A. Brody, Rev. Mex. Fís. **35** (1989) 107 (Chap. 7)
77. "La física computacional", Rev. Mex. Fís. **30** (1984) 513
78. "El cómputo en la investigación". Included in the Supplement dedicated to T.A. Brody, Rev. Mex. Fís. **35** (1989) 121
79. "Software and the dangers to physics from computing" in R. Donaldson and M.N. Kreisler (eds.) *Proc. Symp. Recent Developments in Computing, Processor and Software Research for High-Energy Physics* (Guanajuato, Méx., May 1984) (Fermilab, Chicago) (1984) 355
80. "A random-number generator", Computer Phys. Comm. **34** (1984) 39
†81. "Los determinismos de la física", Revista de la Universidad (UNAM, México) v. 44, n. 463, 23–29 (1989) (Chap. 8)
†82. Resistencia al cambio científico: el caso de la mecánica cuántica (Chap. 21)
*83. "The irrelevance of the Bell inequality" in E.I. Bitsakis and N. Tambakis (eds.) *Determinism in Physics* (Gutenberg, Atenas Grecia) (1985) 243 (Chap. 18)
84. "Where does the Bell inequality lead?", Second International Meeting on Epistemology (Athens, Sep. 1984)
85. "Local realism and the Bell inequality". Included in the Supplement dedicated to T.A. Brody, Rev. Mex. Fís. **35** (1989) 46
*86. "Probability and the way out of the great quantum muddle" in A. van der Merwe, F. Selleri and G. Tarozzi (eds.) *Microphysical Reality and Quantum Formalism* (Urbino, Sep. 1985) (Kluwer Adad. Publ., Dordrecht) (1988) 443 (Chap. 14)
87. "Concerning the Bell puzzle". Included in the Supplement dedicated to T.A. Brody, Rev. Mex. Fís. **35** (1989) 49
88. "Ciencia pura, ciencia aplicada". Included in the Supplement dedicated to T.A. Brody, Rev. Mex. Fís. **35** (1989) 127
89. "Ciencia y democracia", Omnia (Coord. Gral. Estudios Posgrado UNAM) **3**(8), Sep. 1987, p. 19

90. "Actividades, actitudes sociales y políticas de Einstein" in R. Rodríguez and S. Hojman (comp.) *Albert Einstein: Perfiles y perspectivas* (Jornadas Einsteinianas, México, 1979), Nueva Imagen-UNAM (1987) 89

91. "Einstein y la mecánica cuántica" in R. Rodríguez and S. Hojman (comp.) *Albert Einstein: Perfiles y perspectivas* (Jornadas Einsteinianas, México, 1979), Nueva Imagen-UNAM (1987) 155

92. "Einstein y la distribución de Planck" (with J.L. Jiménez and L. de la Peña) in R. Rodríguez and S. Hojman (comp.) *Albert Einstein: Perfiles y perspectivas* (Jornadas Einsteinianas, México, 1979), Nueva Imagen-UNAM (1987) 249

93. "El 'antagonismo' de Ciencias y Humanidades". Included in the Supplement dedicated to T.A. Brody, Rev. Mex. Fís. **35** (1989) 130

94. "Systems, open systems, closed systems" in E.I. Bitsakis (ed.) *Locality and Causality in Microphysics* (Hellenic Physical Society, Athens) (1988)

95. "Closing remarks" in E.I: Bitsakis (ed) *Locality and Causality in Microphysics* (Hellenic Physical Society, Athens) (1988)

96. Conditional independence of probabilities and non-Markov processes (with A.K. Theophilou)

97. "The Suppes-Zanotti theorem and the Bell inequalities", Rev. Mex. Fís. **35** (1989) 170–187

98. "Random-number generation for parallel processors", Computer Phys. Comm. **56** (1989) 147

*99. "The ensemble interpretation of probability" in E.I. Bitsakis and C. Nicolaides (eds.) *The Concept of Probability* (Proc. of the Third Inter. Meeting on Epistemology, Delphi, Greece, Oct. 1987) (Kluwer Acad. Publ., Dordrecht) (1989) (Chap. 10)

100. "Ciencia, universidad e industria", Crítica (Univ. Aut. de Puebla) no. 37, invierno 1988–89, p. 93

*101. "The Bell inequality I: Joint measurability". Included in the Supplement dedicated to T.A. Brody, Rev. Mex. Fís. **35** (1989) 52 (Chap. 16)

*102. "The Bell inequality II: Locality". Included in the Supplement dedicated to T.A. Brody, Rev. Mex. Fís. **35** (1989) 71 (Chap. 17)

*103. "Are hidden variables possible?". Included in the Supplement dedicated to T.A. Brody, Rev. Mex. Fís. **35** (1989) 80 (Chap. 15)

*104. "Algunas ideas sobre el desarrollo del esfuerzo científico en México" (probably written 1970–1975). Included in the Supplement dedicated to T.A. Brody, Rev. Mex. Fís. **35** (1989) 117

105. The implications of the Bell inequalities [undated manuscript]

106. "Filosofía de la Física". (Course delivered at Colegio de Filosofía UAP, March 1979) (Item 100 corresponds to Chap. 1 of this book) (UAP, México 1992)

107. "The Philosophy Behind Physics" [this volume]

* denotes a paper that is reproduced in this volume
† denotes a paper whose English translation is included in this volume

Name Index

Subject Index

Springer-Verlag
and the Environment

We at Springer-Verlag firmly believe that an international science publisher has a special obligation to the environment, and our corporate policies consistently reflect this conviction.

We also expect our business partners – paper mills, printers, packaging manufacturers, etc. – to commit themselves to using environmentally friendly materials and production processes.

The paper in this book is made from low- or no-chlorine pulp and is acid free, in conformance with international standards for paper permanency.